Molecular Biology of Membranes

Structure and Function

Molecular Biology of Membranes

Structure and Function

Howard R. Petty

Wayne State University
Detroit, Michigan

Plenum Press • New York and London

Library of Congress Cataloging-in-Publication Data

Petty, Howard R.
   Molecular biology of membranes : structure and function / Howard
R. Petty.
      p.   cm.
   Includes bibliographical references and index.
   ISBN 0-306-44429-1
   1. Membranes (Biology)--Ultrastructure. 2. Molecular biology.
I. Title.
QH601.P473 1993
574.87'5--dc20   93-12844
                                                          CIP

10 9 8 7 6 5 4 3 2

ISBN 0-306-44429-1

© 1993 Plenum Press, New York
A Division of Plenum Publishing Corporation,
233 Spring Street, New York, N. Y. 10013

Printed in the United States of America

To my family,
Leslie, Aaron, Daniel, and Ian

Preface

This text attempts to introduce the molecular biology of cell membranes to students and professionals of diverse backgrounds. Although several membrane biology books are available, they do not integrate recent knowledge gained using modern molecular tools with more traditional membrane topics. Molecular techniques, such as cDNA cloning and x-ray diffraction, have provided fresh insights into cell membrane structure and function. The great excitement today, which I attempt to convey in this book, is that molecular details are beginning to merge with physiological responses. In other words, we are beginning to understand precisely how membranes work.

This textbook is appropriate for upper-level undergraduate or beginning graduate students. Readers should have previous or concurrent coursework in biochemistry; prior studies in elementary physiology would be helpful. I have found that the presentation of topics in this book is appropriate for students of biology, biochemistry, biophysics and physiology, chemistry, and medicine. This book will be useful in courses focusing on membranes and as a supplementary text in biochemistry courses. Professionals will also find this to be a useful resource book for their personal libraries.

As the title suggests, this book discusses both structural and functional properties of biological membranes. Chapters 2, 3, and 4 are primarily concerned with membrane structure, whereas Chapters 5 through 9 discuss membrane function. Chapters 2 and 3 describe the molecular and macromolecular building blocks of membranes. Chapter 4 discusses how the components presented in Chapters 2 and 3 are assembled to form functional membranes. Chapters 5 and 6 present membrane bioenergetics and transport, respectively. Membrane receptors and signal transduction are discussed in Chapter 7. Chapter 8 describes membrane fusion, biosynthesis, and biosynthetic membrane flow. Chapter 9 focuses on the role of membranes in cancer; these aberrant membranes illustrate the structures and principles presented in Chapters 1 through 8.

My chief concern while preparing this book was always the student reader. I have deliberately avoided the formal, often stilted writing style found in many intermediate and advanced books. I have used a more familiar voice in my writing style. I have also avoided long sentences. Clarity in a textbook's exposition is worth a little choppiness. I have used both the common and IUPAC nomenclatures for several organic molecules, such as fatty acids and lipids. The IUPAC nomenclature, however, is often burdensome and unfamiliar; since many chemical journals don't follow these conventions, they work a hardship on students. Therefore, I teach these conventions but I don't rely on them.

An important feature of this book is its spiral approach to the presentation of material. After the given topic or concept is initially presented, it reappears throughout the text. This facilitates the learning process by reinforcing ideas. For example, bacteriorhodopsin is discussed in chapters on membrane proteins, x-ray and electron diffraction, and bioenergetics. As a student progresses, he or she must continually build upon earlier learning experiences. While studying a given chapter the student is exposed to as many examples of a given phenomenon as is possible within the limited space available. A student can develop a fairly complete background in this fashion.

Biology has traditionally been a descriptive science, much like psychology today. Over the past 30 years biology has evolved into a molecular and mechanism-oriented subject. Bench scientists less frequently describe specimens and catalogue their parts; they instead ask, "How does it work?" Understanding how membranes work at the molecular level is the major goal of this text. For example, in Chapter 5 we follow an electron's path through a photosynthetic reaction center and in Chapter 7 we see how catecholamines bind to adrenergic receptors and relay their message to a cell's interior. Although this is admittedly a reductionist's approach to these problems, it is precisely what the next generation of students will need to learn to be successful.

Unfortunately, a great deal of outstanding research in membrane biology has been omitted to achieve the opposing priorities of balanced, comprehensive coverage in a limited space. I must accept full responsibility for these omissions. I would be happy to receive suggestions from all my readers on improving the balance of material or topics included in this text.

Many people have contributed to the successful completion of this book. I thank Dave Njus and Rob Todd for reviewing one or more chapters, and Al Bochenek, Paul Vergari, and Nahed Jaber for assistance with the illustration program. I also thank my editor at Plenum, Mary Phillips Born, for editing the manuscript and overseeing the completion of this book.

Thanks are due many additional people. The Committee on Higher Degrees in Biophysics at Harvard University can be credited with, or blamed for, steering my research interests toward membrane biology. I would like to thank the Damon Runyon–Walter Winchell Cancer Fund, the Research Corporation, Michigan Heart Association, Children's Leukemia Foundation, NSF, ONR, and especially NIH for supporting my membrane research over the past few years. I am particularly indebted to my wife, Dr. Leslie Isler Petty, for reviewing the entire manuscript. Although a child psychologist by training, she did much to clarify obscure passages in this book. I thank her for her constant support and encouragement while I worked on this project. Without her help I surely would have cut my losses and broken my pencils long ago.

Howard R. Petty

Livonia, Michigan

A Note to the Reader on Nomenclature

It is frequently important to specify a specific side of a bilayer when discussing membrane structure. For example, a plasma membrane has two sides: one faces the extracellular medium and the other faces the cytoplasm. To distinguish between the two sides of a bilayer, the *cis* side is defined as the side that binds (or could be in communication with) membrane-bound ribosomes whereas the *trans* side is the face across a membrane from ribosome attachment sites. Two conditions can introduce some ambiguity in these definitions. First, double-membrane systems such as gram-negative bacterial membranes, nuclear envelopes, and chloroplast and mitochondrial membranes may not have *cis* and *trans* faces by the above definitions. Since only one of the four faces of a double-membrane system could bind ribosomes, what are the other two faces to be called? Second, choloroplasts and mitochondria possess membrane proteins derived both outside and from within the organelle. Therefore, there is no completely unique side based upon protein synthesis. *Cis* and *trans* faces in double-membrane systems will be defined as if the second membrane were not present. For organelles having two sources of protein synthetic activity, the side facing the cytosol (or equivalently protein synthetic activity derived from nuclear mRNA) will be defined as the *cis* face. In the context of freeze-fracture experiments, the P or protoplasmic face is equivalent to the *trans* face.

Abbreviations

ACh, acetylcholine
4-AP, 4-aminopyridine
ATP, adenosine triphosphate
Bchl, bacteriochlorophyll
BiP, binding protein
BPh, bacteriopheophytin
Chl, chlorophyll
CMC, critical micelle concentration
Con A, concanavalin A
DAG, diacylglycerol
DCCD, dicyclohexylcarbodiimide
DGDG, digalactosyl diacylglycerol
DIDS, 4,4′-diisothiocyano-2,2′-stilbene disulfonate
DMPC, dimyristoylphosphatidylcholine
DMSO, dimethyl sulfoxide
DPPC, dipalmitoylphosphatidylcholine
E'_0, standard redox potential
EDTA, ethylenediaminetetraacetic acid
EGF, epidermal growth factor
ER, endoplasmic reticulum
ESR, electron spin resonance
F, Faraday's constant
Fd, Ferredoxin
FMMP, formylmethionylsulfone methyl phosphate
HA, hemagglutinin protein
HEV, high endothelial venule
HPr, heat-resistant protein
hsp, heat shock or heat shock-like protein
IEF, isoelectric focusing
IGF, insulinlike growth factor
IMI, inverted micellar intermediate

IP_3, inositol 1,4,5-trisphosphate
kb, kilobase
kDa, kilodalton
K_d, dissociation constant
LDL, low-density lipoprotein
LHCI, light-harvesting complex I
LHCII, light-harvesting complex II
LPS, lipopolysaccharide
M, molar
MAP, microtubule-associated protein
MGDG, monogalactosyl diacylglycerol
mM, millimolar
M_r, molecular weight
NANA, N-acetylneuraminic acid (or sialic acid)
NCAM, neural cell adhesion molecule
NEM, N-ethylmaleimide
NPG, 4-nitrophenyl-α-D-galactopyranoside
PAS, periodic acid–Schiff
Pc, plastocyanin
PC, phosphatidylcholine
PE, phosphaditylethanolamine
PEG, polyethylene glycol
PEP, phosphoenolpyruvate
PG, phosphatidylglycerol
Pheo, pheophytin a
PI, phosphatidylinositol
PIP, phosphatidylinositol phosphate
PIP_2, phosphatidylinositol bisphosphate
PKA, protein kinase A
PKC, protein kinase C
PQ, plastoquinone
PS, phosphatidylserine
PSI, photosystem I
PSII, photosystem II
PTS, phosphotransferase system
Q, ubiquinone
R, gas constant
SDS-PAGE, sodium dodecyl sulfate polyacrylamide gel electrophoresis
SITS, 4-acetamido-4′-isothiocyano-2,2′-stilbendisulfonate
SM, sphingomyelin
sn, stereospecific numbering
SRP, signal recognition particle
t, time
T, temperature
VSV, vesicular stomatitis virus

$\Delta G°$, standard free energy change
Ψ, transmembrane electrical potential
β-ARK, β-adrenergic receptor kinase
β_2m, β_2-microglobulin
μm, micron
μM, micromolar

Symbols for Amino Acids

	Single-letter	Three-letter		Single-letter	Three-letter
Alanine	A	Ala	Leucine	L	Leu
Arginine	R	Arg	Lysine	K	Lys
Asparagine	N	Asn	Methionine	M	Met
Aspartic acid	D	Asp	Phenylalanine	F	Phe
Cysteine	C	Cys	Proline	P	Pro
Glutamic acid	E	Glu	Serine	S	Ser
Glutamine	Q	Gln	Threonine	T	Thr
Glycine	G	Gly	Tryptophan	W	Trp
Histidine	H	His	Tyrosine	Y	Tyr
Isoleucine	I	Ile	Valine	V	Val

Contents

Chapter 3

Carbohydrates and Cytoskeletal Components

Chapter 4

Supramolecular Membrane Structure

Chapter 5

Bioenergetics: Putting Membranes to Work

Chapter 6

Transport Across Membranes

Chapter 7

Receptors and Responses

Chapter 8

Membrane Fusion, Formation, and Flow

Chapter 9

Membranes in Cancer

Chapter 1

An Introduction
to Biological Membranes

A membrane defined the outer boundary of the first living cell nearly 4 billion years ago. Since then, membranes have acquired additional functions through evolutionary selection. Modern membranes participate in many essential cell activities including barrier functions, transmembrane signaling, forming a locus for metabolic reactions, energy transduction, cell compartmentalization, and intercellular recognition. In this chapter we will explore some of the basic functions of biological membranes. We will also discuss some of the structural features of modern membranes. We will close this chapter with a brief discussion of a few research milestones that have contributed to our knowledge of what membranes are and how they work.

1.1. WHAT MEMBRANES DO

When life began, Earth's atmosphere contained nitrogen, ammonia, hydrogen, and carbon dioxide but very little oxygen or ozone. Consequently, the earliest life forms were anaerobes. Moreover, life could only exist several feet below water since there was no atmospheric ozone layer to filter harmful ultraviolet radiation from sunlight. From this environment, the first living cell emerged. The great leap from complex organic compounds to a living specimen occurred when "several different self-reproducing polymers" were enclosed "within a semi-permeable membrane" (Haldane, 1954). This illustrates how membranes solved life's first problem: defining self from nonself.

Table 1.1 lists some of the early and acquired roles of membranes in cellular life. As mentioned above, membranes define the boundary of a living cell. This semipermeable membrane allows nutrients to enter but prevents the entry of noxious substances. Transmembrane signaling is a primitive and ubiquitous function of membranes. Fossil evidence shows that cyanobacteria are the oldest known living cells, yet they are able to respond to their environment. Examples of biomolecules participating in transmembrane signaling in higher organisms include hormones, neurotransmitters, and growth factors.

1

Table 1.1.

Membrane Functions

Barrier	Membranes distinguish self from nonself, they allow nutrients to enter but keep many harmful substances out of a cell
Transmembrane signaling	Membranes relay information from the extracellular environment to the inside of a cell
Locus for metabolic reactions	Membranes provide a stable two-dimensional surface which allows enzymatic machinery to operate in a more organized assembly line fashion
Energy transduction	Membranes allow light or chemical energy to be converted into more usable forms
Compartmentalization of cell functions	In eukaryotes, membranes segregate distinct functions into different organelles thereby increasing efficiency
Intercellular recognition	Membrane-to-membrane interactions allow cells to recognize one another and to coordinate their activities

Membranes provide a two-dimensional surface that enhances the catalytic rate of enzymes. Since the product of one enzyme is often the substrate for another enzyme, metabolic reaction rates are enhanced by clustering the enzymes into a multisubunit complex and/or by restricting their location to two dimensions, instead of three. In addition to metabolic pathways, this reduction in dimensionality is also used in energy transduction in bacterial, mitochondrial, and chloroplast membranes. By restricting the electron carriers to two dimensions, these reactions occur more efficiently. Furthermore, the sealed structure of membranes allow them to store energy in the form of a proton gradient. The energy stored in a proton gradient can be tapped by a cell to perform work (i.e., movement) or to provide energy for biosynthetic pathways.

Eukaryotes are considerably larger and more complex than prokaryotes. To increase the efficiency of cell functions, eukaryotes developed a compartmentalized cytoplasm. Functional aspects of eukaryotic cells are compartmentalized into distinct membrane-bound organelles.

Intercellular communication and recognition is another important function of membranes. Cells must recognize and adhere to one another to form a tissue. During development cells must take cues from other cells which guide their movement, positioning, and pattern formation. Another good example of intercellular communication is the electrical links between the membranes of heart muscle cells. This electrical communication allows heart muscles to contract in a coordinated way to pump blood. One important goal of this book is to understand how membranes handle these important functions.

1.2. THE DIVERSITY OF MODERN BIOLOGICAL MEMBRANES

Membranes can take many forms (Figure 1.1). For example, cells have evolved several structural strategies to maintain their permeability barrier. Let's begin by considering the simplest form of sealed membranes, lipid vesicles. Lipid vesicles are

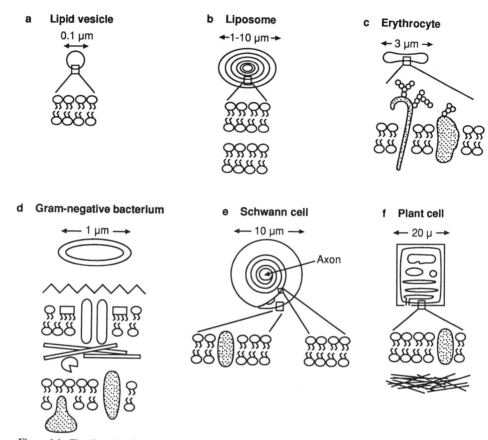

Figure 1.1. The diversity of membranes. Several model and biological membranes are shown schematically. These examples of membranes are: (a) a lipid vesicle, (b) a liposome, (c) an erythrocyte, (d) a gram-negative bacterium, (e) a Schwann cell, and (f) a plant cell.

thin lipid bilayer membranes that assume a spherical shape. Although they do not occur in nature, they are good models of biological membranes since: (1) they maintain a permeability barrier and (2) functional components can be incorporated into these model membranes. In comparison, an erythrocyte's membrane contains lipid, protein, and carbohydrate components. These three types of molecules are found in most mammalian plasma membranes. Schwann cells have a rather distinct type of membrane due to repeated bilayer structures. Multiple bilayers provide electrical insulation for an axon. Bacteria and plants utilize a cell wall in addition to a plasma membrane. Cell walls provide resistance to osmotic pressure differences and protection for some cells. Gram-negative bacteria have a double membrane system whereas gram-positive bacteria almost always have only a single bilayer. The common structural feature among all biological interfaces is the presence of a bilayer membrane.

1.3. A BRIEF HISTORICAL SKETCH OF MEMBRANE RESEARCH

To understand where contemporary membrane biology is headed, one must appreciate a bit of its history. The modern study of biological membranes began in 1925 by Fricke and, independently, by Gorter and Grendel. Fricke (1925) measured the electrical capacitance of erythrocytes in suspension. This work suggested that cells are surrounded by a hydrocarbon layer approximately 5 nm thick. Gorter and Grendel (1925) combined lipid extraction and monolayer techniques to determine the area per molecule. Today we know that the lipid extraction was incomplete, that the calculation of erythrocyte surface area was not entirely correct, and that they made a few assumptions regarding lipid packing. Notwithstanding these difficulties, they were absolutely correct in their conclusions. Gorter and Grendel calculated that there was sufficient lipid to encircle the erythrocyte twice, suggesting a bilayer structure. Davson and Danielli (1935) proposed that membranes are composed of a lipoid core whose polar groups faced outward, although they did not insist on a bilayer structure. These polar groups, in turn, were covered by a protein monolayer. In the mid-1950s the new world of a cell could now be seen with electron microscopes. Membranes were seen as trilamellar structures with a fairly uniform size of 5 nm. Building on the earlier work of others, Robertson (1959) suggested that membranes were made up of a bilayer surrounded by protein in the β-sheet conformation. The conceptual evolution of membrane models is shown in Figure 1.2.

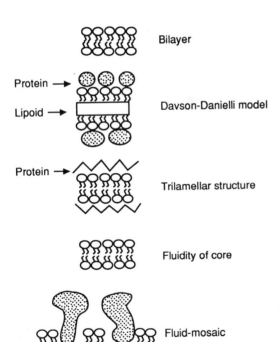

Figure 1.2. The evolution of membrane models. Membrane models and concepts from the early work on the hydrocarbon and bilayer nature of membranes to the modern fluid-mosaic model are shown.

Until the mid- to late 1960s, membranes were considered to be static waxlike structures, similar to bricks in a wall. Evidence to the contrary came from magnetic resonance data, initially from McConnell's laboratory. In 1970 Frey and Edidin discovered that membrane proteins possess lateral mobility, further emphasizing the dynamic or fluid state of cell membranes. Singer and Nicolson (1972) considered these and many other findings in their celebrated review article on the fluid-mosaic model of membrane structure. This timely article called the attention of biologists to the importance of physical properties of membranes in regulating their physiological activities.

During the 1980s molecular biological tools, such as DNA cloning, were widely applied in membrane studies. Early experiments were largely descriptive studies of deduced membrane protein sequences. Mechanism-oriented molecular studies involving site-directed mutagenesis, transfection, and gene knock-out experiments are now being used to test the physiological significance of descriptive molecular observations. Throughout this book we will strive to understand how membranes are built and work from a molecular point of view.

REFERENCES AND FURTHER READING

Danielli, J. F., and Davson, H. 1935. A contribution to the theory of permeability of thin films. *J. Cell. Comp. Physiol.* **5**:495–508.

Fricke, H. 1925. The electrical capacity of suspensions of red blood corpuscles of a dog. *Physical Rev.* **26**:682–687.

Frye, L. D., And Edidin, M. 1970. The rapid intermixing of cell surface antigens and after formation of mouse–human heterokaryons. *J. Cell Sci.* **7**:319–335.

Gorter, E., and Grendel, F. 1925. On bimolecular layers of lipoids on the chromocytes of the blood. *J. Exp. Med.* **41**:439–443.

Haldane, J. B. S. 1954. The origins of life. In: *New Biology* (M. L. Johnson, M. Abercrombie, and G. E. Fogg eds.), Vol. 16, pp. 12–27, Penguin Press, London.

Marchesi, V. T., *et. al.* 1976. The red cell membrane. *Annu. Rev. Biochem.* **45**:667–698.

Robertson, J. D. 1959. The ultrastructure of cell membranes and their derivatives. *Biochem. Soc. Symp.* **16**:3–43.

Singer, S. J., and Nicolson, G. L. 1972. The fluid mosaic model of the structure of cell membranes. *Science* **175**:720–731.

Steck, T. L. 1974. The organization of proteins in the human red cell membrane. *J. Cell Biol.* **62**:1–19.

Chapter 2

Lipids, Oligomers, and Proteins

Biological membranes are composed of lipids, proteins, and carbohydrates, although their relative quantities vary widely. Membranes from various sources are composed of 20 to 60% protein, 30 to 80% lipid and 0 to 10% carbohydrate by weight. In this and the following chapter we discuss the molecules and macromolecules that assemble to form membranes. This chapter focuses on membrane-associated lipids, oligomers, and proteins. Chapter 3 discusses membrane carbohydrates and skeletons. We shall begin by considering lipids, a fundamental component of every membrane.

2.1. MEMBRANE LIPIDS

Since lipids are water-insoluble organic molecules, they are generally extracted from membranes by treatment with nonpolar solvents such as chloroform, ether, or benzene. Lipids are amphipathic molecules, i.e., they possess both polar and nonpolar regions. Most lipids can be categorized as glycerophosphatides, sphingolipids, and sterols. In this section we will study the chemical structures of these molecules and how their amphipathic properties contribute to the formation of membranes.

2.1.1. FATTY ACIDS

Fatty acids are associated with most lipids. Although fatty acids occur throughout nature, they are almost always linked to some other molecule such as glycerol, sphingosine, or cholesterol. Over 100 species of fatty acids have been identified. Fatty acids are characterized by a long hydrophobic tail and a terminal carboxylic acid group. Some fatty acids such as oleic, palmitoleic, and linoleic possess double bonds. These unsaturated fatty acids are structurally heterogeneous and vary in the position, type (*cis* or *trans*), and number of double bonds. Table 2.1 lists the names and shorthand notations for several common fatty acids. Palmitic acid's designation 16:0 indicates that there are 16 carbon atoms with no double bonds. The shorthand notation for oleic acid is $18:1^{\Delta 9}$, which says that this molecule has 18 carbon atoms with one double bond

Table 2.1.

Common Fatty Acids

Notation	Trivial name	IUPAC name
14:0	Myristic acid	n-tetradecanoic
16:0	Palmitic acid	n-hexadecanoic
16:1$^{\Delta 9}$	Palmitoleic acid	cis-9-hexadecenoic
18:0	Stearic acid	n-octadecanoic
18:1$^{\Delta 9}$	Oleic acid	cis-9-octadecenoic
18:1$^{\Delta 9 trans}$	Elaidic acid	$trans$-9-octadecenoic
18:1$^{\Delta 11}$	Vaccenic acid	cis-11-octadecenoic
18:2$^{\Delta 9,12}$	Linoleic acid	cis-cis-9,12-octadecadienoic
18:3:2$^{\Delta 9,12,15}$	α-Linolenic acid	all-cis-9,12,15-octadecatrienoic
20:0	Arachidic acid	n-eicosanoic
20:4$^{\Delta 5,8,11,14}$	Arachidonic acid	all-cis-5,-8,-11,-14-eicosatetraenoic
22:0	Behenic acid	n-docosanoic
24:0	Lignoceric	n-tetracosanoic

beginning at carbon atom number 9. The double bond is assumed to be *cis* unless specifically indicated as *trans*. Due to their biosynthetic mechanisms, most fatty acids have an even number of carbon atoms between 14 and 22 (Table 2.1). Fatty acids containing 16 or 18 carbon atoms occur most frequently in biological membranes. Unsaturated double bonds are often found between carbon atoms 9 and 10. Polyunsaturated fatty acids rarely possess conjugated double bonds; for example, a fatty acid with double bonds at carbon atoms 9 and 12 could be found in cells but not $\Delta 9,11$ since the latter possesses conjugated double bonds.

In addition to the typical fatty acids discussed above, there are several fatty acids with unique structural features. These include branched, β-hydroxy, and cyclopropane fatty acids (compounds to **1–3**, respectively):

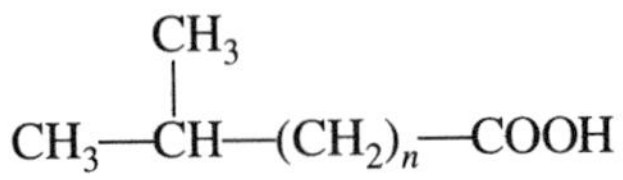

$$CH_3-\overset{\overset{\displaystyle CH_3}{|}}{CH}-(CH_2)_n-COOH$$

(1)

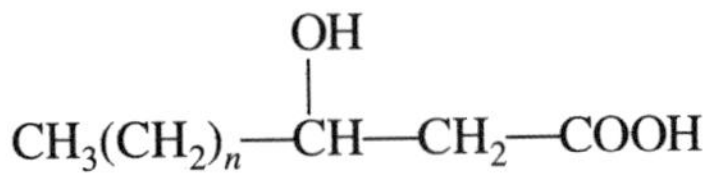

$$CH_3(CH_2)_n-\overset{\overset{\displaystyle OH}{|}}{CH}-CH_2-COOH$$

(2)

$$CH_3-(CH_2)_n-\overset{\overset{\displaystyle CH_2}{\diagup\diagdown}}{HC-CH}-(CH_2)_9-COOH$$

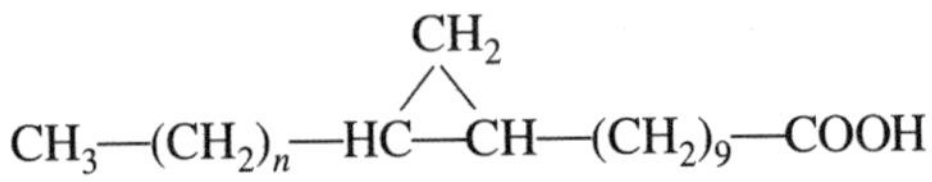

(3)

Branched fatty acid chains are common among gram-positive bacteria. The branch point occurs at various positions along the acyl chain. Since bacteria are generally devoid of unsaturated fatty acids, they apparently utilize branched fatty acids to increase the fluidity of their membranes. Cyclopropane fatty acids have also been observed in bacteria. Their presence is enhanced during conditions of oxygen deprivation.

Figure 2.1 shows the minimal energy conformations of typical saturated and unsaturated fatty acids. Saturated fatty acids can adopt many conformations due to rotation about single bonds. Monounsaturated *cis* fatty acids have one rigid angle of 30° within their hydrocarbon chain. This structure disorganizes bilayers and increases their fluidity. The physical properties of *trans* double bonds are similar to those of saturated fatty acids due to their overall similarity in shape.

2.1.2. Glycerophosphatides

Glycerophosphatides (or phosphoglycerides) are the most abundant class of membrane lipid. They are usually referred to generically as phospholipids, but strictly speaking, the term phospholipid encompasses a rather broad range of molecules including glycerophosphatides. Glycerophosphatides are derived from the glycerol backbone. One of the primary hydroxyl groups of glycerol [position 3 according to the stereospecific numbering convention (*sn*)] is esterified to phosphoric acid whereas the two remaining hydroxyl groups (positions 1 and 2) are esterified to fatty acids. In addition, various alcohols may be linked to the phosphate group leading to considerable structural diversity. In the absence of a second alcohol, the simple glycerophosphatide phosphatidic acid (PA or PtdOH) is obtained. Phosphatidylcholine (PC or PtdCho) is a common glycerophosphatide. In this case the choline head group is esterified to the phosphate moiety (Figure 2.2). The several glycerophosphatides shown in Figure 2.2 illustrate structural diversity created by various alcohols.

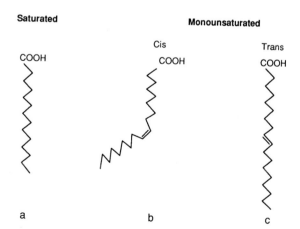

Figure 2.1. Minimal energy conformations of three fatty acids. (a) The saturated fatty acid myristic acid is shown. (b, c) The structures of the unsaturated fatty acids oleic (b) and elaidic (c) acids are illustrated.

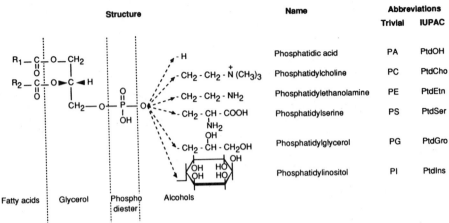

Structure				Name	Abbreviations	
					Trivial	IUPAC
			- H	Phosphatidic acid	PA	PtdOH
			- CH$_2$ - CH$_2$ - N$^+$(CH$_3$)$_3$	Phosphatidylcholine	PC	PtdCho
			- CH$_2$ - CH$_2$ - NH$_2$	Phosphatidylethanolamine	PE	PtdEtn
			- CH$_2$ - CH - COOH NH$_2$	Phosphatidylserine	PS	PtdSer
			- CH$_2$ - CH - CH$_2$OH OH	Phosphatidylglycerol	PG	PtdGro
			(inositol ring)	Phosphatidylinositol	PI	PtdIns
Fatty acids	Glycerol	Phospho diester	Alcohols			

Figure 2.2. Glycerophosphatides. The structures, names, and abbreviations of six common glycerophosphatides are shown. R1 and R2 are hydrocarbon chains of two fatty acids esterified to positions 1 and 2 of the glycerol backbone.

Lipids derived from the glycerol backbone form a major component of all biological membranes. The structural relationships among these molecules are shown in Figure 2.3. Phosphatidylcholine, phosphatidylethanolamine (PE or PtdEtn), and glycosyl diacylglycerols are the predominant glycerolipids of animal, bacterial, and plant membranes, respectively. Generally, an organism's tissues and organelles contain qualitatively similar lipids.

2.1.3. Glycosyl Diacylglycerols

Glycosyl diacylglycerols are glycolipids built upon the glycerol backbone. They are most frequently encountered in bacteria and plants. The carbohydrate moiety is glycosidically attached to the *sn*-3 position of glycerol whereas fatty acids are esteri-

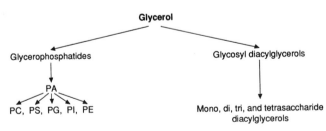

Figure 2.3. Structural relationships among glycerolipids. Both glycerophosphatides and glycosyl diacylglycerols are shown.

fied to positions 1 and 2. Mono- and disaccharide derivatives of glycerol occur most frequently in nature, although certain bacteria possess tetraglycosyl diacylglycerols. Figure 2.4 shows the digalactosyl diacylglycerol 1,2-diacyl-3[α-D-galactosyl-(1′→6′) β-D-galactosyl-(1′→3)]-*sn*-glycerol. It is easily isolated from gram-positive bacteria and chloroplast membranes. In naming this glycolipid, the use of the 3 is optional, since positions 1 and 2 have been specified. Each carbohydrate moiety is preceded by a Greek and an Arabic letter. For example, α-D-galactosyl indicates that the α anomer of the D-galactose is present. The stereochemistry of the galactosyl residue would be indicated as L for L-galactose and X or R for unknown or racemic mixtures, respectively. Only D-galactose is found in biologically derived glycolipids. If necessary, the carbohydrate ring structures of pyranose (6) or furanose (5) are indicated. Since there are many potential sites for covalent linkage on each carbohydrate residue, these sites must be identified. The (1′→6′) specifies that carbon atom number 1 of the β-D-galactosyl residue is linked to carbon number 6 of the second β-D-galactosyl moiety. In this fashion one may chemically specify any glycosyl diacylgylcerol.

2.1.4. Sphingolipids

Sphingolipids are lipids derived from sphingosine (4-sphingenine) or related structures (Figure 2.5). Sphingolipids are a major component of animal cells and are also

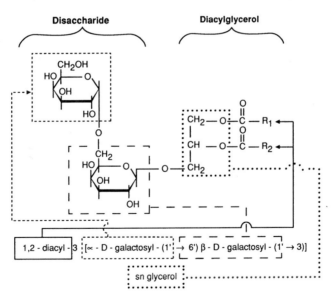

Figure 2.4. Structure and systematic naming of a glycosyl diacylglycerol. The fatty acid, glycerol, and carbohydrate portions are indicated. The relevant structural entities are noted for each component of the systematic name.

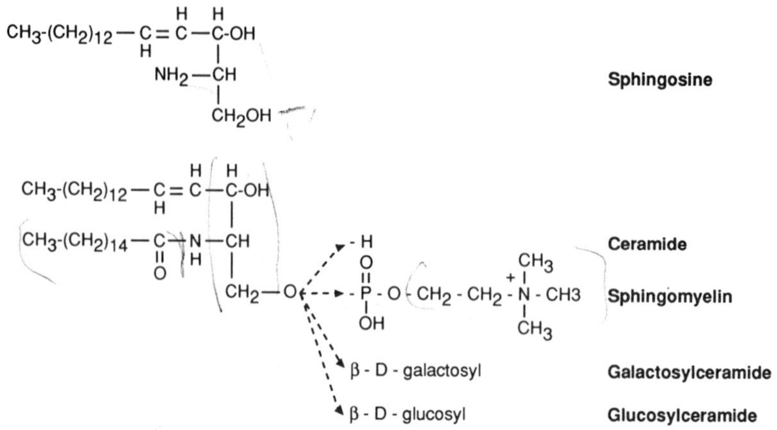

Figure 2.5. Chemical structures of several sphingolipids. The sphingosine molecule is shown at the top of the figure. Four types of sphingolipids are shown in the lower portion.

found in plant membranes and certain microorganisms. Almost all mammalian glycolipids are glycosphingolipids. Glycolipids are only found at the *trans* face of biological membranes. In addition to the 4-sphingenine molecule, phytosphingosine (4-D-hydroxysphingamine; **4**) and sphinganine (dihydrosphingosine; **5**)

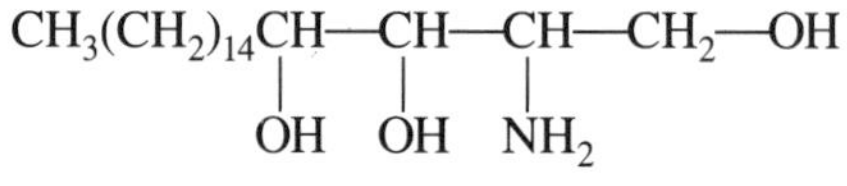

$$CH_3(CH_2)_{14}CH—CH—CH—CH_2—OH$$
$$\underset{OH}{|}\quad\underset{OH}{|}\quad\underset{NH_2}{|}$$

(4)

$$CH_3(CH_2)_{14}CH—CH—CH_2—OH$$
$$\underset{OH}{|}\quad\underset{NH_2}{|}$$

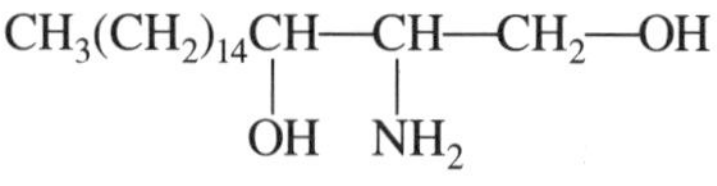

(5)

serve as sphingolipid backbone structures. Phytosphingosine and sphinganine are found in plants and bacteria, respectively. To allow the formation of stable bilayer structures, a fatty acid is linked to the amino group of a sphingoid base via an amide bond. Typical fatty acids are stearic, behenic, lignoceric, and oleic whereas polyunsaturated fatty acids are not found. The simplest membrane-associated sphingolipid is ceramide (Cer); it is a sphingosine molecule with an attached fatty acyl chain (Figure 2.5). Figure 2.6 shows the structural relationships among several simple sphingolipids. Sphingomyelin (SM or CerPCho; Figure 2.5) is a sphingolipid with a phosphocholine headgroup (see phosphatidylcholine, Figure 2.2). Cerebrosides are glycosphingolipids with a single glucose or galactose residue glycosidically linked ($\beta1{\rightarrow}1'$) to the sphingosine base (Figure 2.5). These two glycosphingolipids form the glucosylceramide and galactosyl-

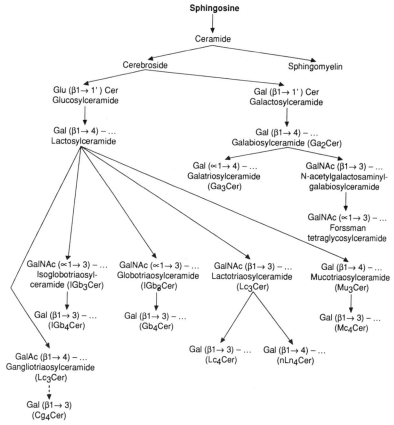

Figure 2.6. Structural relationships among several common sphingolipids. The relationships among sphingo-lipids are illustrated beginning with sphingosine. New molecular components are added to this structure in a stepwise fashion throughout the figure. Each item within the figure shows only the new component added to the preceding structure.

ceramide families of glycosphingolipids (Figure 2.6). The second carbohydrate residue for both groups of glycosphingolipids is Gal($\beta1\rightarrow4$). These two families are further subdivided based on the third saccharide as shown in Figure 2.6. The sequential carbohydrate building blocks, accepted names, and abbreviations of these several glycosphingolipids are given. These subdivisions are of considerable utility in specifying the many complex glycosphingolipids found in nature.

Gangliosides are glycosphingolipids that have one or more negatively charged N-acetylneuraminic acid (NANA) or sialic acid) residues coupled to their oligosaccharide chains. Although gangliosides are derived from both glucosylceramide and galactosylceramide families, most are associated with the ganglio (Gg_3Cer) series (Figure 2.6). Traditionally, gangliosides have been specified by the Svennerholm (1963) nomenclature. In this system a ganglioside is specified by a letter (M, D, T) indicating a mono-, di-, or trisialo derivative and a number. The number is arbitrarily assigned with no correlation to molecular structure. The IUPAC-approved system relies on the definitions given in Figure 2.6. The locations of substituents, such as NANA, are specified by a Roman numeral indicating the position of the modified carbohydrate moiety of the parent oligosaccharide chain. The numbering begins at the ceramide and proceeds to the terminal saccharide. A superscript indicates which hydroxyl group has been modified. For example, the glycosphingolipid

$$Gal(\beta 1 \rightarrow 4)GalNAc(\beta 1 \rightarrow 4)Gal(\beta 1 \rightarrow 4)Glu(\beta 1 \rightarrow 1')Cer$$
$$3$$
$$\uparrow$$
$$2\alpha NANA$$

(6)

is chemically specified as II^3-NANA-Gg_4Cer in the IUPAC system. This ganglioside is designated as GM_1 using the Svennerholm nomenclature. A comparison of various gangliosides shows that similar Svennerholm nomenclatures are given to distinctly different chemical species. For example, GM_3 is in the glucosylceramide family whereas GM_4 is a member of the galactosylceramide family.

2.1.5. Sterols

The third major class of lipids is sterols. Cholesterol and related sterols are biosynthesized by animal, plant, and fungal cells. In general, prokaryotes do not synthesize sterols. One of the rare exceptions to this rule is the microorganism *Methylococcus capsulatus*, which produces 4-methyl sterols. Figure 2.7 shows the structural formulas of cholesterol, stigmasterol, sitosterol, and ergosterol. The cyclic portion of a sterol imparts structural rigidity to the molecule whereas the branched alkyl chain is more flexible. In a lipid bilayer the hydroxyl group resides near the membrane–water interface. The alkyl group is situated near a bilayer's center. Sitosterol and stigmasterol are the major sterols found in plant membranes whereas cholesterol and ergosterol are found in small quantities. The primary sterol of yeast and other fungi is ergosterol. Several sterol derivatives are found in nature. Fatty acids can be esterified to the hydroxyl group of cholesterol, yielding cholesteryl esters. Furthermore, monosaccharides glycosidically linked to the hydroxyl group form steryl glycosides, which are generally associated with plants.

Sterols have profound influences on membrane structure and function. Small quantities of sterols may regulate metabolic pathways as shown by studies with sterol auxotrophs (Dahl *et al.*, 1980). Larger quantities of sterols (i.e., sterol/phospholipid ratio of 0.1 to 1) influence a membrane's physical state. Cholesterol's rigid and flexible regions create unique properties. Cholesterol has the ability to decrease the fluidity of fluid membranes and increase the fluidity of solid membranes, at least in model membrane systems. This is due to its ability to: (1) restrict the motion of fatty acyl chains

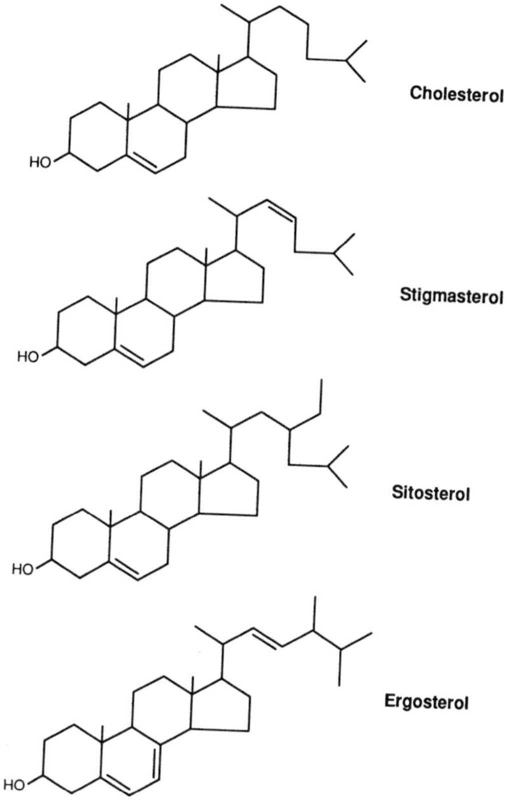

Figure 2.7. Sterols. The chemical structures of four naturally occurring sterols are shown.

in fluid membranes and (2) increase mobility and reduce lateral organization in solid membranes. Studies with artificial membranes and living cells show that cholesterol also decreases membrane permeability to small atoms and molecules. Cholesterol depletion of animal cells affects osmotic and functional properties of cell membranes.

2.1.6. Lipopolysaccharide

As shown in Figure 1.1, gram-negative bacteria have a complex double-membrane system to serve as a permeability barrier. The outer leaflet (*trans* face) of the outer membrane is rich in lipopolysaccharide (LPS) but devoid of glycerophosphatides. LPSs are found in gram-negative bacteria and blue-green algae. They are solubilized from intact cells by both harsh conditions (organic solvents) and mild conditions (salt solutions and EDTA). In contrast, glycerophosphatides in the outer leaflet of other cells require harsh extraction conditions. LPSs express the O-somatic antigens of gram-negative bacteria, which account for antigenic differences among related bacteria. To infect bacteria, bacteriophages often bind to LPS. LPS is responsible for the endotoxin activity of gram-negative bacteria (e.g., they induce fever). LPS interacts with the mammalian immune system to: (1) act as a B-cell mitogen, (2) enhance T-cell helper activity, and (3) activate complement. The functional attributes and solubility properties of LPS can be understood by examining the structure of this unique class of molecules.

The LPS molecule (M_r = 4000) consists of three distinct domains: lipid A, core carbohydrate, and O-antigen structures. The lipid A portion is composed of repeating (β1→6) D-glucosamine disaccharides with fatty acyl residues attached via amide and ester linkages. The amino groups of glucosamine are always linked to β-hydroxy-myristic acid whereas lauric, myristic, palmitic, and β-hydroxymyristic acids are found in ester linkages (Figure 2.8). Furthermore, the β-hydroxyl moiety of β-hydroxy-myristic acid is frequently esterified to another fatty acid. The structure of a representa-tive lipid A molecule is shown in Figure 2.8. The hydrophobic fatty acyl residues are tightly associated with a bilayer.

The core carbohydrate is attached to carbon atom number 6 of lipid A as indicated in Figure 2.8. The core structure consists of roughly 10 carbohydrate groups including ketodeoxyoctonic acid and L-glycero-D-mannoheptose and, to a lesser extent, hexoses and amino sugars. Ketodeoxyoctonic acid occurs as a trisaccharide located between the core structure and lipid A portions of LPS. The distal end of the core region is

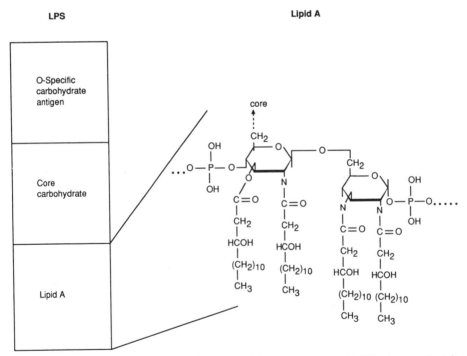

Figure 2.8. Lipopolysaccharide. A schematic diagram of the overall structure of LPS is shown on the left. The detail on the right shows the molecular structure of the lipid A component of LPS.

glycosidically linked to the O-specific side chain. This side chain has repeating carbohydrate components which determine the molecule's antigenic specificity or serotype.

2.1.7. The Disposition of Lipids in Membranes

Lipids form the bilayer structure of model and biological membranes. Lipids are not randomly distributed within a bilayer: both transverse and lateral asymmetries are found. Transverse lipid asymmetries represent distinct lipid compositions at the *trans* versus the *cis* membrane faces. Lateral lipid asymmetries refer to inhomogeneities at one face or leaflet of a membrane. Although these asymmetries have been clearly demonstrated for certain cells or membranes, their general applicability to all membranes is not proven.

Lipid asymmetry was first reported for erythrocyte membranes by Bretscher (1972). The transverse asymmetry of red blood cell phospholipids was demonstrated using the reagent formylmethionylsulfone methyl phosphate (FMMP). Since FMMP reacts with primary amines, it labels phosphatidylserine, phosphatidylethanolamine, and the amino groups of proteins. Its negative charge renders FMMP impermeable or very slowly permeable to membranes. Therefore, FMMP only reacts with components at a membrane's *trans* face. The initial reaction rate of FMMP with erythrocytes has been measured. Although intact RBCs yielded a very slow reaction rate, leaky RBC ghosts were found to react rapidly with FMMP. These data suggest that the aminophospholipids PE and PS are preferentially located at the membrane's *cis* face. To account for 100% of the membrane lipids, the other major components SM and PC must be enriched at the bilayer's *trans* face. Amino lipids are therefore enriched at the erythrocyte membrane's *cis* face whereas choline lipids are enriched at its *trans* face. These results have been confirmed using a wide variety of methods including chemical, enzymatic, and magnetic resonance techniques. Although lipid asymmetry has been established in the case of erythrocytes, the asymmetry of other membrane systems is not well established. Viral, bacterial, organelle, and plasma membranes of plants and animals have been studied (Op den Kamp, 1979). Considerable variability in lipid asymmetry has been found. Nonetheless, in many cases PE and PS are preferentially associated with the *cis* face of biological membranes. Cholesterol may be found associated with both faces of a membrane, although some preference for the *trans* face has been noted in RBC and myelin membranes.

The fatty acyl composition of particular lipids is equivalent at the *cis* and *trans* faces. For example, the small amount of PE at the *trans* face of an RBC's membrane has the same fatty acyl composition as the PE at the *cis* face. However, each lipid may show enrichment in certain fatty acids. Amino lipids generally contain more unsaturated fatty acids than do choline lipids.

In addition to transverse asymmetry, lipids may also exhibit lateral inhomogeneities in membranes. For example, under appropriate conditions model membranes containing a mixture of lipid species can undergo lateral phase separations. During lateral phase separations, lipids are segregated into distinct domains within a membrane. For instance, "fluid" and "solid" glycerophosphatides form separate patches or pools within a bilayer. Similar phenomena have been reported to occur in biological membranes. One of the better examples of lateral asymmetries in biological membranes is the thylakoid membrane of chloroplasts. They contain PC, PG, mono- and digalactosyl diacylglycerols, cholorphylls, and other lipids. Thylakoid membranes can be appressed against another thylakoid membrane (stacked) or not (unstacked). The stacked and unstacked membrane regions possess unique protein compositions. Unique lipid compositions have also been observed for these membrane regions. Stacked regions are enriched in PC and PG and depleted of monogalactosyl diacylglycerols. It is inferred that unstacked regions are enriched in galactolipids and depleted of PC and PG. The lateral inhomogeneities of biological membranes may have important functional roles.

2.1.8. Lipid Structures

The amphipathic nature of lipids is responsible for their spontaneous assembly into organized structures such as membranes. The polar and nonpolar regions of lipid molecules are stable in hydrophilic and hydrophobic environments, respectively. The hydrophobic effect is the most important driving force in the formation of membranes. Lipid hydrocarbon side chains form the hydrophobic core of membranes. The mutual attraction of hydrocarbon chains due to van der Waals forces plays an unimportant role in the formation of lipid structures. The hydrophobic effect, a manifestation of attractive forces among water molecules, is primarily responsible for the spontaneous organization of lipids. Ionic or polar molecules form hydrogen bonds with water. However, nonpolar molecules such as hydrocarbon side chains cannot form hydrogen bonds with water. Water molecules must reorient in the vicinity of hydrophobic substances to maximize their number of hydrogen bonds. The organization of water molecules about hydrocarbon chains leads to an overall decrease in entropy or "randomness" of water molecules.

To illustrate the role of entropy in the spontaneous assembly of micelles and membranes, consider the following thought experiment. Two hydrophobic substances are dissolved in water. Water molecules must reorient around each hydrophobic molecule. However, if both hydrophobic molecules are sequestered into the same cavity, an equivalent volume is contained within less surface area. One consequence of the sequestered hydrophobic molecules is that fewer water molecules will become organized at the surface of one large cavity in comparison to two smaller cavities. There is a net increase in the system's entropy. Therefore, the entropy of a system is maximized by minimizing the number of unfavorable interactions between hydrophobic molecules and water.

A variety of organized structures formed by amphipathic molecules are illustrated in Figure 2.9. These structures include monolayers, micelles, and bilayers. Glycerophosphatides spread at an air–water interface to form a unimolecular layer. The hydrophilic head groups are in contact with water while the methyl ends of the fatty acyl chains point toward the air. Supported lipid monolayers are prepared by passing a

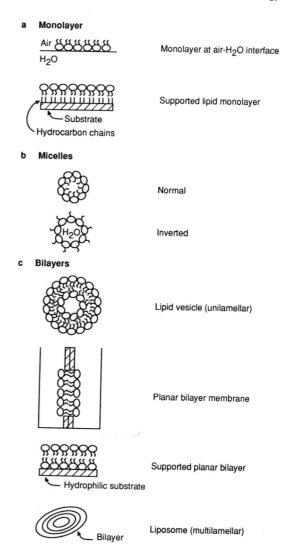

Figure 2.9. Organized lipid structures. Three classes of organized lipid structures are shown: (a) monolayers, (b) micelles, and (c) bilayers. Both monolayers and bilayers serve as useful models of biological membranes.

hydrophobic support through an air–water interface. Glycerophosphatides form a unimolecular layer next to the hydrophobic support.

Lysophospholipids, gangliosides, certain plant lipids, and detergents form micelles. The number of monomers that make up each micelle is called the aggregation number, which varies from 2 to 200. Micelles can only form above the critical micelle concentration (CMC). In normal micelles a molecule's hydrophobic portion is pointed toward the micelle's center, away from water. Inverted micelles sequester water in an internal space with the hydrophilic structures in contact with water. Hydrophobic moieties point outward into a hydrophobic environment.

Glycerophosphatides tend to form monolayer or bilayer structures, not micelles. The bilayers are planar or spherical in shape. Planar or black lipid membranes are formed by drying a solution of lipids on a small orifice. If an appropriate amount of lipid is used, a lipid bilayer is formed at the orifice. Aqueous compartments on either side provide an opportunity to study transmembrane movements of molecules and electrical properties of membranes. Supported lipid bilayers are formed by passing a hydrophilic substrate through an air–water interface twice. Supported bilayers are often used to study cell membrane-to-model membrane interactions.

Glycerophosphatides also form spherical bilayer structures such as lipid vesicles and liposomes. One method of forming lipid vesicles is to bubble a solution of phospholipids in ether into an aqueous medium at 65°C. The ether vaporizes on contact with the hot buffer. Unilamellar lipid vesicles spontaneously form under these conditions. Liposomes are generally defined a multimellar bilayer membranes (Figure 1.1), although some authors do not distinguish between lipid vesicles and liposomes. A liposome's concentric bilayer shells are much like the concentric layers of an onion. Small aqueous spaces separate each adjacent bilayer in a liposome. They are formed by drying a phospholipid solution on the bottom of a beaker. An aqueous buffer is added followed by vigorous agitation. As the lipid peels away from the beaker, liposomes are formed. Certain permeability properties, such as H_2O passage, of lipid vesicles and liposomes resemble those of intact cells. Substances can be entrapped within the aqueous confines of these structures by including them in the buffer. Studies of model membranes offer many insights into the workings of living membranes.

2.2. Oligomeric Membrane Ionophores

Ionophores are molecules or macromolecules capable of conducting ions across a bilayer membrane. This increase in membrane permeability is mediated by either a carrier or a pore-forming mechanism (Figure 2.10). Some ionophores are synthetic whereas others are naturally occurring. Many naturally occurring ionophores are antibiotics of microbial origin. The ionophores discussed in this section are between 200 and 2000 Da in molecular mass. Most ionophores are composed of a few similar subunits, i.e., they are oligomeric in nature. In contrast, polymeric membrane channels formed by proteins will be discussed in the following section. Membrane-associated oligomeric ionophores are particularly useful to study because: (1) they are biologically and medically important reagents and (2) they provide simple models of membrane transport and voltage-gated membrane channels.

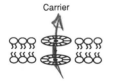

Carrier

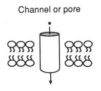

Channel or pore

Figure 2.10. Mechanisms of ion translocation. Ion movement across bilayer membranes can be classified as carrier- or channel-mediated. Carriers shuttle ions across a membrane whereas no transverse movement is required for channel activity.

2.2.1. Ion Carriers

Ion carrying ionophores are classified into two major families: the valinomycin and nigericin groups. The first group includes valinomycin, enniatins, and the macro-tetrolides nonactin and monactin. The nigericin group contains nigericin, monensin, griseorexin, X-206, and X-537A. These two ionophore families are distinguished by their structure and charge.

Valinomycin is an uncharged cyclic compound composed of alternating amino acids and α-hydroxy acids joined by amide and ester linkages (Figure 2.11). Upon interaction with one cation, a positively charged complex is formed. Valinomycin is highly specific for potassium. This cation is sequestered at the center of a valinomycin molecule where carbonyl oxygen atoms form a polar cavity. The exterior of the complex is a hydrophobic shell that is soluble in a membrane's hydrophobic core. Potassium ions are bound and released at membrane surfaces; this results in a net potassium flux across a membrane thereby dissipating a potassium gradient. The molecules of the nigericin family contain several covalently linked heterocyclic groups with carboxylic acid and hydroxyl moieties at the ends. Although nigericin is not a covalently closed circle, a circular structure is formed in a membrane environment by hydrogen bonding between the terminal carboxylic acid and hydroxyl groups. Upon cation binding the negatively charged molecule becomes neutral. Due to its charge, it can only cross the membrane with a bound cation.

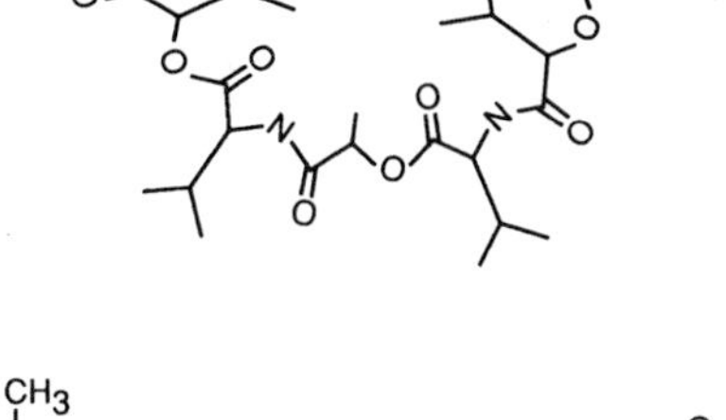

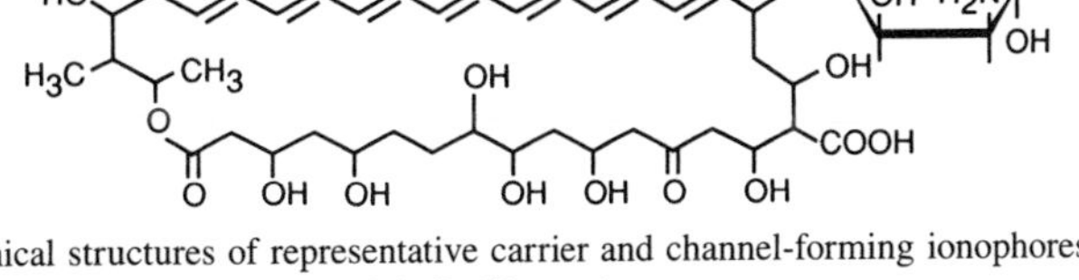

Figure 2.11. Chemical structures of representative carrier and channel-forming ionophores. The molecular structures of valinomycin (a) and amphotericin B (b) are shown.

2.2.2. Pore-Forming Molecules

In the preceding paragraphs we discussed carrier-mediated ionophores. These mobile ionophores shuttle ions between the two faces of a membrane (Figure 2.10). Another strategy to move ions across a bilayer is to form a relatively stationary channel or pore which will allow ions to pass. Pore-forming ionophores fall into two broad classes: those that obey Ohm's law and those that do not.

2.2.2.1. Ohmic Channels

The cyclic polyene antibiotics nystatin, amphotericin B, and filipin form ohmic membrane channels. These molecules require sterols (Section 2.1.5) for their assembly into membrane channels. Ergosterol is most effective although cholesterol is also active. All of these cyclic polyene antibiotics can be employed clinically to treat topical fungal infections. Filipin is also used to localize cholesterol in freeze-fracture electron microscopy (Section 4.2.6). Several structural features are common among these antibiotics (see Figure 2.11 for the covalent structure of amphotericin B). This amphipathic molecule possesses a lactone ring structure. One side of amphotericin B is rich in hydroxyl residues whereas the opposite side has a series of conjugated double bonds. Its carbohydrate and carboxylic acid moieties are found at one side suggesting that this area is in contact with the aqueous environment. In a membrane the series of hydroxyl groups form a hydrophilic channel whereas the polyene portion is positioned near a bilayer's fatty acyl residues.

The conductance of sterol-containing bilayers is greatly increased in the presence of amphotericin B. The measured transmembrane conductivity of antibiotic- and sterol-containing bilayers increases as the 5th to 10th power of the antibiotic's concentration; this indicates that several polyene antibiotic molecules aggregate to form one channel. Therefore, the membrane channel is believed to be formed by 5 to 10 alternating antibiotic and sterol molecules as illustrated in Figure 2.12.

As Figure 2.12 suggests, lateral antibiotic–sterol interactions form one-half of a complete transmembrane channel. Two half-channels at each membrane face form one complete pore. This membrane channel will allow molecules of up to 0.4 nm in size (including univalent cations, anions, water, and nonelectrolytes) to pass across a membrane.

Gramicidins are another class of ohmic membrane channel. They are linear pentadecapeptides made up of alternating L- and D-amino acids. Their constituent amino acids are all hydrophobic. The alternating L- and D-amino acids allow these molecules to form helical three-dimensional structures. The hydrophobic groups are present at the outer surface of the helix whereas the polar peptide bonds point toward the interior of the helix. Within a membrane, gramicidin is believed to form a helical dimer which allows small monovalent cations and water molecules to cross a membrane.

Studies with planar lipid membranes (Figure 2.9) have shown that their ionic conductance depends on the square of gramicidin's concentration. This indicates that dimers are required for the formation of transmembrane channels. When *very* low concentrations of gramicidins are added to a bilayer, the opening and closing of

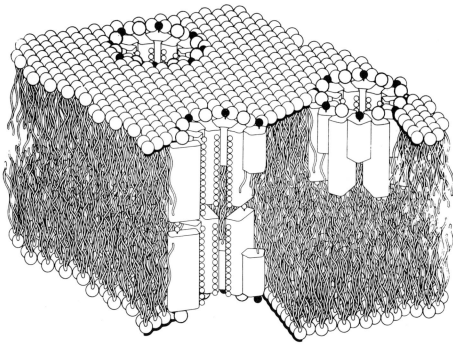

Figure 2.12. Model of sterol–amphotericin B complexes in membranes. A schematic model of channels formed by amphotericin B and cholesterol is shown. A cut-away view of an entire pore is shown near the center. The right-hand side illustrates a half-pore. The small circles at the centers of the pores represent hydrophilic groups. When two half-channels come into confluence, a complete pore is formed. (From de Kruiff and Demel, 1974, *Biochim. Biophys. Acta* **339**:57.)

individual channels can be measured. Channel formation is observed as conductance changes across the bilayer. Figure 2.13 shows a typical conductance profile of gramicidin-treated membranes. When two gramicidin half-channels come into confluence, one complete channel yielding one unit of conductance is formed (Figure 2.13). As the two gramicidin molecules part, the channel closes and conductance stops. If two complete channels form from four half-channels, the conductance is twice the unit value.

The transport of ions across gramicidin-treated membranes is much faster than the rates obtained for carrier molecules such as valinomycin. Carriers must shuttle from one membrane face to the other whereas channels are essentially "holes" in a bilayer. The ability of carriers to move in a membrane is affected by temperature. In solid membranes (below the phase transition temperature), carrier-mediated transport is abolished. However, gramicidin-mediated transport is only slightly diminished.

2.2.2.2. Nonohmic Channels

All of our previous examples of ionophores have demonstrated ohmic membrane electrical properties. Membrane voltage, current, and resistance are related by the

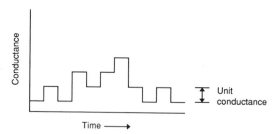

Figure 2.13. Time dependence of conductance. A hypothetical plot illustrating the time dependence of the conductance of gramicidin A-treated planar lipid bilayers is shown. One unit of conductance represents one open gramicidin A channel. Zero, one, two, three, etc. units of conductance are observed.

equation $V = IR$. During nonohmic behavior, membrane resistance is a function of membrane potential, $R(V^n)$; this is in contrast to ohmic systems, in which R is a constant. Nonohmic membrane properties are of considerable importance since they are a requisite for electrical excitability of membranes. In other words, membrane ionic permeability can be controlled by membrane potential.

Nonohmic ionophores were first observed by Paul Mueller and his colleagues during studies of a contaminant in a batch of egg albumin. The contaminant was called EIM (excitability-inducing material). It is now known to be a ribonucleoprotein of microbial origin. Several molecules with similar properties have also been found: alamethicin, monazomycin, and DJ400B. The advantage of these more recently discovered materials is that they are simpler molecules.

Alamethicin is a representative nonohmic membrane ionophore. It is a polypeptide of 18 amino acids produced by *Trichoderma viride*. Alamethicin's conductance properties are shown in Figure 2.14. The current–voltage (*I–V*) curve is very similar to those of neural membranes. At low membrane potentials, the conductance is very low (high resistance). As the potential is increased, the conductance dramatically increases with an inflection point at 60 mV.

The conductance of alamethicin-treated bilayers is dependent on alamethicin's concentration. The conductance increases according to the 5th to 10th power of

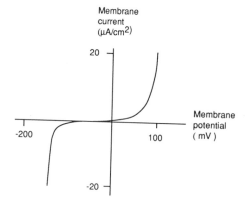

Figure 2.14. Current–voltage curve of alamethicin. The membrane current and voltage are shown along the ordinate and abscissa, respectively. Experiments were conducted using alamethicin in planar lipid bilayers. (Redrawn with permission from *Nature* **217**:713, copyright © 1968 by Macmillan Magazines Ltd.)

alamethicin concentration. This is similar to the behavior previously described for amphotericin B. Several alamethicin molecules are therefore required for the assembly of each alamethicin channel.

The insertion–aggregation model has been proposed by Mueller to account for the voltage-dependent formation of alamethicin channels (Figure 2.15). During resting conditions the monomers are associated with a membrane's hydrophilic face. When a potential is applied to a membrane, alamethicin molecules are driven into the membrane. The molecule's long axis orients perpendicular to the membrane surface. Since alamethicin has a net dipole moment (one end is electronegative whereas the other is electropositive), its electropositive end is forced into a membrane toward an inside-negative membrane potential. Concomitantly, several additional polar moieties are also driven into the hydrocarbon region. Lateral interactions among several alamethicin molecules sequester the polar residues away from the hydrocarbon environment. A polar transmembrane channel is thereby formed. When the transmembrane potential is removed, the alamethicin molecules relax to their original positions.

Recent studies of both ohmic and nonohmic ionophores have focused on constructing oligomers from unnatural amino acids. With only a few rare exceptions, such as gramicidins, oligopeptides and polypeptides are synthesized from L-amino acids. For example, Wade *et al*. (1990) have synthesized channel-forming antibiotics from D-amino acids. Although they were mirror images of their naturally occurring counterparts, they retained complete membrane channel-forming ability. This suggests that channel forma-

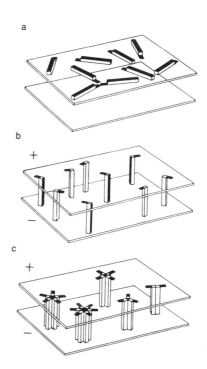

Figure 2.15. Insertion–aggregation model of alamethicin action. (a) Alamethicin molecules are associated with the periphery of a membrane at rest. (b) Application of a membrane potential drives the molecules into the membrane. (c) Alamethicin molecules interact laterally to form stable channels. (Reprinted with permission from Baumann and Mueller, *J. Supramol. Struct.* **2**:538, copyright © 1975 by Wiley–Liss, a division of John Wiley & Sons, Inc.)

tion does not require direct molecular interactions with chiral membrane components, i.e., a hydrophobic membrane is sufficient. Antibiotics constructed from L-amino acids might be clinically useful since they may not be recognized by naturally occurring (D-amino acid-containing) hydrolytic enzymes, thereby potentially increasing their lifetime and potency *in vivo*.

2.3. MEMBRANE PROTEINS

While lipids form the structural barrier of membranes, proteins generally contribute to their functional attributes. Membrane proteins act as enzymes, receptors, pumps, and channels in biological membranes. Some membrane proteins, particularly those facing the extracellular milieu, are glycosylated. Most membranes have a protein-to-lipid weight ratio of 1:1 to 3:1. For example, plasma membranes have a protein to lipid ratio of 1:1 whereas mitochondrial membranes have a ratio of 3:1.

2.3.1. The Diversity of Membrane Proteins

Membrane proteins are attached to lipid bilayers in a variety of ways. These associations are operationally defined as integral, peripheral, or glycophospholipid-linked. These classes of membrane proteins are illustrated in Figure 2.16. Peripheral (or extrinsic) membrane proteins are removed from bilayers by relatively gentle conditions. Treatment of membranes with hypotonic solvents, high-ionic-strength media, or removal of divalent cations is often sufficient to release peripheral membrane proteins. Cytoskeletal proteins are a major class of peripheral membrane proteins; they will be discussed in Section 3.2. On the other hand, integral (or intrinsic) proteins require harsh conditions for their extraction from membranes. Amphipathic molecules such as detergents are used to disrupt or solubilize a bilayer. Many integral membrane proteins span the bilayer. Glycophospholipid-linked membrane proteins were discovered by Low and

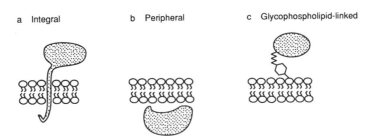

Figure 2.16. Types of membrane proteins. Membrane proteins can be classified into three types based on operational criteria. (a) Integral membrane proteins require membrane disruption for their extraction. (b) Peripheral proteins can be dislodged from a membrane by relatively mild treatments. (c) Glycophospholipid-linked proteins can be released from a membrane by exposure to specific phospholipases.

others using a phosphatidylinositol-specific phospholipase C. Several membrane proteins are released from their membranes by this enzyme (Table 2.2). Glycophospholipid-linked proteins behave as though they are integral proteins yet the polypeptide chain is not in direct contact with a bilayer. Due to their unique association with membranes, glycophospholipid-linked proteins merit a separate classification. All membrane proteins are asymmetrically oriented in native biological membranes, but not necessarily in reconstituted membranes. Carbohydrate residues are always associated with the *trans* (or outer) face of a bilayer. Peripheral membrane proteins are always associated with just one face of the membrane (frequently the *cis* or inner face). The orientation of an individual integral membrane protein's NH_2- and COOH-termini in biological membranes is always the same.

Table 2.3 lists many of the currently sequenced membrane proteins. These data are derived from protein and/or DNA sequencing experiments. The known or suggested positions of NH_2- and COOH-termini are given. The NH_2- and COOH-termini of membrane proteins can be present on either side of a membrane. The proposed transmembrane dispositions of several selected membrane proteins are schematically shown in Figure 2.17.

Membrane transport proteins have multiple membrane-spanning domains. Membrane channels are formed by several properly aligned transmembrane domains. Membrane transport proteins generally have hydrophilic, often charged, amino acid residues buried within a bilayer. Theses hydrophilic residues may form part of an active site and/or a polar transmembrane cavity. Typically, membrane channels have several subunits, each spanning a membrane several times. For example, gap junctions have six identical subunits. Each subunit spans a membrane four times yielding 24 transmembrane domains per channel. The large number of membrane-spanning domains likely accounts for the wide (1.6 nm) diameter of each channel. Molecules up to 1000 Da can pass through these pores. Smaller channels have fewer transmembrane domains. Enzymatic pumps that transport electrons or protons across a bilayer typically have 10 to 12 transmembrane segments. However, enzymes that act on a substrate at only one membrane face may possess one or zero transmembrane domains. Some membrane proteins that transduce information across a bilayer, such as hormone receptors, have just one transmembrane domain in monomeric form. However, others such as the adrenergic receptors have seven transmembrane domains. Conformational changes are

Table 2.2.

Examples of Lipid-Containing Membrane Proteins

Glycophospholipid-linked proteins	Integral proteins with covalently associated lipids
Thy-1	Lipoprotein
Acetylcholinesterase	HLA
Variable surface glycoprotein	Lipophilin
p63	VSV G protein
5'-nucleotidase	Sindbis virus envelope glycoprotein
Alkaline phosphatase	Transferrin receptor
	Insulin receptor

Table 2.3.
Summary of Many Sequenced Membrane Proteins[a,b]

	$M_r{}^c$	Trans-membrane domains	Terminus[d] N	C	Literature[e]
Structural proteins					
M-13 coat protein	5,260	1	t	c	PNAS **73:**1159
M2 protein (influenza virus)	15,000	1	t	c	Cell **40:**627
E. coli lipoprotein	7,000		t	c	PNAS **69:**970
Glycophorin	29,000	1	t	c	PNAS **72:**2964
Myelin P_0 protein	28,000	1	t	c	Cell **40:**501
Bacteriochlorophyll a binding protein	4,900				FEBS Lett. **191:**34
Chlorophyll a/b-binding protein (pAB90)	15,000		t	nd	JBC **258:**1399
Lens fiber major intrinsic protein	26,000	6	c	c	Cell **39:**49
Myelin basic protein (rat, small sub-unit)	14,000	0	c	c	Cell **34:**799
Myelin-associated glycoprotein	100,000	1?			PNAS **84:**600
Semliki Forest virus					
E1 glycoprotein	49,000	0	t	t	Nature **288:**236
E2 glycoprotein	52,000	1	t	c	Nature **288:**236
Sendai virus fusion protein					J. Gen. Virol **66:**317
Simian virus fusion protein					Cell **48:**441
VSV G protein	70,000	1	t	c	PNAS **77:**3884
Integrin	89,000	1	t	c	Cell **46:**271
17-kDa protein of cytochrome b_6–f complex	17,000	3	c	t	PNAS **81:**674
Heymann nephritis antigen	330,000	?			Science **244:**1163
Amyloid A4 precursor	91,500	1	t	c	EMBO J. **7:**949
VAMP-1	13,000	1	t	c	PNAS **85:**4538
Pore					
Matrix porin	36,500	8	c	c	PNAS **76:**5014
α-toxin, Staphylococcus aureus		1			Immunology **46:**615
Nuclear pore gp210	204,205	2	c	t	JCB **108:**2083
Transport					
Bacteriorhodopsin	26,000	7	t	c	FEBS Lett. **100:**219
Lac carrier	46,500	12	c	c	Nature **283:**541
Melibiose carrier	52,029	10			JBC **259:**4320
EII mannitol permease	67,896	7	t	c	JBC **258:**10761
Band 3 (murine)	95,000	12	c	c	Nature **316:**234
Ca^{2+}-ATPase (rabbit muscle)		10	c	c	Nature **316:**696
Na^+/K^+-ATPase (sheep kidney)					
α	100,000	8			Nature **316:**691
β	55,000				
Na^+/H^+ antiporter	38,683	10		?	JBC **263:**10408
Na^+/Ca^{2+} exchanger	108,000	12	c	c	Science **250:**562
Sodium channel	250,000	?		?	Science **237:**744
Potassium channel	70,200	7	c?	t	Science **237:**770
Slow voltage-dependent K^+ current	14,698	1	t	c	Science **242:**1042
Ars B protein	15,030	7	c	t	JBC **261:**15030
Synaptophysin	38,000	4	c	c	Science **238:**1142

Table 2.3.

(*Continued*)

	$M_r{}^c$	Trans-membrane domains	Terminus[d] N	C	Literature[e]
Glucose transporter (murine RBC)	55,000	12	*t*	*t*	*Science* **229**:941
Gap junction protein	27,000	4			*JCB* **103**:123
ADP/ATP translocator (yeast mito-chondria)					*Mol. Cell Biol.* **6**:626
Chromaffin granule H^+-ATPase	31,495	1			*JBC* **263**:17638
Histidine transport complex					
P protein	28,738	0?			*Nature* **298**:723
Q protein	24,573	3			*Nature* **298**:723
M protein	26,484	3			*Nature* **298**:723
Dopamine transporter		12	*c*	*c*	*Science* **254**:578
Serotonin transporter	73,000	12	*c*	*c*	*Science* **254**:579
Noradrenaline transporter	69,000	12	*c*	*c*	*Nature* **350**:350
GABA transporter	67,000	12	*c*	*c*	*Science* **249**:1303
Enzyme/energy transduction					
H^+-ATPase (*E. coli*)[f]					
F_0, DCCD-binding protein c	8,000	2	nd		*ARBB* **11**:445
F_0, protein a	30,260				*ARB* **52**:801
F_0, protein b	17,220	1			*ARB* **52**:801
Cytochrome b (yeast)	42,000	9			*PNAS* **81**:674
Cytochrome b_6					
Cytochrome b_5 (porcine liver)	16,000	0	*c*	*c*	*PNAS* **74**:3725
Cytochrome c oxidase (yeast)					
Subunit 1	56,000				*JBC* **255**:11927
Subunit 2	28,480				*JBC* **254**:9324
Subunit 3	30,340				*JBC* **255**:6173
Subunit 4	14,570		?		*EMBO J.* **3**:2831
Subunit 5	14,858	1			*JBC* **260**:9513
Subunit 7	6,603	1			*JBC* **261**:9206
Subunit 7a	6,303	1			*JBC* **261**:17183
Subunit 8	5,364				*JBC* **259**:6571
Coenzyme QH_2–cytochrome c reductase					
Core protein (yeast)	44,000	0?			*JBC* **261**:17163
14k protein	14,000	1?			*EJB* **138**:169
Cytochrome b	42,000	9			*PNAS* **81**:674
NADH dehydrogenase (*E. coli*)	47,300				*EJB* **116**:165
Cytochrome f (pea)	39,000	1	*c*	*t*	*Cell* **36**:555
3-Hydroxy-3-methylglutaryl coenzyme A reductase (hamster)	97,000	7	*t*	*c*	*JBC* **260**:522
Pro-sucrase-isomaltase	260,000	1	*c*	*t*	*Cell* **46**:227
Leader peptidase	35,994	1	*c*	*t*	*JBC* **258**:12073
VSG lipase	40,760	?			*PNAS* **85**:8914
A1 protein of PSI (P_{700} chlorophyll a)	83,200	11			*JBC* **260**:1413
A2 protein of PSI (P_{700} chlorophyll a)	82,500	11			*JBC* **260**:1413

Continued

Table 2.3.
(*Continued*)

	$M_r{}^c$	Trans-membrane domains	Terminus[d] N	Terminus[d] C	Literature[e]
Cytochrome b-559 of PSII					
Subunit 1	9,400	1			*FEBS Lett.* **176**:239
Subunit 2	4,400	1			
HB protein of PSII	38,900	7	*c*	*t*	
D2 protein of PSII	39,500	7	*c*	*t*	*NAR* **12**:8819
44 k protein of PSII	51,800	7	*c*	*t*	*NAR* **12**:8819
51 k protein of PSII	56,000	7	*c*	*t*	*NAR* **12**:8819
Ribulose 1,5-bisphosphate carboxylase small subunit	15,000				*JBC* **258**:1399
Penicillinase (membrane-bound)	32,000				*PNAS* **78**:3506
UDP-glucuronosyltransferase	52,000	1	*t*	*c*	*Biochem. J.* **242**:581
β-1,4-galactosyltransferase	44,416	1	*c*	*t*	*JBC* **263**:10420
Dolicholphosphate mannose synthase	29,962	1 or 2	?		*JBC* **263**:17499
α-Glutamyltranspeptidase					
Heavy chain	41,650	1	*c*	*t*	*PNAS* **85**:8840
Light chain	19,750	0	*t*	*t*	*PNAS* **85**:8840
Adenylyl cyclase	120,000	6	*c*	*c*	*Science* **244**:1558
Neuraminidase (influenza virus)	50,087	1			*Nature* **290**:213
Cytochrome b-561	30,061	6	*c*	*c*	*EMBO J.* **7**:2693
Cytochrome b-563		5	*c*	*t*	
Receptor/recognition					
Hemagglutinin (influenza virus)	224,640	1	*t*	*c*	*Cell* **19**:683
Rhodopsin (bovine)	38,000	7	*t*	*c*	*Cell* **34**:807
LDL receptor	160,000	1	*t*	*c*	*Cell* **37**:577
Insulin receptor	120,000	1	*t*	*c*	*Cell* **40**:747
Insulinlike growth factor	250,000	1	*t*	*c*	*Nature* **329**:301
Platelet-derived growth factor type 2 receptor	120,000	1	*t*	*c*	*Science* **243**:800
Basic fibroblast growth factor receptor		1	*t*	*c*	*Science* **245**:57
Erythropoietin	55,000	1	*t*	*c*	*Cell* **57**:227
Prolactin receptor	41,000	1	*t*	*c*	*Cell* **53**:69
Endothelin receptor	48,516	7	*t*	*c*	*Nature* **348**:730
HLA antigen					
Class I	66,000	1	*t*	*c*	*PNAS* **76**:4395
Class II	62,000	1			
Leukocyte common antigen	200,000	1	*t*	*c*	*Cell* **41**:83
Carcinoembryonic antigen	180,000	0	*t*	*t*	*PNAS* **84**:2965
Thy-1 antigen	13,500	0	*t*	*t*	
CD2 (T11)	50,000	1	*t*	*c*	*PNAS* **84**:2941
CD4	51,000	1	*t*	*c*	*Cell* **42**:93
CD8 antigen					
37-kDa chain (rat)	37,000	1	*t*	*c*	*Nature* **323**:74
32-kDa chain (rat)	32,000	1	*t*	*c*	*Nature* **323**:74
CD19	51,800	1	*t*	*c*	*JEM* **168**:1205
CD20 (B1)	33,000	3?	*t*?	*c*	*PNAS* **85**:208
CD28	23,000	1	*t*	*c*	*PNAS* **84**:8573

Table 2.3.
(*Continued*)

	$M_r{}^c$	Trans-membrane domains	Terminus[d] N	C	Literature[e]
Fc receptor for IgG2b/1	30,040	1			*Science* **234**:718
Fc receptor, type 3 (human)	26,000	0	*t*	*t*	*Nature* **333**:568
Fc receptor for IgE	25,000	1			*Cell* **47**:657
Interleukin-1 receptor	64,598	1	*t*	*c*	*Science* **241**:585
Interleukin-2 receptor	55,000	1	*t*	*c*	*Nature* **311**:631
B chain	58,000	1	*t*	*c*	*Science* **244**:551
Interleukin-3 receptor	94,723	1	*t*	*c*	*Science* **247**:324
Interleukin-4 receptor	140,000	1	*t*	*c*	*Cell* **59**:335
Lymphocyte homing receptor	90,000	1	*t*	*c*	*Science* **243**:1165
Tumor necrosis factor	26,000	1			*Cell* **53**:45
Epidermal growth factor receptor	132,000	1	*t*	*c*	*Nature* **309**:418
Nicotinic acetylcholine receptor					
α	40,000	4	*t*		*Nature* **299**:793
β	50,000	4	*t*		*Nature* **301**:251
γ	60,000	4	*t*		*Nature* **301**:251
δ	65,000	4	*t*		
Muscarinic acetylcholine receptor	51,670	7	*t*	*c*	*FEBS Lett.* **209**:367
Muscarinic type M2 receptor	51,700	7	*t*	*c*	*Science* **236**:600
β_2-adrenergic receptor	64,000	7	*t*	*c*	*PNAS* **84**:46
N-methyl-D-aspartate receptor	105,500	4			
Thyrotropin receptor	82,000	7	*t*	*c*	*Science* **246**:1620
Lutropin-choriogonadotropin receptor	75,000	7	*t*	*c*	*Science* **245**:494
Substance K receptor	43,066	7	*t*	*c*	*Nature* **329**:836
Serotonin 1c receptor	51,899	7	*t*	*c*	*Science* **241**:558
Dopamine D2 receptor		7	*t*		*PNAS* **86**:9762
cAMP	44,243	7	*t*	*c*	*Science* **244**:1558
Fasciclin II	97,000	1	*t*	*c*	*Science* **242**:700
α-factor	47,000	7			*PNAS* **85**:3855
a-factor		7			*PNAS* **83**:1418
Natriuretic peptide (guanylyl cyclase B)	115,085	1	*t*	*c*	*Cell* **58**:1155
IgM membrane-bound	86,000	1	*t*	*c*	*Cell* **20**:303
Ly-1	67,000	1	*t*		*PNAS* **84**:204
Poly Ig receptor	82,041	1	*t*	*c*	*Nature* **308**:37
T cell receptor	35,000	1	*t*	*c*	*Nature* **308**:149
Asialoglycoprotein receptor (RHL-1)	36,000	1	*c*	*t*	*JBC* **259**:770
Scavenger receptor type 1	49,975	1	*c*	*t*	*Nature* **343**:531
Scavenger receptor type 2	38,388	1	*c*	*t*	*Nature* **343**:570
Transferrin receptor	90,000	1	*c*	*t*	*Nature* **311**:675
Chicken hepatic lectin	26,000	1	*c*	*t*	*JBC* **256**:5827
Thrombomodulin	100,000	1	*t*	*c*	*JCB* **103**:1635
N-cell adhesion molecule (major, 1d)	160,000	1	*t*	*c*	*PNAS* **83**:3037
L-cell adhesion molecule	124,000	1	*t*	*c*	*PNAS* **84**:2808
Endothelial leukocyte adhesion molecule	64,000	1	*t*	*c*	*Science* **243**:1160

Continued

Table 2.3.

(Continued)

	$M_r{}^c$	Trans-membrane domains	Terminus[d] N	C	Literature[e]
Nerve growth factor receptor (fast)	83,000	1	t	c	*Nature* **325**:593
Contactin	130,000	1	t	c	*JCB* **107**:1561
Bile/canaliculus glycoprotein	110,000	1	t	c	*PNAS* **84**:7962
Ryanodine	425,000	4	t	t	*Nature* **339**:439
Cholera toxin transcriptional activator protein (Tox R protein)	32,527	1	c	t	*Cell* **48**:271
dna B protein (*B. subtilis*)	55,000			t?	*PNAS* **84**:657
lamp A	120,000	1	t	c	*PNAS* **85**:3743
Nod C	46,800	1	t	c	*EMBO J.* **7**:583
Mot B	60,000	1?	c	t	*Science* **239**:276
Zipper	56,000	1	t	c	*EMBO J.* **7**:1115
Sevenless gene product	232,000	1	t	c	*Science* **236**:55

*a*Only one representative example of each protein is listed. In many cases a particular protein has been sequenced in several species.

*b*Sequence information regarding oncogenes and secretory pathway proteins can be found in Chapters 9 and 8, respectively.

cM_r was determined using the deduced amino acid sequence or protein sequence data.

*d*N, NH$_2$-terminus; C, COOH-terminus; t, *trans* face; c, *cis* face; nd, not determined.

*e*Journal abbreviations: ARB, *Annu. Rev. Biochem.*; ARBB, *Annu. Rev. Biophys. Bioeng.*; EJB, *Eur. J. Biochem.*; JBC, *J. Biol. Chem.*; JCB, *J. Cell Biol.*; JEM, *J. Exp. Med.*; NAR, *Nucleic Acids Res.*; PNAS, *Proc. Natl. Acad. Sci. USA.*

*f*Additional sequences of peripheral membrane proteins associated with the H$^+$-ATPase can be found in *Annu. Rev. Biophys. Bioeng.* **11**:445.

presumably important in the transfer of information. Some membrane receptor proteins, such as acetylcholine receptors, have 20 transmembrane domains. In this case the information is carried across a membrane in the form of an ion flux, thus accounting for its many domains.

In addition to primary sequences, the diversity of membrane proteins is also expressed by their covalent chemical modifications. Included among the frequent covalent modifications are: glycosylation, phosphorylation, methylation, acetylation, proteolytic processing, and fatty acyl linkage. Covalent modifications are frequently performed by enzymes that recognize specific amino acid sequences within the primary structure of a membrane protein (e.g., Table 2.4). Glycosylation will be considered in Section 3.1. We shall return to phosphorylation of membrane proteins in our discussions of energy transduction and transmembrane signaling. In general, proteolytic processing of membrane proteins involves at least the removal of its signal sequence, a series of NH$_2$-terminal amino acids that specify a protein's delivery to a membrane (Section 8.2). Fatty acyl groups have been found in association with many membrane proteins; this includes both glycophospholipid-linked and integral membrane proteins. Integral membrane proteins can possess both a hydrophobic polypeptide domain and a hydrophobic fatty acyl residue (Table 2.2). The fatty acyl group of these proteins may be a signal for their incorporation into membranes (Schlesinger *et al.*, 1980) and/or insulate the bilayer from the polypeptide chain.

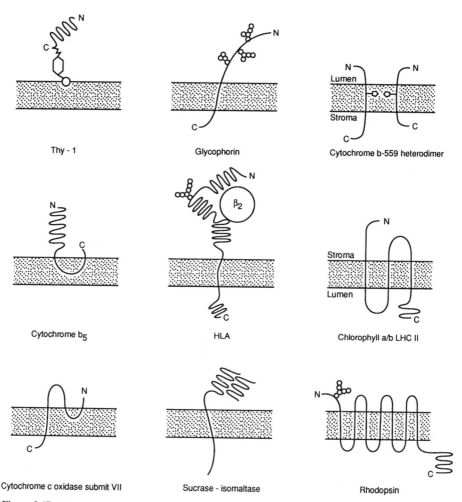

Figure 2.17. A gallery of models for the disposition of proteins in membranes. Nine membrane proteins with various structural and functional properties are illustrated.

2.3.2. Analysis of Membrane Protein Sequences

It has become quite popular to display the hydropathic character of membrane proteins using hydropathy plots (Kyte and Doolittle, 1982). In this method every amino acid is assigned a value of hydrophobicity. At each point along the sequence an average hydropathy is calculated for a predetermined interval of amino acids in the vicinity of that point. This process is illustrated in Figure 2.18. This procedure is repeated in stepwise fashion throughout a protein's entire length. The resulting plot has peaks and troughs corresponding to regions of high hydrophobicity and hydrophilicity, respec-

Table 2.4.
Possible Peptide Signals and Modification Sites

Sequence[a]	Function
20 hydrophobic residues	Membrane-spanning domain
Basic residues at *cis* face	Stop-transfer and/or membrane anchor
Arg-X-Ser (or Thr)	*N*-linked carbohydrate
Ser or Thr	*O*-linked carbohydrate
Ser, Thr, or Tyr	Phosphorylation
NH$_2$-terminal or ϵ-amino	Acetylation
Lys-Arg-X-X-Ser-X or Arg-Arg-X-Ser-Y	Protein kinase
Leu-Ala-Gly-Cys	Glyceride-Cys
gly-X-Gly-X-X-Gly . . . X$_{17}$ . . . Lys	ATP binding
Leu-X-X-X-X-X-Leu . . .	Leucine zipper

[a]X represents any amino acid.

tively. This procedure aids in predicting protein regions possessing transmembrane domains.

The major hydrophobic domain of proteins possessing one transmembrane domain is generally easy to recognize since they have one very hydrophobic segment 20 to 25 amino acids in length. However, additional functional groups such as quinones, hemes, chlorophylls, and iron–sulfur groups associated with transmembrane domains influence their hydrophobicity. Furthermore, proteins with multiple membrane-spanning segments often possess charged residues buried within the bilayer. One consequence of

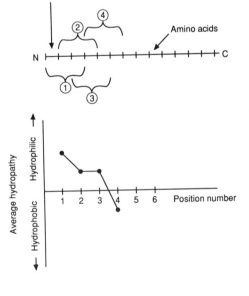

Figure 2.18. Calculating a hydropathy plot. The procedures employed to determine a hydropathy plot are illustrated. Each amino acid is assigned a value of hydropathy based on its physical properties. An average hydropathy for an interval of amino acids (three in this figure) is calculated (upper panel, position 1) and plotted (lower panel). The process is repeated after moving one amino acid toward the COOH-terminus (position 2). The hydropathy is again plotted. The procedure is repeated until the entire polypeptide has been analyzed.

the varied sequence positions of hydrophobic and hydrophilic residues is that the number and positions of transmembrane domains can be in error. For example, the number of transmembrane domains of the nicotinic acetylcholine receptor has been contested. The presence of extensive intramembrane β-structure further complicates an analysis (Section 2.3.3). Nucleotide data analysis may also be flawed by the assignment of nucleotide sequences corresponding to an immature protein to its mature counterpart. For example, a COOH-terminal sequence of Thy-1, a glycophospholipid-linked membrane protein, was thought to be a hydrophobic transmembrane domain, although it is not associated with the mature protein (Tse *et al.*, 1985).

2.3.3. Intramembrane Conformations of Membrane Proteins

We will now consider the secondary structures of polypeptide chains residing *within* a bilayer's hydrophobic core. As we have already learned, the thermodynamically favored location of hydrophilic molecules is in an aqueous medium whereas hydrophobic molecules are sequestered into a hydrocarbon environment. Frequently, but not always, transmembrane domains are rich in hydrophobic amino acid residues. In addition to side-chain residues, a protein's polar peptide bonds also contribute to its hydropathic character. Peptide bonds are sequestered from the hydrophobic core by formation of hydrogen bonds. An α-helix is formed by hydrogen bonding within one polypeptide strand (Figure 2.19). Roughly 20 amino acids are required to cross a membrane's hydrophobic core. About 30 residues (approximately 4 kDa) are needed to transverse an entire membrane. As suggested by sequencing data and direct physical measurements, α-helices are the predominant transmembrane conformation of proteins.

Although α-helices are the rule, there are several illuminating exceptions. It has been reported that β-structures are found in association with M-13 coat protein and porin. The data regarding porin are especially strong; we shall return to this protein in Section 2.3.4.2. β-sheets need only nine amino acid residues to cross a bilayer's

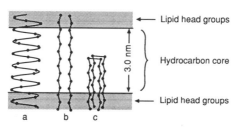

Figure 2.19. Possible intramembrane conformations of proteins. Possible conformations of membrane proteins traversing (a, b) and penetrating (c) a bilayer membrane are shown. The solid circles represent α-carbon atoms of constituent amino acids. The conformations are: (a) α-helix, (b) antiparallel β-structure, and (c) β-structure with a reverse turn. Four or more strands of β-structure are required to satisfy all hydrogen bonding requirements. (Redrawn with permission from Tanford, *The Hydrophobic Effect*, copyright © 1980 by John Wiley & Sons, Inc.)

hydrocarbon region once (Figure 2.19). Since β-sheets require lateral hydrogen bonding between polypeptide chains, a protein must cross a membrane several times to form stable β-sheets.

The preponderance of the 20-amino-acid transmembrane domain and thermodynamic arguments have led to the suggestion that 20 amino acids are required to cross a bilayer. However, this supposed consensus is incorrect. Adams and Rose (1985) studied the transmembrane domain of vesicular stomatitis virus (VSV) G protein. The G protein gene was cloned with plasmids. Using site-directed mutagenesis, the DNA coding for the transmembrane domain was altered. Nucleotides corresponding to 2, 4, 6, 8, 12, and all 20 transmembrane amino acids were deleted in separate plasmids. Cells were transfected with a vector containing modified DNA. The expression of G protein in fibroblasts was studied with anti-G protein antibodies. Deletion of up to 6 transmembrane amino acids had no effect on the protein's cell surface expression. When 8 to 12 amino acids were deleted, the protein was not transported to the cell surface, although it remained in a transmembrane configuration. When the entire transmembrane domain was deleted, VSV G protein behaved as a secretory protein. Only 8 amino acids were capable of forming a transmembrane domain. Several candidate explanations can be considered: (1) the transmembrane domain adopted a random coil or β-sheet structure, (2) lipid packing became severely distorted near the protein to form an unusually thin bilayer, and (3) hydrophilic residues were forced into the bilayer. Whatever the correct explanation(s) might be, it is fair to say that 20 amino acids are simply not a physical requirement to establish a protein in a transmembrane configuration.

2.3.4. *E. coli* Membrane Proteins as a Prokaryote Model System

Prokaryotes have no internal membrane systems. As a consequence, many membrane-associated functions are restricted to the cytoplasmic or inner membrane of bacteria. For example, oxidative phosphorylation, lipid biosynthesis, membrane transport, and peptidoglycan synthesis are restricted to the inner membrane of gram-negative bacteria. Diffusion pores and LPS molecules are found in the outer membranes of gram-negative bacteria. Prokaryotes such as *E. coli* are advantageous experimental subjects because: (1) they are simple cells, (2) bacteria are easy to grow and manipulate, (3) genetic techniques are easily applied, and (4) their membranes are easy to isolate.

2.3.4.1. Lipoprotein

The first reported primary sequence of a membrane protein was that of *E. coli* lipoprotein (Braun and Bosch, 1972). Its name is rather unfortunate since it can be confused with serum lipoproteins, such as low-density lipoprotein, of higher organisms. Lipoprotein is exclusively found in the outer membrane of gram-negative bacteria ($M_r = 7000$). It is an abundant protein with roughly 7×10^6 copies per cell. Two-thirds of these molecules are not covalently linked to peptidoglycan. The remainder are linked via the COOH-terminal lysine's ϵ-amino group to a carboxyl group of *meso*-diaminopimelic acid of peptidoglycan. The NH_2-terminal amino acid, glycerylcysteine, is located at the *trans* face of outer membranes. Two fatty acids are esterified to this residue. Lipoprotein's NH_2-terminus is blocked by an amide-linked fatty acid. The most commonly occurring fatty acids of lipoprotein are palmitic, palmitoleic, *cis*-vaccenic, cyclopropylenehexadecanoic, and cyclopropyleneoctadecanoic acid. The NH_2-terminal cysteine residue of the mature protein is the 21st residue of its precursor protein, prolipo-protein. The sequence Leu-Ala-Gly-Cys (positions 18–21) of prolipoprotein is believed to be recognized by enzymes participating in the formation of glyceride-Cys (Table 2.4; Lai *et al.*, 1981).

Lipoprotein is rich in α-helical content as indicated by both infrared spectroscopy and circular dichroism. The sequence positions of hydrophobic amino acids within lipoprotein form a repeating pattern every three or four residues. Since each turn of a right-handed α-helix possesses 3.6 residues, two hydrophobic faces are formed (Inouye, 1974), which may be in contact with the outer membrane's hydrophobic core. Chemical cross-linking studies have indicated that lipoprotein exists as a dimer or trimer *in situ*. Lateral interactions between lipoprotein molecules may sequester polar sites from hydrophobic portions of the outer membrane. Lipoprotein molecules are thought to interact with porin. Lipoprotein likely plays an important role in stabilizing outer membrane structure. *E. coli* mutants lacking lipoprotein are very sensitive to low Mg^{2+} concentrations, possess leaky outer membranes, and demonstrate morphological defects.

2.3.4.2. Porin

Porin is a large transmembrane protein that permits the flow of small molecules (<900 Da) across the outer membrane of *E. coli*. There are 1.1×10^6 molecules of porin per cell. Porin, or matrix protein as it is also known, has 336 amino acid residues and a molecular mass of 36.5 kDa. Two forms of this protein, designated Ia and Ib, have been identified by electrophoretic mobility and peptide fragmentation studies. Porin is tightly but noncovalently bound to peptidoglycan. The porin–peptidoglycan association is disrupted by high salt concentration or head denaturation.

Porin's amino acid sequence has been determined (Table 2.3). The amino acid sequence is rather exceptional for an integral membrane protein because there is no contiguous series of 20 hydrophobic amino acids. This suggests that there are no putative transmembrane α-helices. However, β-structure has been observed by biophysical techniques and confirmed by x-ray diffraction (Weiss *et al.*, 1991). Figure 2.20 shows a proposed model of an intramembrane portion of porin. This model suggests that porin's transmembrane segment is made up of eight transmembrane β-strands. Although porin is assembled as a trimer, each monomer is believed to possess channel activity. Porin monomers may also interact with lipoprotein molecules.

Porin can be reconstituted into liposomes, lipid vesicles, and planar bilayer membranes. Reconstitution is generally carried out in the presence of phospholipid and LPS. Porin greatly increases the permeability of lipid vesicles to sucrose and other low-

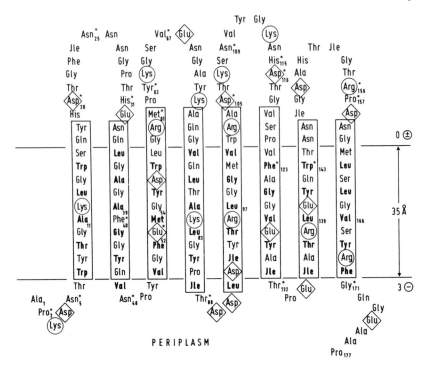

Figure 2.20. Structural model of a portion of porin. A porin fragment from residues 1 to 117 is shown. The large rectangles are transmembrane β-structures. Asterisks designate β-turns, circles and diamonds denote positively and negatively charged residues, respectively. [Reprinted with permission from Jahnig, in: *Prediction of Protein Structure and the Principles of Protein Conformation* (G. Fasman, ed.), copyright © 1989 by Plenum Press.]

molecular-weight compounds. The morphology of reconstituted membranes is most similar to the outer membrane of *E. coli* when lipoprotein is included. When reconstituted into planar lipid bilayer membranes, no increase in transmembrane conductance was observed. However, when an applied transmembrane potential exceeds a threshold value (V_i), a transmembrane current is established. The initiation process is irreversible; once a current is established, potentials below V_i can elicit a current. Figure 2.21 shows a current–voltage curve of porin in planar bilayer membranes after initiation. The response is linear between −140 and +140 mV. The current declines below −140 or above +140 mV suggesting that the channels close.

2.3.5. Erythrocyte Membrane Proteins as a Eurkaryote Model System

In general, eukaryotic cells possess numerous intracellular membranes. However, most types of mature erythrocytes possess no internal membranes. This makes

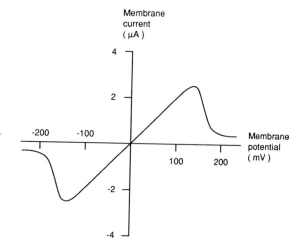

Figure 2.21. Current–voltage curve of porin. Membrane current and voltage are shown along the ordinate and abscissa, respectively. Porin was reconstituted into planar lipid bilayer membranes. (Redrawn from Schindler and Rosenbusch, 1978, *Proc. Natl. Acad. Sci. USA* **75**:3751.)

erythrocytes particularly amenable to experimental study. We will now discuss erythrocyte membrane proteins with particular emphasis on glycophorin and band 3.

Figure 2.22 shows sodium dodecyl sulfate–polyacrylamide gel electrophoresis (SDS-PAGE) profiles of red blood cell membrane proteins. Proteins are visualized with Coomassie blue staining whereas carbohydrates are located with periodic acid–Schiff (PAS) staining (see Section 3.1.1). The erythrocyte's peripheral membrane proteins are all associated with the bilayer's *cis* face. Bands 1 and 2 are subunits of spectrin, a major component of the erythrocyte's membrane skeleton. Actin, band 5, is another peripheral membrane protein participating in cytoskeletal structure. Band 6 is a subunit of glyceraldehyde 3-phosphate dehydrogenase. This peripheral protein participates in glycolysis and binds to band 3's NH_2-terminal domain. Band 2.1 is ankyrin; it binds to band 3 and acts as a bridge between band 3 and the cytoskeleton. Band 3 is an integral membrane protein that participates in anion transport. Band 4.5 is the glucose transporter. PAS-1 is an integral membrane protein called glycophorin. The locations of additional PAS bands are identified in Figure 2.22. The glycoprotein glycoconnectin, not visualized in this gel, may play a role in membrane–cytoskeleton interaction. The functions of the other proteins are not certain. Furthermore, there may be additional proteins not detectable by these methods.

2.3.5.1. Glycophorin

Glycophorin was the first eukaryotic membrane protein to be sequenced (Tomita and Marchesi, 1975). The basic concepts gleaned from these early data have been observed in almost all biological membranes. Glycophorin is composed of three major portions: extracellular, intramembrane, and intracellular. The NH_2-terminal domain is located at the *trans* face whereas the COOH-terminal region is at the *cis* face. Glycophorin is over 60% carbohydrate by weight. All of the approximately 100 carbohydrate

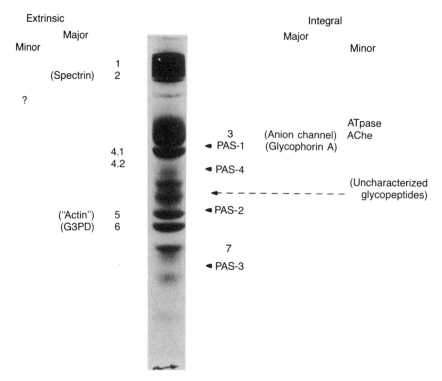

Figure 2.22. Gel electrophoresis of erythrocyte membrane proteins. SDS-PAGE analysis of human erythrocyte membrane proteins is shown. Membrane proteins were labeled with Coomassie blue. The locations of glycoproteins are revealed by treatment with PAS reagent, which is specific for carbohydrates. The identities of extrinsic membrane proteins are listed on the left whereas intrinsic membrane proteins are listed on the right. (Reprinted with permission from *Annual Review of Biochemistry* Vol. 45, copyright © 1976 Annual Reviews, Inc.)

residues are present in the NH_2-terminal domain. Carbohydrates in the NH_2-terminal portion contain ABO and MN blood group antigens and provide binding sites for influenza viruses.

Glycophorin's intramembrane portion is a sequence of 20 largely nonpolar amino acids. As mentioned above, this sequence likely forms a transmembrane α-helix. Immediately upon exiting a membrane's *cis* face, a series of four positively charged amino acids are found. This series of positively charged amino acids may interact with negatively charged phospholipids such as PS. This interaction may minimize transverse displacement of the protein in a membrane, thus serving as an anchor. The anchoring function may serve a second role as a "stop-transfer" sequence during glycophorin biosynthesis. The "stop-transfer" sequence is proposed to inhibit further translocation of a nascent polypeptide chain across a bilayer.

The first definitive evidence that membrane proteins could span a bilayer was provided by Bretscher (1971a,b) during studies of glycophorin and band 3. Intact erythrocytes and ruptured ghosts were radiolabeled with the reagent FMMP. When intact erythrocytes were labeled with FMMP, only two major proteins residing at the *trans* face were found using SDS-PAGE. However, 20 labeled membrane proteins were observed in broken ghosts since both *trans* and *cis* faces were accessible to FMMP. Clearly, more membrane proteins reside at the cytosolic face. After SDS-PAGE the band corresponding to glycophorin was physically cut from the polyacrylamide gel. Appropriate controls were performed to show that any contaminating proteins cut from the gel with glycophorin did not affect the results. Glycophorin was then digested with a protease. The small radioactive peptides were separated in a two-dimensional system which provides a fingerprint or map of glycophorin. Maps of glycophorin labeled from the *trans* or both faces of the membrane were compared. These experiments showed that unique peptide fragments of glycophorin were associated with the *cis* and *trans* faces of a bilayer. Therefore, since one protein can be labeled from either face of a membrane, the protein must extend through the membrane. Cotmore *et al.* (1977) employed the antigenic properties of glycophorin to define its asymmetric orientation and transmembrane configuration in erythrocyte membranes. In this study rabbits were immunized with the COOH-terminal 51 amino acids of human glycophorin A. This polyclonal antibody preparation did not react with intact erythrocytes, pronase-treated erythrocytes, or other mammalian cells. However, the antibodies were found to react with the COOH-terminal domain of glycophorin between amino acid residues 102 and 118. When linked to the electron microscopic label ferritin, the antibodies were found to exclusively label an erythrocyte's *cis* membrane face. Unconjugated antibodies compete with the ferritin-conjugated antibodies for binding sites; therefore, the binding is a property of the antibody, not the ferritin. Similarly, ferritin-conjugated concanavalin A binds only to an erythrocyte membrane's *trans* face. Since glycophorin is know to bind this lectin, it must be asymmetrically positioned in a bilayer.

2.3.5.2. Band 3

Band 3 is an erythrocyte integral membrane protein with a molecular mass of 95 kDa. It represents almost one-third of the erythrocyte membrane protein, there being over one million copies of band 3 in each RBC. Band 3 has a relatively low carbohydrate content, 8% by weight. Experiments utilizing covalent labeling and protease treatment indicate that band 3 is exposed at both membrane faces. It possesses two major domains: an NH_2-terminal region and a membrane-bound COOH-terminal domain. A proposed model of band 3 illustrating these domains is shown in Figure 2.23. Band 3's NH_2-terminal region is located at the cytoplasmic face of erythrocyte membranes. Its mass of 41 kDa is distributed throughout a highly extended structure. This structure contains at least three types of binding sites. Binding sites for hemoglobin and glycolytic enzymes are located within the first 11 amino acid residues. The phosphorylation state of Tyr-8 may regulate the association of glycolytic enzymes with human band 3. The binding site for the cytoskeletal protein ankyrin is located in the vicinity of residues 170 to 180.

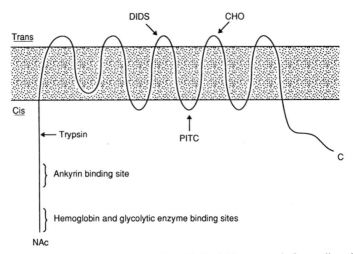

Figure 2.23. Postulated membrane topography of band 3. Band 3 is composed of cytosolic and membrane-bound domains. The NH_2-terminus is acetylated (Ac). The following sites are indicated: DIDS, stilbene disulfonate binding site; PITC, phenyl isothiocyanate reactive site; CHO, carbohydrate linkage site.

The membrane-associated region of band 3 is a 52-kDa fragment that contains the COOH-terminus. It is rich in α-helical structure. Various models based on cDNA sequencing studies have suggested 10, 12, or 14 transmembrane segments (e.g., Kopito and Lodish, 1985a,b; Jay and Cantley, 1986). The α-helices are probably arranged in the form of a cylinder. Models of band 3 suggest that each face possesses multiple positively charged amino acids. These charges likely affect the distribution of ions at both faces of a membrane: cations are repelled and anions are attracted. This would be expected to enhance the rate of anion transport in both directions.

Band 3 participates in erythrocyte anion exchange as shown in Figure 2.24. In this example, carbon dioxide permeates the membrane of erythrocytes within a tissue's capillaries by simple diffusion. Carbonic anhydrase catalyzes the production of bicarbonate and a proton. The released proton binds to hemoglobin, thus facilitating the release of oxygen to tissues. Band 3 exchanges a bicarbonate ion for an extracellular chloride ion. This reaction pathway is reversible in pulmonary capillary beds where the level of oxygen rises and carbon dioxide falls. Reagents such as stilbene disulfonates inhibit erythrocyte anion exchange by covalently labeling band 3. These reagents bind to the membrane-bound portion of band 3. This domain accounts for all of band 3's anion translocation ability.

The *in situ* structural organization of band 3 molecules has not been unambiguously established. However, evidence from cross-linking experiments suggests the presence of band 3 dimers and higher oligomers. Furthermore, dimers are found following solubilization of erythrocyte membranes with mild detergents. Current data indicate that anion translocation may be mediated by band 3 in any form whereas water permeation via band 3 may require the dimeric form.

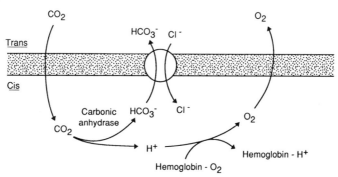

Figure 2.24. Band 3-catalyzed anion exchange. The molecular physiology surrounding the band 3-mediated exchange of bicarbonate and chloride ions is shown. These events take place in peripheral tissues. During conditions of high oxygen tension, the directions of the arrowheads are reversed. (Elements of this diagram are redrawn with permission from Kopito and Lodish, *J. Cell. Biochem.* **29:**1, copyright © 1985 by Wiley–Liss, a division of John Wiley & Sons, Inc.)

2.3.6. The LDL Receptor as a Model Recognition Structure

The cholesterol of animal cell membranes originates from foodstuffs or *de novo* synthesis by the liver. Because of its hydrophobic character, cholesterol is not soluble in aqueous bodily fluids. The serum protein low-density lipoprotein (LDL) transports cholesterol as cholesteryl esters to tissues throughout the body. The LDL receptor is a cell surface integral membrane glycoprotein that recognizes LDL and stimulates its endocytosis. The LDL is one of several cell surface molecules that trigger ingestion via coated pits.

The LDL receptor is a transmembrane glycoprotein with an apparent molecular mass of 175 kDa. The calculated molecular mass based on sequence data is 115 kDa. Yamamoto *et al.* (1984) have sequenced the cDNA coding for the human LDL receptor. A model of this receptor based on its primary sequence is shown in Figure 2.25. On the basis of antibody binding experiments, the NH_2-terminus has been localized to the plasma membrane's *trans* face. The deduced amino acid sequence suggests the presence of five major domains.

The LDL receptor's NH_2-terminal region is rich in cysteine residues. One in seven amino acids in this domain is cysteine. Most of these residues participate in the formation of disulfide bridges. The disulfide bonds are expected to hold the NH_2-terminus in a highly folded structure, as indicated in Figure 2.25. The eight loops represent the eight very similar amino acid sequence repeats of roughly 40 residues each in the NH_2-terminal domain. Seven of these loops contain short negatively charged segments. These segments likely bind to complementary positively charged regions of LDL.

The LDL receptor's second domain consisting of 350 amino acids possess considerable sequence homology with the epidermal growth factor precursor protein. These two proteins may have evolved from a common ancestral gene. The third region of the LDL

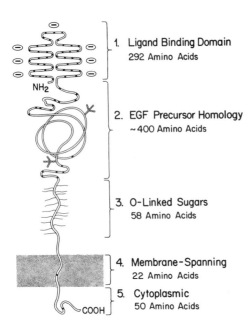

1. **Ligand Binding Domain**
   292 Amino Acids

2. **EGF Precursor Homology**
   ~400 Amino Acids

3. **O-Linked Sugars**
   58 Amino Acids

4. **Membrane-Spanning**
   22 Amino Acids

5. **Cytoplasmic**
   50 Amino Acids

Figure 2.25. Structural model of the LDL receptor. The five-domain structure of the LDL receptor is shown. The identity of each domain and its number of constituent amino acids is listed on the right. (Courtesy of Dr. M. Brown.)

receptor is rich in carbohydrate. This region likely includes two *N*-linked oligosaccharides and 9–18 *O*-linked carbohydrate chains. The receptor's fourth region is composed of 22 hydrophobic amino acid residues; these likely cross the membrane as an α-helix. The transmembrane disposition of this region has been confirmed by proteolysis experiments. The intracellular COOH-terminal portion of the LDL receptor consists of 50 amino acid residues. This domain interacts with clathrin molecules in coated pits.

2.3.7. Bacteriorhodopsin: An Example of a Membrane Protein Participating in Energy Transduction

Transmembrane electrochemical gradients participate in many biological processes. For example, in eukaryotes a proton gradient across the inner mitochondrial membrane drives ATP synthesis. The bacterial membrane protein bacteriorhodopsin is one of the most thoroughly studied proteins involved in membrane energy transduction. Bacteriorhodopsin is a light-driven proton pump isolated from the purple membrane of *Halobacterium halobium*. This important biological system was discovered by Stoeckenius in 1967. *H. halobium* possesses one bilayer membrane separating its cytoplasm from the cell wall. During oxygen starvation, bacteriorhodopsin is synthesized and accumulates within the membrane as purple patches. Bacteriorhodopsin converts light energy into a proton gradient across the membrane. The gradient is used to drive the synthesis of ATP. The purple membrane is 75% protein by weight. Bacteriorhodopsin, the sole protein of the purple membrane, forms a two-dimensional crystal. These native crystals have allowed extensive supramolecular studies of purple membranes.

Bacteriorhodopsin, a model membrane protein, consists of a single polypeptide chain of 26 kDa. Biophysical studies indicate that its secondary structure is largely α-helical (70–80%). Most of its mass is associated with seven transmembrane α-helices (Figure 2.26). Proteolysis experiments confirm that only several peptides extend beyond the membrane surface. As might be expected, the amino acid sequence demonstrates that most charged residues are outside the membrane. However, nine charged residues are apparently associated with the hydrophobic membrane core. At least some of these may form ion pairs within a membrane, thus minimizing unfavorable interactions. Figure 2.27 shows how such an ion pair might form along an intramembrane sequence of bacteriorhodopsin. The spacing of three or four main acids between positive and negative residues is optimal for intrahelix ion pair formation. Other charged residues may form interhelix ion pairs. Retinal is attached to the ε-amino group of an intramembrane lysine where it participates in proton pumping. We will discuss how bacteriorhodopsin pumps protons in Chapter 5.

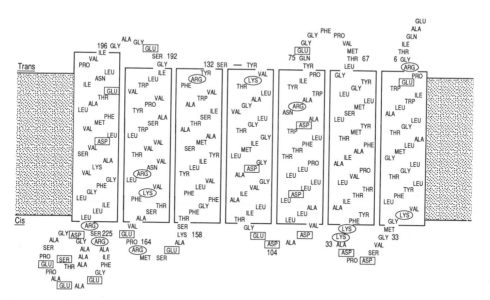

Figure 2.26. A proposed spatial arrangement of the polypeptide chain of bacteriorhodopsin. The seven α-helical transmembrane domains of bacteriorhodopsin may be arranged as shown. Positively and negatively charged amino acids are indicated by ellipses and boxes, respectively. Residue numbering begins at the NH₂-terminus. (Redrawn from Engelman *et al.*, 1980, *Proc. Natl. Acad. Sci. USA* **77**:2023.)

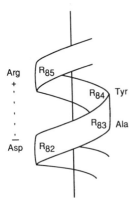

Arg
+
.
.
.
.
Asp

Figure 2.27. A proposed intramembrane ion pair of bacteriorhodopsin. Charged amino acids within the hydrophobic core of a membrane may be stabilized by forming ion pairs. A potential ion pair within bacteriorhodopsin is shown.

REFERENCES AND FURTHER READING

Lipids

Bretscher, M. S. 1972. Asymmetrical lipid bilayer structure for biological membranes. *Nature* **236:**11–12.

Buldt, G., and Wohlgemuth, R. 1981. The headgroup conformation of phospholipids in membranes. *J. Membr. Biol.* **58:**81–100.

Dahl, J. *et al.* 1980. Sterols in membranes: Growth characteristics and membrane properties of *Mycoplasma capricolum* cultured on cholesterol and lanoserol. *Biochemistry* **19:**1467–1472.

Demel, R. A., *et al.* 1977. The preferential interaction of cholesterol with different classes of phospholipids. *Biochim. Biophys. Acta* **465:**1–10.

Engleman, D. M., and Rothman, J. E. 1972. The planar organization of lecithin–cholesterol bilayers. *J. Biol. Chem.* **247:**3694–3697.

Hannun, Y. A., and Bell, R. M. 1989. Functions of sphingolipids and sphingolipid breakdown products in cellular regulation. *Science* **243:**500–507.

Johnston, N. C., and Goldfine, H. 1985. Phospholipid aliphatic chain composition modulates lipid class composition, but not lipid asymmetry in *Clostridium butyricum. Biochim. Biophys. Acta* **813:**10–18.

Karnovsky, M. J., *et al.* 1982. The concept of lipid domains in membranes. *J. Cell Biol.* **94:**1–6.

McConnell, H. M., *et al.* 1986. Supported planar membranes in studies of cell–cell recognition in the immune system. *Biochim. Biophys. Acta* **864:**95–106.

Miljanich, G. P., *et al.* 1981. The asymmetric transmembrane distribution of phosphatidylethanolamine, phosphatidylserine, and fatty acids of the bovine retinal rod outer segment disk membrane. *J. Membr. Biol.* **60:**249–255.

Montal, M., and Mueller, P. 1972. Formation of biomolecular membranes from lipid monolayers and a study of their electrical properties. *Proc. Natl. Acad. Sci. USA* **69:**3561–3566.

Op den Kamp, J. A. F. 1979. Lipid asymmetry in membranes. *Annu. Rev. Biochem.* **48:**47–71.

Rilfors, L. 1985. Difference in packing properties between iso and anteiso methyl-branched fatty acids as revealed by incorporation into the membrane lipids of *Acholeplasma laidlawii* strain A. *Biochim. Biophys. Acta* **813:**151–160.

Rogers, J., *et al.* 1979. The organization of cholesterol and ergosterol in lipid bilayers based on studies using non-perturbing fluorescent sterol probes. *Biochim. Biophys. Acta* **552**:23–37.

Rujanavech, C., and Silbert, D. F. 1986. LM cell growth and membrane lipid adaptation to sterol structure. *J. Biol. Chem.* **261**:7196–7203.

Somerville, C., and Browse, J. 1991. Plant lipids: Metabolism, mutants, and membranes. *Science* **252**:80–87.

Svennerholm, L. 1963. Chromatographic separation of human brain gangliosides. *J. Neurochem.* **10**:613–623.

Tanford, C. 1980. *The Hydrophobic Effect*. Wiley, New York.

Thompson, T. E., and Tillack, T. W. 1985. Organization of glycosphingolipids in bilayers and plasma membranes of mammalian cells. *Annu. Rev. Biophysics Biophys. Chem.* **14**:361–386.

Vance, D. E., and Vance, J. E. 1985. *Biochemistry of Lipids and Membranes*. Benjamin–Cummings, Menlo Park, Calif.

Oligomers

Anderson, O. S. 1984. Gramicidin channels. *Annu. Rev. Physiol.* **46**:531–548.

Bamberg, E., *et al.* 1977. Structure of the gramicidin A channel. Distinction between $\pi_{L,D}$ the β helix by electrical measurements with lipid bilayer membranes. *Proc. Natl. Acad. Sci. USA* **74**:2402–2406.

Baumann, G., and Mueller, P. 1975. A molecular model of electrical excitability. *J. Supramol. Struct.* **2**:538–557.

Christensen, H. N. 1975. *Biological Transport*. Benjamin, Reading, Mass.

Lear, J. D., *et al.* 1988. Synthetic amphiphilic peptide models for protein ion channels. *Science* **240**:1177–1181.

Morrow, J. S., *et al.* 1979. Transmembrane channel activity of gramicidin A analogs: Effects of modification and deletion of the amino terminal residue. *J. Mol. Biol.* **192**:733–738.

Toro, M., *et al.* 1987. Formation of ion-translocating ligomers by nigericin. *J. Membr. Biol.* **95**:1–8.

Urry, D. W., *et al.* 1971. The gramicidin A transmembrane channel: Characteristics of head-to-head dimerized $\pi(L,D)$ helices. *Proc. Natl. Acad. Sci. USA* **68**:1907–1911.

Wade, D., *et al.* 1990. All-D amino acid-containing channel-forming antibiotic peptides. *Proc. Natl. Acad. Sci. USA* **87**:4761–4765.

Wallace, B. A., and Ravikumar, K. 1988. The gramicidin pore: Crystal structure of a cesium complex. *Science* **241**:182–187.

Proteins

Adams, G. A., and Rose, J. K. 1985. Structural requirements of a membrane-spanning domain for protein anchoring and cell surface transport. *Cell* **41**:1007–1015.

Bordier, C., *et al.* 1986. *Leishmania* and *Trypanosoma* surface glycoproteins have a common glycophospholipid membrane anchor. *Proc. Natl. Acad. Sci. USA* **83**:5988–5991.

Braun, V., and Bosch, V. 1972. Repetitive sequences in the murein-lipoprotein of the cell wall of *Escherichia coli*. *Proc. Natl. Acad. Sci. USA* **69**:970–974.

Bretscher, M. S. 1971a. A major protein which spans the human erythrocyte membrane. *J. Mol. Biol.* **59**:351–357.

Bretscher, M. S. 1971b. Major human erythrocyte glycoprotein spans the cell membrane. *Nature New Biol.* **231**:229–232.

Chamberlain, B. K., *et al.* 1978. Association of the major coat protein of fd bacteriophage with phospholipid vesicles. *Biochim. Biophys. Acta* **510**:18–37.

Chen, R., *et al.* 1979. Primary structure of major outer membrane protein I of *Escherichia coli* B/r. *Proc. Natl. Acad. Sci. USA* **76**:5014–5017.

Ching, G., and Inuoye, M. 1986. Expression of the *Proteus mirabilis* lipoprotein gene in *Escherichia coli*. Existence of tandem promoters. *J. Biol. Chem.* **261**:4600–4606.

Cotmore, S., *et al.* 1977. Immunochemical evidence for a transmembrane orientation of glycophorin A: Localization of ferritin–antibody conjugates in intact cells. *J. Mol. Biol.* **113**:539–553.

Engelman, D. M., *et al.* 1980. Path of the polypeptide in bacteriorhodopsin. *Proc. Natl. Acad. Sci. USA* **77**:2023–2027.

Ferguson, M. A. J., *et al.* 1988. Glycosyl-phosphatidylinositol moiety that anchors *Trypanosoma brucei* variant surface glycoprotein to the membrane. *Science* **239**:753–759.

Gerber, G. E. 1977. Orientation of bacteriorhodopsin in *Halobacterium halobium* as studied by selective proteolysis. *Proc. Natl. Acad. Sci. USA* **74**:5426–5430.

Henderson, R. 1977. The purple membrane from *Halobacterium halobium*. *Annu. Rev. Biophys. Bioeng.* **6**:87–109.

Henning, U., *et al.* 1979. Cloning of the structural gene (omp A) for an integral outer membrane protein of *Escherichia coli* K-12. *Proc. Natl. Acad. Sci. USA* **76**:4360–4364.

Inuouye, M. 1974. A three-dimensional molecular assembly model of a lipoprotein from the *Escherichia coli* outer membrane. *Proc. Natl. Acad. Sci. USA* **71**:2396–2400.

Jay, D., and Cantley, L. 1986. Structural aspects of the red cell anion exchange protein. *Annu. Rev. Biochem.* **55**:511–538.

Jennings, M. L., *et al.* 1984. Peptides of human band 3 protein produced by extracellular papain cleavage. *J. Biol. Chem.* **259**:4652–4660.

Kopito. R. R., and Lodish, H. F. 1985a. Primary structure and transmembrane orientation of the murine anion exchange protein. *Nature* **316**:234–238.

Kopito. R. R., and Lodish, H. F. 1985b. Structure of the murine anion exchange protein. *J. Cell. Biochem.* **29**:1–17.

Kyte, J., and Doolittle, R. F. 1982. A simple method for displaying the hydropathic character of a protein. *J. Mol. Biol.* **157**:105–132.

Lai, J.-S., *et al.* 1981. *Bacillus licheniformis* penicillinase synthesized in *Escherichia coli* contains covalently linked fatty acid and glyceride. *Proc. Natl. Acad. Sci. USA* **78**:3506–3510.

Low, M. G., and Saltiel, A. R. 1988. Structural and functional roles of glycosyl-phosphatidylinositol in membranes. *Science* **239**:268–275.

Maelicke, A. 1988. Structural similarities between ion channel proteins. *Trends Biochem. Sci.* **13**:199–202.

Maniol, C., and Beckwith, J. 1986. A genetic approach to analyzing membrane protein topology. *Science* **233**:1403–1408.

Mueckler, M., *et al.* 1985. Sequence and structure of a human glucose transporter. *Science* **229**:941–945.

Russell, D. W., *et al.* 1985. Domain map of the LDL receptor: Sequence homology with the epidermal growth factor precursor. *Cell* **37**:577–585.

Schindler, H., and Rosenbusch, J. P. 1978. Matrix protein from *Escherichia coli* outer membranes forms voltage-controlled channels in lipid bilayers. *Proc. Natl. Acad. Sci. USA* **75**:3751–3755.

Schindler H., and Rosenbusch, J. P. 1981. Matrix protein in planar membranes: Clusters of channels in a native membrane environment and their functional reassembly. *Proc. Natl. Acad. Sci. USA* **78**:2302–2306.

Schindler, M., and Rosenbusch, J. P. 1982. Chemical modification of matrix porin from *Escherichia coli*: Probing the pore topology of a transmembrane protein. *J. Cell Biol.* **92:**742–746.

Schlesinger, M. J., *et al.* 1980. Fatty acid acylation of proteins in cultured cells. *J. Biol. Chem.* **255:** 10021–10024.

Tomita, M., and Marchesi, V. T. 1975. Amino acid sequence and oligosaccharide attachment sites of human erythrocyte glycophorin. *Proc. Natl. Acad. Sci. USA* **72:**2964–2968.

Tse, A. G. D., *et al.* 1985. A glycophospholipid tail at the carboxyl terminus of the Thy-1 glycoprotein of neurons and thymocytes. *Science* **230:**1003–1008.

Walder, J. A., *et al.* 1984. The interaction of hemoglobin with the cytoplasmic domain of band 3 of the human erythrocyte membrane. *J. Biol. Chem.* **259:**10238–10246.

Weiss, M. S., *et al.* 1991. Molecular architecture and electrostatic properties of bacterial porin. *Science* **254:**6127–6130.

Yamamoto, T., *et al.* 1984. The human LDL receptor: A cysteine-rich protein with multiple Alu sequences in its mRNA. *Cell* **39:**27–38.

Chapter 3

Carbohydrates and Cytoskeletal Components

The previous chapter dealt with membrane components that frequently, but not always, have an amphipathic structure. For example, phospholipids, integral membrane proteins, and ionophores possess both hydrophobic and hydrophilic domains. This chapter examines hydrophilic building blocks of biological membranes. These structures are generally localized to the aqueous environment at either side of a membrane.

3.1. MEMBRANE CARBOHYDRATES

Many constituents of biological membranes can be glycosylated. About 5% by weight of a cell's total lipid is found in the form of glycolipid. These glycolipids were discussed in Section 2.1. Certain pore-forming ionophores such as amphotericin B express a carbohydrate residue as an integral structural feature. Almost all membrane proteins at a plasma membrane's *trans* or outer face are glycosylated. As a rule, carbohydrates are located exclusively at the *trans* face of biological membranes.

3.1.1. Microscopic Detection of Carbohydrates

Carbohydrates were first observed at cell surfaces using light microscopy and histochemical staining. The periodic acid–Schiff reagent (PAS stain) was employed to detect carbohydrates. This procedure is also employed to identify glycoproteins in polyacrylamide gels (Figure 2.22). Periodic acid oxidizes vicinal hydroxyl groups of carbohydrates to aldehydes. Acid fuchsin reacts with dialdehydes to form a Schiff base. The reaction product appears reddish purple by light microscopy. It has been observed at the external surface of all cells. The staining is rather specific for carbohydrates since vicinal hydroxyl groups are required. This is confirmed by the observation that glycosidase treatment of cells diminishes staining.

Electron microscopy is also used to detect cell surface carbohydrates. Some staining methods such as ruthenium red/OsO_4, cationized ferritin, and colloidal iron/thorium hydroxide are fairly nonspecific, i.e., they react with many carbohydrates instead of just one or two. All cell surface carbohydrates are believed to be labeled by ruthenium red/OsO_4. Cationized ferritin binds to anionic sites on membranes. Since N-acetylneuraminic acid (NANA) is negatively charged, it binds to cationized ferritin. However, it is not completely specific for NANA since other membrane components are also negatively charged.

3.1.2. General Structural Properties of Carbohydrates

Although over 50 carbohydrates have been identified in nature, only 9 account for the vast majority of monosaccharides found on membranes. Figure 3.1 shows Haworth structural projections of commonly occurring monosaccharides. The D or L preceding each name identifies the stereochemical configuration of the asymmetric carbon atom. If two or more asymmetric carbons are present in a monosaccharide, the asymmetric carbon farthest from the carbonyl carbon is used to specify the configuration.

Many proteins expressed at a plasma membrane's *trans* face are glycosylated. Due to the nature of glycoprotein biosynthesis, certain general structural features can be identified. Figure 3.2 shows an arrangement of monosaccharides of a "typical" membrane protein. Although it does not represent any particular membrane protein, it does present a compilation of frequently encountered spatial relationships among carbohydrates. Oligosaccharides are only attached to asparagine, serine, and threonine residues of proteins. Asparagine residues are most frequently glycosylated. The amide group of asparagine is linked to N-acetylglucosamine via N-glycosidic linkages (Figure 3.3). The second amino acid following the asparagine residue is always serine or threonine; this is the so-called Asn-X-Ser(Thr) rule. This is a general characteristic of glycoproteins including membrane-bound glycoproteins. Both serine and threonine residues provide hydroxyl groups that are capable of hydrogen bonding with the carbonyl carbon of N-glycosidic bonds. N-glycosidic bonds are only formed between asparagine and N-acetylglucosamine. The linkage is relatively stable to treatment with alkali. The center of many oligosaccharide chains contains mannose. Fucose and NANA are frequently terminal carbohydrate residues. NANA is generally linked to galactose.

Membrane oligosaccharides are also attached to proteins via O-glycosidic linkages. In this configuration a carbohydrate is linked to serine or threonine (Figure 2.3). The O-glycosidic linkage is sensitive to alkali. In most cases the amino acid is linked to N-acetylglucosamine. However, there are a few uncommon O-glycosidic linkages in cell walls of plants and microorganisms.

3.1.3. Oligosaccharide Chains of Glycophorin and Band 3

The erythrocyte membrane glycoproteins glycophorin and band 3 illustrate both the common features and structural diversity of membrane carbohydrates. Human

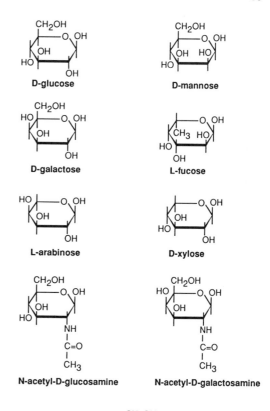

Figure 3.1. Chemical structures of common membrane carbohydrates. Haworth structural projections of nine frequently encountered carbohydrates are shown.

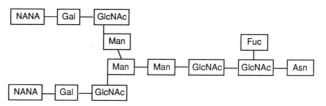

Figure 3.2. A representative spatial arrangement of carbohydrates in an oligosaccharide chain. An *N*-linked oligosaccharide chain is shown. The general spatial relationships among these monosaccharides are found in many oligosaccharides.

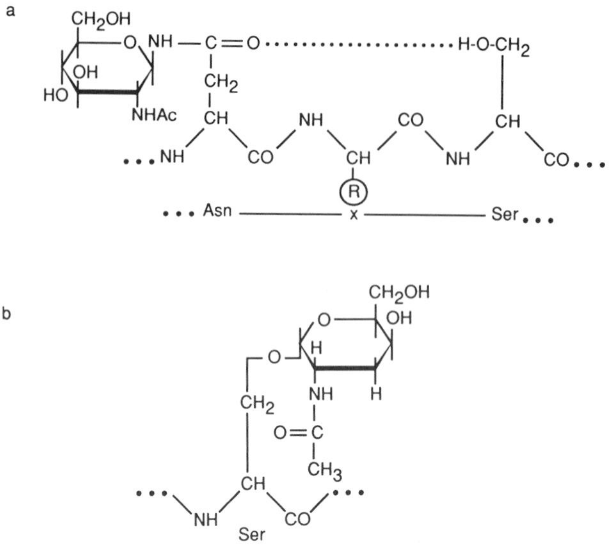

Figure 3.3. The *N*-glycosidic and *O*-glycosidic linkages between peptides and carbohydrates. (a) The *N*-glycosidic bond between *N*-acetylglucosamine and asparagine is shown. The Asn-X-Ser(Thr) rule is also illustrated. (b) An example of an *O*-glycosidic linkage between *N*-acetylgalactosamine and serine is shown.

glycophorin A possesses 15 *O*-linked oligosaccharide chains and one *N*-linked chain. Figure 3.4 shows the structures of these carbohydrate chains. The tetrasaccharide shown in panel a represents the majority of *O*-linked sugar chains associated with glycophorin A. When attached to glycophorin A this carbohydrate sequence possesses blood group MN antigenic activity.

Panel b of Figure 3.4 illustrates the *N*-linked oligosaccharide chain of glycophorin A (Yoshima *et al.*, 1980; Irimura *et al.*, 1981). This carbohydrate sequence represents the principal structure encountered for this oligosaccharide. Other species of this oligosaccharide chain have one or both NANA residues missing.

Band 3 migrates as a diffuse or broad molecular weight band during SDS-PAGE (Figure 2.22). This can be accounted for, at least in part, by heterogeneous glycosylation. Band 3 possesses one (Jay, 1986), possibly two (Fukuda *et al.*, 1984b), glycosylation sites. Two distinct oligosaccharide side chains are associated with band 3 via *N*-glycosidic bonds. These are the lactosaminoglycan and complex-type oligosaccharides (Figure 3.4). Although these two carbohydrate chains are found in equimolar amounts, the lactosaminoglycan accounts for most of the band 3 carbohydrate because of its large size (7000 to 11,000 Da). The complex-type side chain is almost identical to glycophorin A's *N*-linked chain. Some variability of the terminal galactose residues of band 3's complex-type oligosaccharide has been noted (Tsuji *et al.*, 1981). The core structure of the complex-type chain is also shared with lactosaminoglycan. Oligolactosaminoglycan (. . . Galβ1→4GlcNAcβ1→ . . .) side chains are attached to the core structure as shown in Figure 3.4. One chain consists of 10–12 lactosaminyl groups

a   NANA ∝ 2 → 3Gal β1 → 3GalNAc – Ser (Thr)
                        ↑∝2-6
                     NANA

b   NANA2 → 6Gal β1 → 4GlcNAc β1 → 2Man
                              ↓∝1→6
                  GlcNAc β1 → 4Man β1 → 4GlcNAc β1 → 4GlcNAc – Asn
                            ↑∝1→3
     NANA2 → 6Gal β1 → 4GlcNAc β1 → 2Man

c   Gal β1 → 4GlcNAc β1 → 2Man
                      ↓∝1→6
            GlcNAc β1 → 4Man β1 → 4GlcNAc β1 → 4GlcNAc – Asn
                  ↑∝1→3   ↑∝1→6
     Gal β1 → 4GlcNAc β1 → 2Man   Fuc

d   [10-12 N-acetyllactosaminyl] ··· Gal β1 → 4GlcNAc β1 → 2Man
                                 ↓∝1→6
        [±Glc β1 → 4̅ GlcNAc β1 → 4Man β1 → 4GlcNAc β1 → 4GlcNAc – Asn
                             ↑∝1→3   ↑∝1→6
     [5-6 N-acetyllactosaminyl] ··· Gal β1 → 4GlcNAc β1 → 2Man   Fuc

Figure 3.4. Carbohydrate sequences of erythrocyte glycophorin and band 3. The oligosaccharides shown are: (a) the *O*-glycosidic carbohydrate of glycophorin A, (b) the *N*-linked moiety of glycophorin A, (c) the complex-type carbohydrate chain of band 3, and (d) the lactosaminoglycan of band 3.

whereas the second possesses 5–6 groups. In adult erythrocytes the lactosaminyl chains are branched; this structure is manifest *in vivo* as the I blood group antigen. The lactosaminoglycan of fetal erythrocyte band 3 is unbranched; it is the i antigen. In addition to the Ii blood group, the ABO(H) blood group can also be attached to this carbohydrate chain. Fucose and NANA may be associated with the nonreducing ends of lactosaminyl side chains. The presence or absence of fucose and NANA constitute two types of microheterogeneity within these oligosaccharide chains.

3.1.4. Structural Features of Cell Wall Carbohydrates

Carbohydrate polymers such as cellulose and peptidoglycan of plants and bacteria, respectively, are peripheral membrane components. For example, peptidoglycan is. associated with the integral protein lipoprotein in the outer membrane of *E. coli*. Cellulose is synthesized by enzymes of plant cell plasma membranes. Both polymers are closely associated with membrane structure and function. We will now discuss saccharide components of cell walls.

As we have previously discussed, gram-negative bacteria possess an inner membrane, a peptidoglycan layer, teichoic acids, and an outer membrane (Figure 1.1). In some cases bacteria may also possess a capsule. Lipopolysaccharide, one carbohydrate-containing component of outer membranes, was discussed in Section 2.1.6. Two

additional carbohydrate-rich components of bacterial envelopes are the peptidoglycan
layer and capsule. As its name implies, peptidoglycan is composed of both peptide and
saccharide components; it is found in almost all prokaryotes. In gram-negative bacteria
peptidoglycan is found as a single thin layer whereas gram-positive bacteria possess
multiple peptidoglycan layers. In both cases peptidoglycan forms a cross-linked matrix
that surrounds a cell and contributes to its shape. The glycan chains are composed of
10 to 150 disaccharide units. Each disaccharide generally contains N-acetylglucosamine
and N-acetylmuramic acid. The glycan polymers lie parallel to cell surfaces. The
carboxyl group of N-acetylmuramic acid is linked to L-alanine, as shown in Figure 3.5.
The peptide chains are cross-linked in a variety of fashions, one of which is shown in the
figure. The polymeric matrix forms a sturdy but porous network.

In some cases a bacterial envelope may possess an outer capsule or gel-like coat.
These capsules are generally polysaccharide in nature. Although the sugars may vary, a
frequent component is glucuronic acid (GlcUA), a negatively charged saccharide. A
repeating and branched pattern of a small number of saccharide residues is most
frequently encountered. A typical branched sequence is:

$$GlcUA$$
$$|$$
$$—(Gal\text{-}Man\text{-}Gal)_x—$$

where X is a large integer. The presence of a capsule is associated with the virulence of
pathogenic bacteria such as those responsible for anthrax and pneumonia. The polysaccharide capsule plays a vital role in evasion of a host's defense mechanisms.

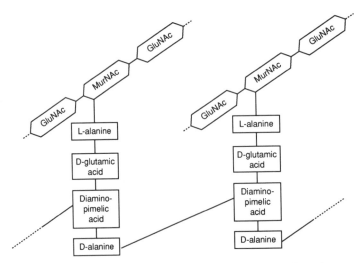

Figure 3.5. The structure of peptidoglycan. Peptidoglycan is composed of both peptide and carbohydrate
components. The peptide chains are cross-linked as shown. The glycan portion contains a repeating
disaccharide composed of N-acetylglucosamine (GluNAc) and N-acetylmuramic acid (MurNAc). The long
carbohydrate chains can be linked to multiple L-alanine residues via different MurNAc groups.

In addition to microorganisms, organized cell wall structures are found in plants. Cell walls provide protection from osmotic and mechanical damage while conferring shape to plant cells. A principal component of plant cell walls is cellulose. Cellulose is a linear, unbranched chain of $\beta1\rightarrow4$-linked glucose molecules. Roughly 1000 glucose monomers make up one cellulose molecule. Long ribbonlike chains are formed. Hydrogen bonding within and between the 40–60 separate cellulose molecules forms one microfibril with a diameter of 10 nm. Many microfibrils are assembled to form a cell wall. The microfibrils are embedded within a gluelike matrix composed of polysaccharides such as hemicellulose and pectin. Together these molecules are responsible for the highly rigid structural properties of plant cells.

3.1.5. Lectins

Lectins are proteins or glycoproteins that noncovalently bind to specific carbohydrates. Carbohydrates are not chemically altered by lectin binding. Lectins can be isolated from a great variety of sources, particularly plants. In plants lectins may serve as a mechanism of nonself recognition, just as antibodies recognize foreign substances in mammals. More recent evidence suggests that plants may utilize lectins to further stabilize cell walls by forming additional cross-linkages. Many lectins with differing carbohydrate specificities have been identified (Table 3.1).

One of the most thoroughly studied lectins is concanavalin A (Con A) which is isolated from jack beans. Con A is a tetramer composed of identical subunits. Each subunit has one monosaccharide binding site specific for D-manose and D-glucose. When multiple binding sites are occupied, surface carbohydrate chains become cross-linked. Cells respond to Con A cross-linkage of their surface oligosaccharides in a

Table 3.1.
Selected Lectins

Scientific name	Common name	M_r	Subunits	Carbohydrate specificity
Arachis hypogaea	Peanut	120,000	4	β-D-Gal(1→3)-D-GalNAc
Canavalia ensiformis	Concanavalin A	102,000	4	α-D-Man, α-D-Glc
Lens culinaris	Lentil	49,000	2	α-D-Man
Limulus polyhemus	Horseshoe crab	400,000	18	NANA
Phaseolus vulgaris	Kidney bean	128,000	4	Oligosaccharide
Phytolacca americana	Pokeweed	32,000	—	(D-GlcNAc)₃
Ricinus communis				
Toxin RCA60		60,000	2	D-GalNAc, β-D-Gal
Agglutinin, RCA120		120,000	4	β-D-Gal
Tetragonolobus purpureas	Lotus	120,000	4	α-L-Fuc
Triticum vulgaris	Wheat germ	36,000	2	(D-GlcNAc)₂, NANA
Vigna radiata	Mung bean	160,000	4	α-D-Gal
Viscum album	Mistletoe	115,000	4	β-D-Gal

variety of fashions. Patches of lectin and cross-linked carbohydrates form at cell surfaces. By an energy-dependent process, a cell can collect its patches into a large cap. Cells also respond to Con A by stimulation of DNA synthesis and the formation of endocytic vesicles, as though they receive a physiological signal.

One application of lectins is to locate specific carbohydrates in a complex biological setting. Con A can be covalently coupled to fluorescent probes or ferritin. Fluorescent molecules are visualized by fluorescence microscopy. The electron-dense protein ferritin is observed by transmission electron microscopy. Figure 3.6 shows an example of fluorescence microscopy of Con A bound to cell surfaces. Although electron microscopy possesses greater resolution, fluorescence microscopy can be used to observe living cells as they carry out specific activities.

3.1.6. Mapping Glycosylation Sites of Membrane Proteins

Although carbohydrate components of membrane glycoproteins are exclusively associated with a membrane's *trans* face, they may be linked to a polypeptide chain at any one of many potential sites. Jay (1986) has reported a technique to map the distance between the NH_2-terminus of a polypeptide and the closest or first glycosylation site. Multiple glycosylation sites could be mapped using multiple large and well-characterized proteolytic fragments each possessing a unique NH_2-terminus. The NH_2-terminus is labeled with radioactive phenylisothiocyanate. Relatively small fragments of a protein are prepared by exposure to various proteolytic enzymes. The proteolytic fragments are then separated by SDS-PAGE using 18% polyacrylamide gels. The resulting pattern of bands provides a "fingerprint" of the protein. In a second experiment the proteolytic fragments are subjected to affinity chromatography prior to electrophoresis. For example, Con A covalently linked to a solid support provides a useful affinity chromatography matrix. Those peptides bearing an appropriate carbohydrate bind to the matrix. Peptides without the appropriate glycosylation site are washed away from the bound peptides. A sugar that competes for the Con A binding site is used to elute bound peptides. These

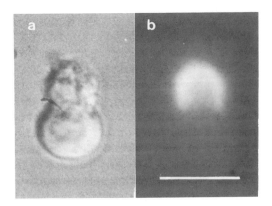

Figure 3.6. Microscopy of concanavalin A binding sites. Panel a shows a differential interference contrast micrograph of a lymphocyte after capping by exposure to Con A. Panel b is a corresponding fluorescence photomicrograph showing the location of the capped fluorescent Con A molecules. Bar = 5 μm.

peptides are then separated by electrophoresis. In this experiment only those peptides containing *both* the glycosylation or lectin-binding site and the radioactive NH_2-terminus are visualized. The smallest proteolytic fragment retained by the affinity column represents the longest possible distance between the NH_2-terminus and the carbohydrate moiety. The largest proteolytic fragment *not* retained by the column (i.e., washed away) defines the shortest possible distance between the NH_2-terminus and the first carbohydrate chain. This is schematically illustrated for the membrane protein HLA-B7 in Figure 3.7. The two molecular sizes identified by electrophoresis define the upper and lower limits of the glycosylation site as shown in Figure 3.7. Assuming a mass of 110 Da per amino acid, a glycosylation site is identified at residue 90 ± 10. From other data, this has been found to be residue 86, thus supporting this method's validity. Using this approach, Jay (1986) determined that the glycosylation site of band 3 is at $28,000 \pm 3000$ Da from the COOH-terminus.

3.1.7. Conformations of Carbohydrates

Carbohydrates play important roles in membrane events such as immune recognition and cell–cell interactions. Membrane-to-membrane recognition phenomena include both protein–carbohydrate and carbohydrate–carbohydrate binding interactions (Turley and Roth, 1980). Many physiological activities could be reflected in the conformations adopted by membrane carbohydrates. Unfortunately, very little is know about carbohydrate conformations. Polymeric carbohydrates can form ordered three-dimensional structures such as helices and ribbons. These ordered structures doubtlessly play an important role in forming stable cell walls in plants and microorganisms (Preston, 1979). However, membrane-bound carbohydrates of animal cells almost never form such structures due to their aperiodic, branching, and short-chain properties. One possible exception to this rule is the polysialic acid chains of the neural cell adhesion molecule (Section 7.5.2). Carbohydrate chains rapidly rotate at model membrane surfaces,

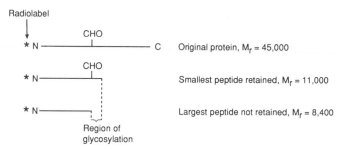

Figure 3.7. Strategy for mapping glycosylation sites. The NH_2-terminus (N) is selectively labeled with a radioactive atom (*). The original protein is cleaved into many distinct fragments. The peptides are separated based on their ability to be retained by a lectin affinity chromatography column. The glycosylation site must be located at a position corresponding to the greater M_r than the largest peptide not retained but smaller than the M_r of the smallest peptide retained.

although free rotation in cell membranes may be hindered by interchain hydrogen bonding—just as we have seen with cellulose but on a much smaller scale.

3.2 MEMBRANE SKELETONS

Three remarkable features of eukaryotic cells are their abilities to assume distinct shapes, crawl along surfaces, and rapidly reorganize or remove surface components. Cytoskeletal structures participate in these and other membrane-related phenomena. One important function of cytoskeletal structures is to act as a cell's skeleton. They are found throughout the cytosol and in contact with organelles and plasma membranes, thereby providing a structural framework for a cell's components. In most cells cytoskeletal assemblies also form a layer beneath plasma membranes. Three types of fibrous cytoskeletal structures have been found: microfilaments, microtubules, and intermediate filaments. These structures are composed of a complex set of proteins that are controlled by poorly understood interactions. Although clathrin-coated membranes and erythrocyte spectrin–actin networks are not "technically" cytoskeletal in nature, they do participate in membrane functions similar to those of a cytoskeleton. Their attachment to membranes and participation in membrane activities indicate that they represent membrane skeletal structures. In this section we will discuss various peripheral proteins that contribute to the formation of membrane skeletons.

3.2.1. An Erythrocyte Membrane Skeleton: A Revealing Model System

The *cis* face of an erythrocyte membrane is covered by a fibrous network of peripheral proteins collectively referred to as a membrane skeleton. The membrane skeleton confers both structural stability and shape upon an erythrocyte membrane. Figure 3.8 shows a current model of an erythrocyte membrane skeleton and enhanced views of two regions (Bennett, 1985; Marchesi, 1985). The properties of the constituent proteins are summarized in Table 3.2. We shall return to this model after considering each protein in more detail.

3.2.1.1. Building Blocks of an Erythrocyte Membrane Skeleton

Spectrin is a heterodimer of α and β subunits of 260 and 225 kDa, respectively. These subunits are very tightly but noncovalently held together at numerous sites along their lengths. The dimer and its subunits are long and flexible rods. The dimeric rods have a length of 100 nm. The major form of spectrin is a tetramer formed by head-to-head contacts between two dimers. There are roughly 100,000 tetramers in each erythrocyte. The deduced amino acid sequences of both spectrin subunits have been determined. Considerable sequence homologies have been found between the α and β subunits and within each individual subunit. Similar structural features are repeated each 106 amino acids (12 kDa). Furthermore, each 106-amino-acid domain may form

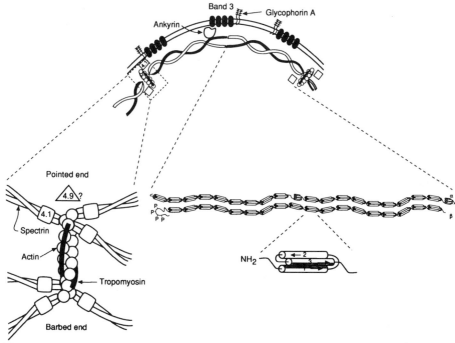

Figure 3.8. A current model of the erythrocyte membrane skeleton. The global structure of the erythrocyte membrane skeleton is shown at the top of the figure. The relative positions of integral membrane proteins, skeletal proteins, and their transmembrane linkages are indicated. The detail at the lower left shows the assembly of the actin/band 4.1 membrane-binding complex. The detail at the right-hand side shows the triple-helix structure of spectrin. [Redrawn with permission from the *Annual Review of Biochemistry* Vol. 54, copyright © 1985 by Annual Reviews, Inc., and the *Annual Review of Cell Biology* Vol. 1, copyright © 1985 by Annual Reviews, Inc. (insert on right-hand side).]

Table 3.2.

Cytoskeletal Proteins of Human Erythrocytes

Protein	M_r	Assembly state	Copies/cell	Shape
Spectrin	α 260,000	$(\alpha\beta)_2$	10^5	Rod
	β 225,000			
Ankyrin	210,000	Monomer	10^5	Globular
Myosin				
Heavy	210,000	Oligomer ($\approx$30/	6000	Rod
Light	25,000	filament)		Globular
Light	19,500			Globular
Band 4.1	78,000	—	2×10^5	
	80,000			
Band 4.9 (acting-binding protein)	45,000	—	5×10^4	
Actin (band 5)	43,000	Oligomer (12–	5×10^5	Globular (oligomer
		17 subunits)		rods)
Tropomyosin	29,000	Dimer	7×10^4	Rod
	27,000			

three α-helical strands. A suggested spatial arrangement of these domains and strands is shown in Figure 3.8. Spectrin acts as a membrane framework. It contains binding sites for ankyrin, band 4.1, actin, and calmodulin.

Spectrin is linked to an erythrocyte membrane by ankyrin. Ankyrin is a monomeric phosphoprotein of 210 kDa which migrates as band 2.1 during SDS-PAGE. As its name implies, ankyrin anchors band 3 to the membrane skeleton (Figure 3.8). It contains binding sites for both the β subunit of spectrin and band 3's COOH-terminal region. Interactions among these three proteins provide one transmembrane linkage that participates in the control of erythrocyte shape.

A second type of transmembrane linkage is mediated by band 4.1. As is the case with band 3, band 4.1 derives its name from SDS-PAGE experiments. Band 4.1 is composed of two almost identical proteins of 78 and 80 kDa. In addition to forming a transmembrane link, it increases the affinity of spectrin–actin interactions. The tail region of spectrin's β subunit binds to band 4.1. In turn, band 4.1 binds to glycophorin A and/or glycoconnectin (Figure 3.8). This binding reaction requires the phospholipid cofactor 4,5-bisphosphate phosphatidylinositol (PIP_2). PIP_2 is a trace phospholipid localized exclusively at the *cis* face of an erythrocyte bilayer. Although PIP_2 is apparently required for membrane interaction, its role in promoting the interaction is unknown. It is interesting, however, because phosphoinositide turnover is believed to be an important mechanism of transmembrane signaling (Section 7.6). Recently, it has been found that a relatively low-affinity band 4.1 binding site is present on band 3 at a position distinct from ankyrin's binding site. Band 4.1 may bind to band 3 during special conditions such as after PIP_2 hydrolysis. Recently, band 4.1 has been found to bind and regulate myosin molecules. Band 4.1 plays a central role in erythrocyte membrane shape and stability.

Actin is a major component of an erythrocyte membrane skeleton (Bennett, 1985; Pollard and Cooper, 1986). There are approximately 500,000 actin monomers per erythrocyte. Most of this actin is in the form of oligomers composed of 12–17 subunits. Actin oligomerizes or polymerizes to form filaments 6 nm in diameter. Oligomeric or polymeric actin is bipolar. For example, when F-actin is labeled with heavy meromyosin, the filament looks like a series of arrowheads with morphologically defined "barbed" and "pointed" ends. Filamentous actin is referred to as F-actin whereas the monomeric or globular form is G-actin (Section 3.2.2.2). There are roughly four spectrin tetramers for each F-actin oligomer. The actin oligomers are stabilized by interactions with tropomyosin and band 4.9, the actin-binding protein (Figure 3.9). Actin is capable of activating myosin ATPase.

Erythrocyte myosin is a bipolar molecule comprised of two heavy chains of 210 kDa each and two pairs of light chains of 25 and 19.5 kDa. Erythrocyte myosin is similar in shape to myosins derived from other sources. It is a rod-shaped molecule 150 nm in length with two globular heads that bind actin. It is a fairly rare protein with only about 6000 copies per cell. Similar to other myosins, erythrocyte myosin possesses calcium-stimulated ATPase activity. Phosphorylation of the 19.5-kDa light chain influences its actin-sensitive ATPase activity. Although myosin's role is unknown, it could participate in erythrocyte shape changes by a sliding filament mechanism as seen in muscle cells.

Tropomyosin, another component of muscle fibers, is also part of an erythrocyte

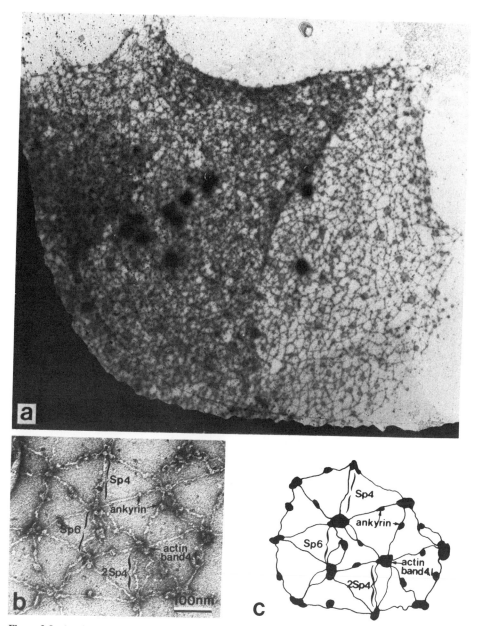

Figure 3.9. An electron microscopic study of the erythrocyte membrane skeleton. Erythrocyte membranes from Triton-X-treated ghosts were spread on an electron microscopy grid, fixed with glutaraldehyde, and negatively stained with uranyl acetate. Panel a shows a survey electron micrograph of a spread detergent-treated RBC ghost. The right-hand side of this panel shows only one (bottom) surface of the ghost's membrane skeleton. (The darker middle and left-hand sides show two superimposed bilayers.) In panel b a high-magnification image (bar = 0.1 μm) of the membrane skeleton's meshwork is given. Panel c shows a schematic interpretation of the image in panel b. Sp4, spectrin tetramer; Sp6, a three-armed spectrin hexamer; 2Sp4, a double spectrin tetramer. (Reproduced from the *Journal of Cell Biology*, 1987, Vol. 104, p. 527, by copyright permission of the Rockefeller University Press.)

membrane skeleton. It is a rod-shaped heterodimer containing subunits of 29 and 27 kDa. There are likely two tropomyosin molecules bound to each actin oligomer. Tropomyosin stabilizes F-actin and regulates actin–spectrin interactions.

3.2.1.2. The Erythrocyte Membrane Skeleton: A Summary View

The erythrocyte transmembrane glycoproteins band 3, glycophorin, and possibly glycoconnectin are tethered to a membrane skeleton by ankyrin and band 4.1 (Figure 3.8). This drawing is somewhat deceptive because spectrin is shown in only one dimension. Spectrin forms a complex and densely packed web beneath a membrane. The meshlike quality is introduced by three or more spectrin molecules meeting at the same point. For example, several spectrin molecules could meet at one actin and band 4.1 complex. In addition to spectrin tetramers, hexamers and octomers can form. Additional spectrin branching could be introduced, for example, by combining three head-to-head interactions (Marchesi, 1985).

Figure 3.9 shows some of the complex interactions observed with negative-stain electron microscopy. Untreated membrane skeletons are far too densely packed to clearly observe membrane skeletal structure. The data of Figure 3.9 were obtained by laterally stretching or "spreading" the skeleton. Panel a shows a representative negative-stain electron microscopy profile of an erythrocyte membrane skeleton. Panel c interprets various portions of the micrograph. A hexagonlike meshwork is observed. Ankyrin is located along the rod-shaped spectrin tetramers whereas actin and band 4.1 are located at the intersections of spectrin tetramers. The combined biochemical and microscopic studies of the erythrocyte membrane skeleton have yielded a detailed map of its constituents. This, in turn, has led to a greater understanding of the control of erythrocyte shape (Elgsaeter et al., 1986).

3.2.2. Eukaryotic Cytoskeletons: The General Case

Most eukaryotic cells generally possess cytoskeletal assemblies of microfilaments, microtubules, and intermediate filaments. These components perform a myriad of functions including plasma membrane-associated activities. The combination of biochemical and cell biological techniques has been particularly rewarding in cytoskeletal studies. We will first consider a few methods employed to localize cytoskeletal proteins and then explore their constituent proteins and activities.

3.2.2.1. Visualizing Cytoskeletons

The cellular locations of cytoskeletal proteins and their movements during physiological processes must be determined to identify relevant interactions and mechanisms. Optical and electron microscopic techniques are used to obtain this information. This section will focus on optical imaging using fluorescence microscopes. The two basic approaches in this type of experiment are immunocytochemistry and fluorescent analogue cytochemistry.

Under appropriate conditions, cytoskeletal proteins are immunogenic. Polyclonal and monoclonal antibodies reactive with specific cytoskeletal antigenic sites have been prepared. To perform immunocytochemistry, antibodies or their fragments are covalently linked to fluorescent dyes. Cells are fixed with formalin or dried in air. Since antibodies are much too large to penetrate plasma membranes, cells must first be disrupted. This is often accomplished by extracting a sample with cold acetone. The cytoskeleton is then directly or indirectly labeled with anticytoskeleton antibodies followed by fluorescence microscopy. Only polymers such as F-actin are visualized since extraction removes monomeric proteins. Examples of immunofluorescence microscopy are provided in Figures 3.13 and 3.14.

Fluorescent analogue cytochemistry is an alternate means of viewing cytoskeletons (Taylor and Wang, 1980). This method does not use antibodies. Individual cytoskeletal proteins are purified to homogeneity and then labeled with a fluorescent molecule. Fluorescent cytoskeletal proteins are microinjected into individual living cells. In general, it is best to include a nonspecific protein, such as bovine serum albumin, conjugated to a different fluorescent tag to serve as a control. Fluorescent cytoskeletal proteins become part of a cell's cytoskeletal protein pool. The proteins are incorporated into functioning cytoskeletal assemblies such as actin filaments (see Figure 3.14). The advantage of fluorescent analogue cytochemistry is that unfixed, living cells are studied as biological events occur.

3.2.2.2. Microfilaments

Actin is an abundant protein accounting for up to 15% by weight of a nonmuscle cell's protein. Several of actin's properties were described above (Section 3.2.1.1). Actin is a spherical protein with a diameter of 5nm and a molecular mass of 43 kDa. The amino acid sequences of several actin molecules from different species have been determined (Pollard and Cooper, 1986). Its primary structure has been highly conserved throughout evolution.

Crude eukaryotic cell extracts prepared by homogenization and centrifugation form a solid gel when warmed to room temperature in a calcium-free buffer containing ATP. As might be expected, a sample's viscosity dramatically increases during gelation. Microfilaments built from actin are the primary component of these gels, although lesser quantities of other participatory proteins are also present. These actin-binding proteins will be discussed below. Figure 3.10 shows actin gelation. Panels b and c illustrate the inability to pour a gel from a cuvette and the increased turbidity and contraction of a gel, respectively. If myosin is present in an extract, its gel state can contract. The fluid or gel state of a cytoplasmic matrix may be an important regulator of cell function.

The polymerized actin filaments form a double-helical strand 6 nm wide, like two intertwined strings of pearl. The helix repeats every 37 nm. Polymerization proceeds in three distinct phases. The slow event in polymer formation is nucleation. During this lag phase several monomers bind together to form a short oligomer. The nucleus is elongated

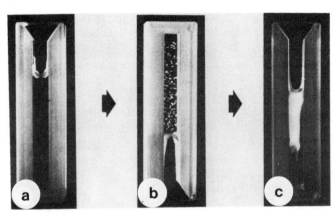

Figure 3.10. Gelation and contraction of cytosolic proteins. A cytosolic fraction of *Acanthamoeba* prepared by homogenization and ultracentrifugation (panel a). Upon warming in an appropriate buffer, the cell extract gels as shown by its inability to be poured from the cuvette (panel b). The contraction and turbidity of the gel are shown in panel c. (Reproduced from the *Journal of Cell Biology*, 1981, Vol. 91, p. 1565, by copyright permission of the Rockefeller University Press.)

by addition of monomers in both directions, although polymerization is faster at a filament's barbed end. Two filaments may subsequently anneal to form one larger filament. The kinetics of nucleation and polymerization are shown in Figure 3.11. Polymerization continues until a critical concentration of actin monomers is reached. Actin concentrations below this level are too dilute to support net polymerization. At this point the assembly rate equals the disassembly rate.

Microfilaments may be attached to a plasma membrane. Careful isolation of plasma membrane fractions has revealed associated actin, although this represents a small fraction of a cell's total actin. In a few instances the barbed end of an actin filament has been reported to be attached to the *cis* face of a plasma membrane. In this case the force generated by the actin–myosin complex would pull perpendicular to the membrane's surface. The sides of actin filaments, including stress fibers, are in contact with membranes at sites called focal contacts or adhesion plaques. Several proteins including vinculin, metavinculin, α-actinin, talin, and the 110-kDa protein of brush border membranes have been identified as possible linkage proteins (Pollard and Cooper, 1986). Figure 3.12 shows a highly schematic model of the transmembrane associations in a region of cell adherence. This represents one of the better, although still incompletely, understood transmembrane cytoskeletal linkages. Microfilaments are linked to integral membrane proteins via talin and/or vinculin. Talin binds directly to integrin protein(s), a family of adherence-promoting membrane proteins. At a bilayer's *trans* face these integral membrane proteins recognize the RGD sequence of fibronectin (or other ligands). We shall return to the physiological properties of adherence proteins in Section 7.5.

The molecular architecture of membrane–cytoskeleton interactions has not yet been clearly mapped. Nonetheless, the overall disposition of microfilaments can be

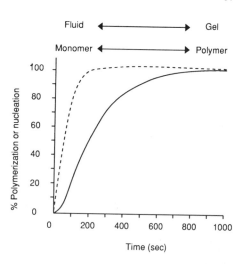

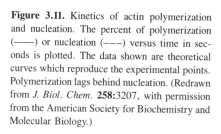

Figure 3.11. Kinetics of actin polymerization and nucleation. The percent of polymerization (——) or nucleation (– – –) versus time in seconds is plotted. The data shown are theoretical curves which reproduce the experimental points. Polymerization lags behind nucleation. (Redrawn from *J. Biol. Chem.* **258**:3207, with permission from the American Society for Biochemistry and Molecular Biology.)

determined by the techniques of immunofluorescence and fluorescent analogue cytochemistry. Figure 3.13 shows the distribution of F-actin in tissue culture cells. In this example large bundles of microfilaments known as stress fibers are present in a cell. Stress fibers are found in surface-attached nonmotile cells. They are capable of forming transmembrane linkages with integral membrane proteins. Figure 3.14 shows double-label immunofluorescence studies of surface-attached fibroblasts. Histocompatibility antigens were cross-linked with an F(ab′)$_2$ antibody fragment. The pattern of labeled antigen (Figure 3.14, panel B) can be superimposed on the stress fiber pattern (Figure 3.14, panel A), suggesting a transmembrane association. A fluorescent analogue cytochemical study of actin in actively motile amoebae is shown in Figure 3.15. Actin is dispersed throughout the cytoplasm of motile cells. In this case, exterior surface cross-

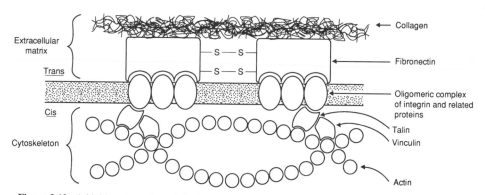

Figure 3.12. A highly schematic model of a possible transmembrane linkage at adherence sites. F-actin is linked to integrin and related proteins via talin and/or vinculin. The integrin proteins bind to fibronectin with the RGD amino acid sequence. Fibronectin, in turn, is associated with the extracellular matrix.

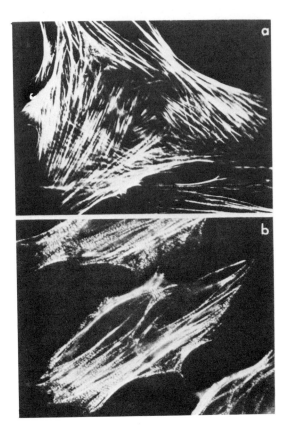

Figure 3.13. The distribution of actin
and myosin in cells as visualized by immunofluorescence. Nonmotile cells were
stained with antiactin (panel a) or anti-
myosin (panel b) antibodies. Arrays of
stress fibers can be seen throughout the
cytoplasm in panel a. A similar distribution of fluorescence is seen for myosin in
panel b. In addition, myosin labeling is
punctate, suggesting that myosin clusters
along stress fibers. (Reproduced from the
Journal of Cell Biology, 1981, Vol. 91,
p. 156s, by copyright permission of the
Rockefeller University Press.)

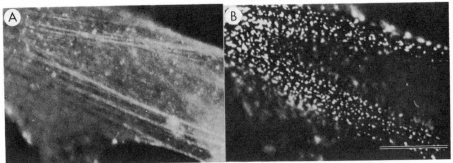

Figure 3.14. Double-labeling study of membrane antigens and stress fibers. Fluorescence microscopy
experiments were conducted on cells to visualize actin (A) or membrane antigens (B). Cells were then treated
with F(ab')$_2$ fragments of anti-β_2-microglobulin antibodies. Patches of membrane antigens were formed. Cells
were then fixed and labeled with a fluorescent derivative of heavy meromyosin to visualize actin. The antigen
clusters are aligned above the stress fibers. Bar = 20 μm. (From Ash *et al.*, 1977, *Proc. Natl. Acad. Sci. USA*
74:5583.)

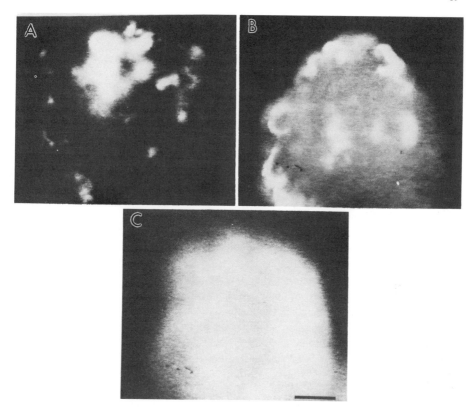

Figure 3.15. Fluorescent analogue cytochemistry of actin. Fluorescent actin or ovalbumin was microinjected into *Chaos carolinensis* cells. Fluorescent Con A was capped at the surface of living cells. (A) A fluorescence photomicrograph of Con A at the cell surface. (B) Distribution of fluorescent actin in a living cell. Actin clusters in the vicinity of Con A caps. (C) Fluorescent ovalbumin is uniformly distributed under similar conditions. Bar = 40 μm. (Reproduced from the *Journal of Cell Biology*, 1980, Vol. 86, p. 580, by copyright permission of the Rockefeller University Press.)

linking by Con A has induced clustering of actin at the plasma membrane's *cis* face. Actin contains binding sites for calcium, ATP, myosin, and a host of accessory proteins. To support polymerization, actin has binding sites for other actin molecules. Cytosolic calcium levels may act as messengers to influence the behavior of actin. Upon actin binding the myosin Mg^{2+}-ATPase becomes activated. These interactions between actin, myosin, and ATP are very likely the primary mechanism responsible for the cytoskeleton's contractile properties. Both the polymerization of actin and the associated contractile events are highly orchestrated phenomena. Sophisticated control mechanisms are required to generate coordinated physiological processes. The actin-binding proteins discussed in the following paragraphs provide regulatory mechanisms.

3.2.2.3. Actin-Binding Proteins

Just as the actin–myosin system of muscles must be carefully controlled to yield movement, the actin–myosin interactions of nonmuscle cells must be regulated to achieve cell shape, endocytosis, and movement. A family of actin-binding proteins participate in regulatory activities (Stossel *et al.*, 1985; Pollard and Cooper, 1986). These activities are categorized into five major groups: monomer binding, cross-linking, severing, capping, and membrane attachment. Examples of membrane-attachment proteins were presented above. Table 3.3 lists representative members of each of these classes of actin-binding proteins. The proteins are listed under their apparent primary function since some proteins may possess multiple functions. Several of these complex functions are illustrated in Figure 3.16. Actin-binding proteins create the great variety of actin structures found in cells. Actin monomer-binding proteins preferentially interact with G-actin. The assembly of actin filaments is thereby diminished. Profilin, a representative actin monomer-binding protein, increases the length of the lag phase and decreases the rate and extent of polymerization. Profilin–actin complexes bind to PIP_2, which promotes their dissociation. This may be one cytoskeletal regulatory pathway.

Capping proteins bind to one end of a filament and inhibit monomer reactions at that point. This would be expected to stabilize filament length. Capping proteins make filaments shorter. They potentiate nucleation leading to more and necessarily shorter filaments. If capping proteins are added to F-actin, intact filaments become shorter. This postassembly shortening may be due to additional nucleation followed by removal of monomers from the normal monomer-to-polymer equilibrium. Alternatively, capping proteins may actively sever F-actin. Some proteins have only been associated with a severing function (Table 3.3). Comparatively little is known about these severing proteins. The capping and severing activities of actin-binding proteins likely control filament length and participate in regulating cytosol viscosity.

Table 3.3.
A Summary of Actin-Binding
Proteins According to Their Functions

Actin monomer binding	Capping
Profilin	Gelsolin
DNase I	Villin
Vitamin D-binding protein	Fragmin/severin
Cross-linking	Capping protein
Filamin	α-actinin
Fodrin	Acumentin
TW 260/2	Membrane attachment
α-actinin	110-kDa protein
Fascin	Vinculin
Fimbrin	Metavinculin
Severing	Talin
Actophorin	24-kDa protein
Depactin	
Actin depolymerizing factor	

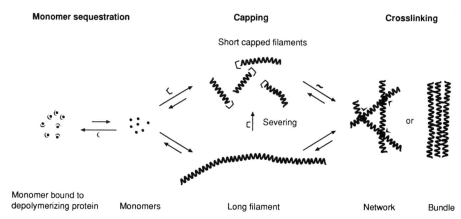

Figure 3.16. Functional activities of actin-binding proteins. The activities of actin-binding proteins are shown. To designate these activities, the following symbols are used: (, monomer-binding protein; [, capping or severing protein; and <, network-forming cross-linking protein. (Redrawn with permission from the *Annual Review of Biochemistry*, Vol. 55, copyright © 1986 by Annual Reviews Inc.)

Certain actin-binding proteins promote cross-linkage of filaments. All of these proteins must possess at least two actin-binding sites. They vary in size, affinity for actin, and calcium sensitivity. Cross-linking proteins increase both the viscosity and the stability of F-actin solutions. This cross-linking occurs in two fashions: (1) side-by-side linkage resulting in the formation of bundles and (2) linkage at various angles resulting in the formation of two-dimensional meshs or networks. Proteins such as filamin and myosin lead to network formation whereas α-actinin and fimbrin create actin bundles like those found in stress fibers.

3.2.2.4. Myosin

Myosin represents another class of actin-binding protein. Although myosin is a major component of muscle cells, it generally constitutes less than 1% of nonmuscle cell protein. The properties of erythrocyte myosin were presented above (Section 3.2.1.1). Figure 3.17 shows electron micrographs and schematic drawings of platelet myosin. Panels A and B illustrate the characteristic features of an individual myosin molecule. Panels C and D show an intact myosin fiber composed of 28 myosin molecules. The overlapping tail regions and the numerous globular heads are seen. Panel b of Figure 3.13 shows an immunofluorescence photomicrograph of myosin within tissue culture cells. The punctate labeling of stress fibers is apparent.

3.2.2.5. Microtubules

Microtubules are found in almost all eukaryotic cells but never in prokaryotes. They are the largest cytoskeletal fiber with a diameter of 25 nm. Microtubules are built from α and β tubulin. Both tubulin molecules are globular proteins of 55 kDa. The

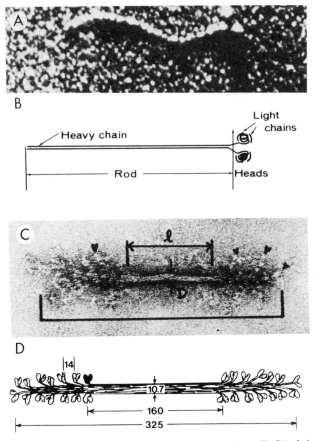

Figure 3.17. Myosin. Electron micrographs (A, C) and schematic drawings (B, D) of platelet myosin are shown. Panels A and B show a single myosin molecule. Panels C and D show a myosin filament. The bare zone length is given by a "1." The dimensions of a filament in nanometers are given in panel d. (Reproduced from the *Journal of Cell Biology*, 1980, Vol. 91, p. 1565, by copyright permission of the Rockefeller University Press.)

amino acid sequences of α and β tubulin are similar, suggesting that they arose from a common ancestral gene. Tubulin heterodimers polymerize to form microtubules in a fashion similar to that described above for actin. Above the critical tubulin concentration, polymerization begins with a lag phase followed by a more rapid elongation phase (see Figure 3.11). Microtubules depolymerize in the presence of calcium, certain drugs such as colchicine (Figure 3.18), and low temperatures. Tubulin polymerizes to form cylindrical structures as shown in Figure 3.19. Thirteen tubulin subunits spiral about a microtubule's hollow center.

Weisenberg (1972) was the first to assemble microtubules *in vitro* using brain extracts. Polymerization requires an absence of calcium ions. In addition to conventional

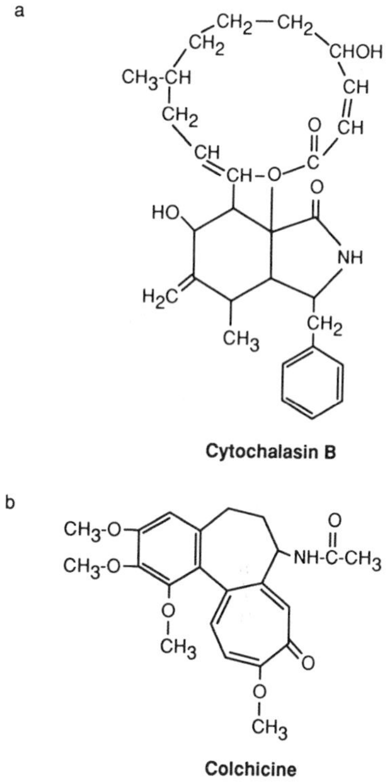

Cytochalasin B

Colchicine

Figure 3.18. Chemical structures of cytochalasin B and colchicine. The drugs cytochalasin B and colchicine promote the depolymerization of microfilaments and microtubules, respectively.

biochemical procedures, tubulin can be purified by a cyclical assembly–disassembly procedure. While the subunits are assembled in the form of a microtubule, non-microtubule contaminants are washed away by centrifugation. By repeating this cycle many times a homogeneous preparation of microtubule-bound proteins is obtained. In addition to tubulin, a collection of proteins known as microtubule-associated proteins (MAPs) are found (Olmsted, 1986; Section 3.2.2.6). These proteins represent up to 20% of a microtubule's protein by weight. They likely play an important role in microtubule assembly and stability.

Microtubules are not static structures; they are continually undergoing subunit flux and change. One manifestation of their dynamics is the phenomenon of treadmilling. Treadmilling is the continual flow of tubulin subunits through a microtubule. It results from a microtubule's polarity. Assembly of new tubulin subunits into microtubules occurs at the + or A end. Microtubule disassembly takes place at the − or D end. Figure 3.20 illustrates the continual assembly and disassembly of microtubules and the resultant treadmilling. Since tubulin binds GTP especially well during assembly, a pulse of radiolabeled GTP can be added to microtubules. The radioisotope is then removed and replaced by unlabeled GTP. The samples are chased for various periods of time. A stripe of radiolabeled tubulin subunits moves along the microtubule's length

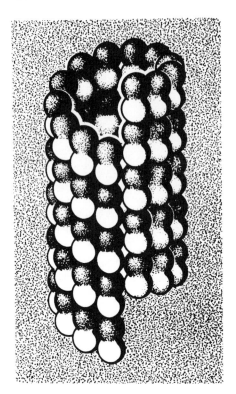

Figure 3.19. A model of microtubule structure. A current model of microtubule structure is shown. Each of the 13 protofilaments is shown vertically. The α-subunits are darkened. (Reproduced from the *Journal of Cell Biology*, Vol. 102, p. 1064, by copyright permission of the Rockefeller University Press.)

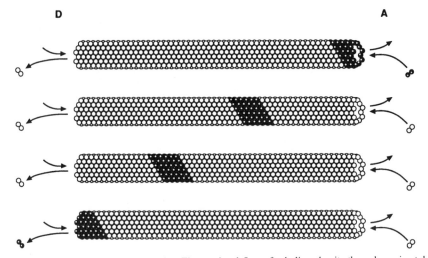

Figure 3.20. Treadmilling of microtubules. The continual flow of tubulin subunits through a microtubule studied *in vitro* is shown. The radiolabeled tubulin subunits are darkened. The upper illustration shows a pulse with radioactive tubulin. This is followed by chase periods of increasing lengths of time. A stripe of radiolabeled tubulin flows from the assembly to disassembly ends. (Redrawn with permission from *Nature*, Vol. 293, p. 705, copyright © 1981 by Macmillan Magazines Ltd.)

(Figure 3.20). The continual flow of tubulin subunits through a microtubule is tread-milling.

The dynamic state of microtubules within living cells has been shown by fluorescent analogue cytochemistry. Fluorescent tubulin microinjected into cells equilibrated with microtubule-bound tubulin with a half-time of 20 min (Salmon *et al.*, 1984). Schulze and Kirschner (1986) provided further evidence indicating that rapid tubulin incorporation was due to elongation from one end of a microtubule, not random insertion along its entire length. Figure 3.21 shows a double-labeling fluorescence microscopy study of a single cell. The cell was microinjected with a specially labeled tubulin molecule. After 50 sec the cell was fixed to inhibit tubulin exchange. This procedure uniquely labels portions of microtubules assembled during the 50-sec incubation period. Panel A shows the results of this experiment. Short labeled microtubule segments are observed. The fixed cells were then labeled by conventional immunofluorescence techniques using a second color fluorescent dye (panel B). Long microtubules that include the segments labeled in the previous panel are observed. This indicates that tubulin subunits must be rapidly (3.6 μm/min or 100 subunits/sec) incorporated at one end of a microtubule. A second class of microtubules representing about 10% of the total exchange very slowly. This study indicates that treadmilling occurs in the cytoplasm of living cells.

In living cells, tubulin's concentration is too low to support its polymerization into microtubules. Proteins that cap the D end of microtubules diminish the loss of tubulin subunits. These may be found, for example, at microtubule-organizing centers within a cell.

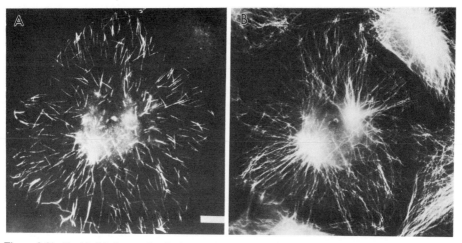

Figure 3.21. Double-labeling study of microtubules. Tubulin was microinjected into the cytosol of living fibroblasts. After 50 sec the cells were fixed. Panel A shows the incorporation of tubulin subunits into microtubules. Short segments of microtubules were labeled. Panel B shows the same cell labeled by immunofluorescence. The entire length of the microtubule can be observed. Bar = 10 μm. (Reproduced from the *Journal of Cell Biology*, Vol. 102, p. 1020, by copyright permission of the Rockefeller University Press.)

3.2.2.6. Microtubule-Associated Proteins

MAPs were first identified using *in vitro* polymerization experiments (see above). MAPs are a heterogeneous collection of proteins that interact with microtubules (Table 3.4). These proteins likely contribute to assembly and regulatory reactions of microtubules.

MAP1 is a high-molecular-weight protein complex composed of three heavy (1A, 1B, and 1C) and two light chains. This protein stimulates microtubule assembly. MAP2 is a distinct protein complex that binds to microtubules. Two very similar isotypes of MAP2, 2A and 2B, have been described on the basis of their electrophoretic migration in gels. MAP2 and 2B are almost identical. MAP2 is a highly asymmetric protein. It possesses two domains: a microtubule-binding domain and a domain that projects from an assembled microtubule. The microtubule-binding domain also binds calmodulin. The projecting domain binds cAMP-dependent protein kinase A (PKA). Both sites provide possible control mechanisms for MAP2–microtubule interactions. Although MAP2 primarily binds to microtubules, it can also interact with microfilaments as shown by its ability to influence the viscosity of actin gels. The actin-binding domain of MAP2 may be similar to spectrin based on immunological cross-reactivity between the proteins.

Several additional MAPs have been identified. Tau proteins have a highly complex structure consisting of three to six proteins depending on species. Phosphorylation sites and calmodulin-binding sites may regulate the activity of tau proteins. Molecular biological, biochemical, and immunological studies indicate that tau proteins are highly conserved. They are capable of binding to F-actin, similar to MAP2.

MAP3 co-purifies with other MAPs, although very little is known about it. MAP4 is composed of three proteins, MAP4A, B, and C. They are related but nonidentical polypeptides. The STOPS (stable tubule-only proteins) may participate in sliding mechanisms along microtubules. Chartins are another class of MAPs which have not been very well characterized.

Fluorescent analogue cytochemistry has shown that MAP2 and MAP4 bind to

Table 3.4.
Summary of Microtubule-Associated Proteins

MAP	M_r ($\times$ 10³)	Postulated function
MAP1		
Heavy chains (1A, 1B, 1C)	350	Stimulate MT formation
Light chains	28, 30	
MAP2 (2A, 2B)	270	Regulatory?
		Binds F-actin
MAP3	180	?
MAP4 (4A, 4B, 4C)	220–240	Similar to ankyrin?
		Binds F-actin
Tau (three to six proteins)	55–62	Regulatory?
STOPS	150, 72, 56	Sliding
Chartins	69, 72, 80	?

microtubules within living cells. Fluorescence photobleaching recovery experiments have shown that MAP2 and MAP4 exchange very rapidly with microtubule-bound proteins. Both microtubules and their associated proteins are very dynamic structures.

3.2.2.7. Intermediate Filaments

Intermediate filaments are the third major class of cytoplasmic filaments. Their name derives from the fact that their diameter (8–11 nm) is intermediate between that of microfilaments and microtubules. Intermediate filaments are a heterogeneous group of proteins possessing similar structures, morphologies, and cellular locations. The five classes of intermediate filaments are: (1) keratin (M_r = 40,0000 to 60,000), (2) desmin (M_r = 50,000), (3) neurofilaments (M_r = 68,000, 150,000 and 200,000), (4) glial filaments (M_r = 51,000), and (5) vimentin (M_r = 52,000). At least one or two intermediate filament classes are found in every higher cell type.

Several properties are common to all classes of intermediate filaments. Intermediate filaments are composed of numerous rod-shaped subunits. These filaments are extremely stable structures; they do not display the dynamic characteristics of microfilaments and microtubules. Reconstitution experiments indicate that intermediate filaments assemble without the aid of other proteins or cofactors. This polmerization reaction is irreversible.

The five classes of intermediate filaments are structurally similar (Figure 3.22, panel a). The amino acid sequences of desmin, neurofilaments, glial filaments, and vimentin are all over 70% homologous. Keratin is the least similar with 25% homology. These proteins contain three regions: NH_2-terminal domain, rod domain, and COOH-terminal domain. The NH_2- and COOH-terminal domains of each protein subunit project from the polymeric intermediate filament. The rod domain is rich in α-helical content. It has four α-helical regions separated by three linker regions as shown in Figure 3.22, panel a. Two intermediate filament proteins wrap around each other forming a cord of two α-helices, or protofilaments, as shown in panel b. Two protofilaments (panel c) form the basic repeating pattern of an intact filament (panel d). Intermediate filaments are distributed throughout the cytoplasmic matrix of cells in complex patterns. The distribution of intermediate filament is dependent on the presence of microtubules. Although colchicine does not bind to intermediate filaments, it can disrupt intermediate filament patterns due to its action on microtubules. Microscopic studies have shown that intermediate filaments frequently cluster about microtubules. Bridges can be observed between microtubules and intermediate filaments. This provides another link among the cytoskeletal systems.

Although intermediate filaments have been the object of intense study, an unambiguous function has not yet been assigned to these structures. Microinjection of anti-intermediate filament antibodies into living cells causes the filaments to clump within cells. Although the distribution of filaments is dramatically changed, there is no effect on cell shape, motility or mitosis. They likely provide structural rigidity for a cell, especially in its tissue environment. This is certainly true for desmosome–tonofilament interactions, which help hold tissues together.

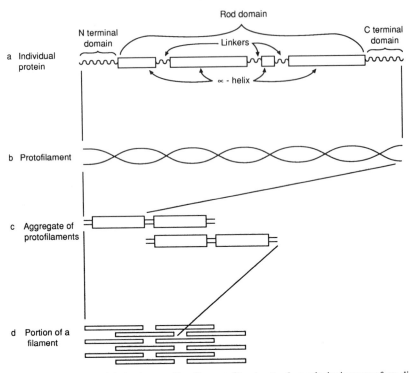

Figure 3.22. A possible model of an intermediate filament. The structural complexity increases from diagram a to d. Diagram a shows one polypeptide with its constituent domains, b shows one coiled protofilament, and c and d show the construction of an intermediate filament from interacting protofilaments.

3.2.2.8. Clathrin

Roughly 2% of a fibroblast's cell surface is covered by clathrin-coated pits. Coated pits participate in one mechanism of receptor-mediated endocytosis and intracellular transport. Coated pits are indentations of a plasma membrane which have specialized compositions at both membrane faces. Certain cell surface receptor activities are associated with the *trans* face of coated pits. The inner face of these specialized membrane domains is covered by a polyhedral coat of the protein clathrin. During endocytosis, the coated pit buds from a plasma membrane to form a coated vesicle. Polyhedral coats of coated vesicles are shown in the electron micrographs of Figure 3.23.

Clathrin coats are composed of clathrin trimers which are also know as triskelions. The three-armed structure of a clathrin trimer is schematized in Figure 3.24, panel a. Panel b is an electron micrograph of a trimer. Each triskelion contains three heavy chains of 180 kDa and three light chains of 20 kDa to 40 kDa. These chains are peripheral membrane proteins since they are easily dissociated from membranes using 2.0 M urea or low-salt buffers. The heavy chains provide the structural framework of a coated pit.

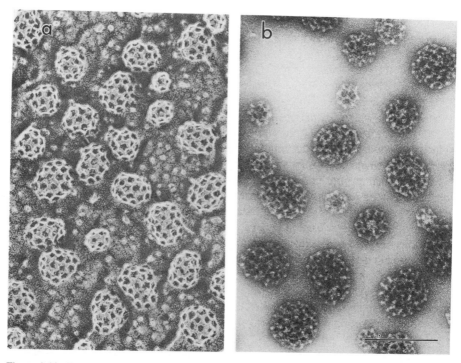

Figure 3.23. Electron microscopic views of clathrin baskets. Freeze-etched (a) and negative-stained (b) electron micrographs of clathrin baskets are shown. Empty baskets were prepared by dialysis of purified clathrin in a polymerization-promoting buffer. Bar = 0.2 μm. (Micrographs produced by Dr. John E. Heuser of Washington University School of Medicine, St. Louis.)

The light chains likely provide regulatory functions since they bind to calmodulin and uncoating enzymes that disassemble clathrin oligomers. Every cell possess two kinds of light chains. These light chains have been sequenced (Jackson *et al.*, 1987). One region of the light chain, rich in α-helical content, demonstrates significant homology with intermediate filaments. But how are clathrin molecules anchored to a membrane?

3.2.2.9. Adaptins Form Transmembrane Links between Membrane Proteins and Clathrin

A layer of assembly proteins (AP) is found between a membrane bilayer and a clathrin network, much like ankyrin links band 3 with the spectrin/actin network beneath erythrocyte membranes. Assembly proteins form complexes of 250 to 300 kDa called HA-1 and HA-2 adaptors. These two adaptors were first separated from one another by hydroxy-apatite chromatography, hence the HA nomenclature. Each adaptor is made from four assembly proteins: two different 100-kDa proteins (called adaptins), one ~50-kDa protein, and one ~20-kDa protein. The assembly proteins that make up

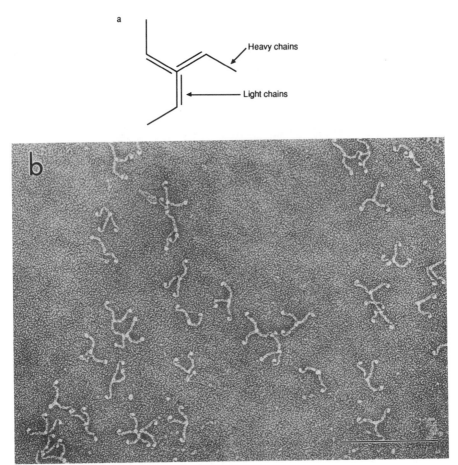

Figure 3.24. Clathrin trimers. A schematic diagram of a clathrin trimer is shown in panel a. Panel b is an electron micrograph of clathrin trimers. Bar = 0.05 μm. (Micrograph produced by Dr. John E. Heuser of Washington University School of Medicine, St. Louis.)

the HA-1 and HA-2 adaptors are different, although they possess some overall similarities. The HA-1 and HA-2 adaptors are found in association with Golgi and plasma membranes, respectively.

Adaptins are classified into three major families (α, β, and γ) which differentially assemble to form adaptors. The deduced amino acid sequences of adaptins revealed three structural regions: a globular NH_2-terminal region, a globular COOH-terminal region, and a stem region connecting the globular domains (Ponnambalam *et al.*, 1990). The stem region is rich in Gly and Pro and likely constitutes a flexible hinge domain. The NH_2-terminal region binds to clathrin while the COOH-terminal region binds to the tails of membrane proteins (Pearse, 1988). In HA-2 adaptors the α adaptin binds to the "hub" or central region of a triskelion while the β adaptin binds to the distal end of an "arm." During our discussion of endocytosis we will consider the molecular signals that adaptins recognize in membrane proteins (Section 6.2.1.3). Adaptins thereby form transmembrane links between membrane proteins and clathrin. The AP50 of HA-2 adaptors is a protein kinase that can be phosphorylated on Ser and Thr residues (Thurieau *et al.*, 1988), although the physiological significance of this phosphorylation is unknown. AP50 and AP17 of HA-2 adaptors and their homologues in HA-1 adaptors are believed to participate in adaptor assembly. Both HA-1 and HA-2 adaptors trigger clathrin coat assembly *in vitro*. The regulation of adaptin function is an active area of research.

REFERENCES AND FURTHER READING

Carbohydrates

Drickamer, K. 1988. Two distinct classes of carbohydrate recognition domains in animal lectins. *J. Biol. Chem.* **263:**9557–9560.

Fukuda, M., *et al.* 1984a. Structure of fetal lactosaminoglycan. The carbohydrate moiety of band 3 isolated from human umbilical cord erythrocytes. *J. Biol. Chem.* **259:**4782–5791.

Fukuda, M., *et al.* 1984b. Structure of branched lactosaminoglycan, the carbohydrate moiety of band 3 isolated from adult human erythrocytes. *J. Biol. Chem.* **259:**8260–8273.

Fukuda, M., *et al.* 1987. Structures of novel sialyated O-linked oligosaccharides isolated from human erythrocyte glycophorins. *J. Biol. Chem.* **262:**11952–11957.

Irimura, T., *et al.* 1981. Structure of a complex-type sugar chain of human glycophorin A. *Biochemistry* **20:**560–566.

Jay, D. G. 1986. Glycosylation site of band 3, the human erythrocyte anion-exchange protein. *Biochemistry* **25:**554–556.

Lis, H., and Sharon, N. 1986. Lectins as molecules and as tools. *Annu. Rev. Biochem.* **55:**35–67.

Pasternack, G. R., and Racusen, R. H. 1989. Erythrocyte protein 4.1 binds and regulates myosin. *Proc. Natl. Acad. Sci. USA* **86:**9712–9716.

Preston, R. D. 1979. Polysaccharide conformation and cell wall function. *Annu. Rev. Plant Physiol.* **30:**55–78.

Rademacher, T. W., Parekh, R. B., and Dwek, R. A. 1988. Glycobiology. *Annu. Rev. Biochem.* **57:** 785–838.

Ryan, C. A. 1987. Oligosaccharide signalling in plants. *Annu. Rev. Cell Biol.* **3:**295–317.

Taylor, R. B., *et al.* 1971. Redistribution and pinocytosis of lymphocyte surface immunoglobulin molecules induced by anti-immunoglobulin antibody. *Nature New Biol.* **233:**225–229.

Tsuji, T. *et al.* 1981. The carbohydrate moiety of band 3 glycoprotein of human erythrocyte membranes. Structures of lower molecular weight oligosaccharides. *J. Biol. Chem.* **256:** 10497–10502.

Turley, E. A., and Roth, S. 1980. Interaction between carbohydrate chains of hyaluronate and chondroitin sulphate. *Nature* **283:**268–271.

Yoshima, H. *et al.* 1980. Structures of asparagine-linked sugar chains of glycophorin A. *J. Biol. Chem.* **255:**9713–9718.

Cytoskeletons

Anderson, R. A., and Lovrein, R. E. 1984. Glycophorin is linked by band 4.1 to the human erythrocyte membrane skeleton. *Nature* **307:**655–657.

Bennett, V. 1985. The membrane skeleton of human erythrocytes and its implications for more complex cells. *Annu. Rev. Biochem.* **54:**273–304.

Branton, D., *et al.* 1981. Interaction of cytoskeletal proteins on the human erythrocyte membrane. *Cell* **24:**24–31.

Elgsaeter, A., *et al.* 1986. The molecular basis of erythrocyte shape. *Science* **234:**1217–1223.

Harrison, S. C., and Kirchhausen, T. 1983. Clathrin, cages, and coated vesicles. *Cell* **33:**650–659.

Horowitz, A., *et al.* 1986. Interaction of plasma membrane fibronectin receptor with talin-A transmembrane linkage. *Nature* **320:**531–533.

Ip, W. *et al.* 1985. Assembly of vimentin in vitro and its implications concerning the structure of intermediate filaments. *J. Mol. Biol.* **183:**365–375.

Jackson, A. P., *et al.* 1987. Clathrin light chains contain brain-specific insertion sequences and a region of homology with intermediate filaments. *Nature* **326:**154–159.

Lazarides, E. 1980. Intermediate filaments as mechanical integrators of cellular space. *Nature* **238:**249–256.

Liu, S. C., *et al.* 1987. Visualization of the hexagonal lattice in the erythrocyte membrane skeleton. *J. Cell Biol.* **104:**527–536.

Marchesi, V. T. 1985. Stabilizing infrastructure of cell membranes. *Annu. Rev. Cell Biol.* **1:**531–561.

Moore, M. S., *et al.* 1987. Assembly of clathrin-coated pits onto purified plasma membranes. *Science* **236:**558–563.

Niggli, V., and Burger, M. M. 1987. Interaction of the cytoskeleton with the plasma membrane. *J. Membr. Biol.* **100:**97–121.

Olmsted, J. B. 1986. Microtubule-associated proteins. *Annu. Rev. Cell Biol.* **2:**421–457.

Pearse, B. M. F. 1988. Receptors compete for adaptors found in plasma membrane coated pits. *EMBO J.* **7:**3331–3336.

Pearse, B. M. F., and Robinson, M. S. 1990. Clathrin, adaptors, and sorting. *Annu. Rev. Cell Biol.* **6:**151–171.

Pollard, T. D., and Cooper, J. A. 1986. Actin and actin-binding proteins. A critical evaluation of mechanisms and functions. *Annu. Rev. Biochem.* **55:**987–1035.

Ponnambalam, S., *et al.* 1990. Conversation and diversity in families of coated vesicle adaptins. *J. Biol. Chem.* **265:**4814–4820.

Salmon, E. D., *et al.* 1984. Diffusion coefficient of fluorescein labeled tubulin in the cytoplasm of embryonic cells of a sea urchin: Video image analysis of the fluorescence redistribution after photobleaching. *J. Cell Biol.* **99:**2157–2164.

Scherson, T., *et al.* 1984. Dynamic interactions of fluorescently labeled microtubule-associated proteins in living cells. *J. Cell Biol.* **99:**425–435.

Schliwa, M., and van Blerkom, J. 1981. Structural interaction of cytoskeletal components. *J. Cell Biol.* **90:**222–231.

Schulze, E., and Kirchner, M. 1986. Microtubule dynamics in interphase cells. *J. Cell Biol.* **102:** 1020–1031.

Steinert, P. M., and Parry, D. A. D. 1985. Intermediate filaments: Conformity and diversity of expression and structure. *Annu. Rev. Cell Biol.* **1:**41–65.

Stossel, T. P., *et al.* 1985. Non-muscle actin-binding proteins. *Annu. Rev. Cell Biol.* **1:**353–402.

Taylor, D. L., and Wang, Y. L. 1980. Fluorescently labeled molecules as probes of the structure and function of living cells. *Nature* **284:**405–410.

Thevenin, B. J. M., *et al.* 1989. The redox state of cysteines 201 and 317 of the erythrocyte anion exchanger is critical for ankyrin binding. *J. Biol. Chem.* **264:**15886–15892.

Thurieau, C., *et al.* 1988. Molecular cloning and complete amino acid sequence of AP50, an assembly protein associated with clathrin-coated vesicles. *DNA* **7:**663–669.

Ungewickell, E., and Branton, D. 1981. Assembly units of clathrin coats. *Nature* **289:**420–423.

Weeds, A. 1982. Actin-binding proteins—Regulators of cell architecture and cell motility. *Nature* **296:**811–813.

Weisenberg, R. 1972. Microtubule formation *in vitro* in solutions containing low calcium concentrations. *Science* **177:**1104–1105.

Willardson, B. M., *et al.* 1989. Localization of the ankyrin-binding site on erythrocyte membrane protein, band 3. *J. Biol. Chem.* **264:**15893–15899.

Chapter 4

Supramolecular Membrane Structure

To this point we have been concerned with the molecular and macromolecular building blocks of membranes. This chapter will focus on how these building blocks are assembled to become functional membranes.

4.1. X-RAY AND ELECTRON DIFFRACTION AND IMAGE PROCESSING

The pioneering research of Kendrew and Perutz in the 1950s led to high-resolution three-dimensional structures of myoglobin and hemoglobin. During the past quarter century, high-resolution structures of many proteins, including a few very complex multisubunit proteins, have been obtained using x-ray diffraction. Studies of membrane-bound proteins have not fared so well. Complete high-resolution structures of a bacterial photosynthetic reaction center and porin have been obtained (Deisenhofer *et al.*, 1984; Weiss *et al.*, 1991). In addition, crystal structures of the extracellular regions of influenza virus hemagglutinin, the variable surface coat protein of trypanosomes, HLA antigens, and the aspartate receptor have been determined (Wilson *et al.*, 1981; Bjorkman *et al.*, 1987a; Milburn *et al.*, 1991). Although three-dimensional crystals of many membrane proteins and their fragments are now available, the photosynthetic reaction center and porin provide the only high-resolution structures of transmembrane domains. X-ray diffraction has also been useful in the study of lipids and ionophores (Table 4.1). Medium-resolution structures of the membrane-bound domains of bacteriorhodopsin and porin have been obtained from two-dimensional crystals using electron diffraction (Henderson and Unwin, 1975; Jap *et al.*, 1991). Unfortunately, two-dimensional crystals with sufficiently long range order for electron diffraction have been difficult to obtain. Negative-stain electron microscopy in combination with image processing has been employed to study numerous membrane proteins, although it provides the lowest resolution. In this section we will discuss the information that molecular structural techniques provide regarding the supramolecular organization of membranes and ex-

Table 4.1.
Summary of Diffraction Studies on Membrane Structures

	Form	Reference[a]
Lipids		
DMPC	3-D crystal	*Nature* **281**:499
Dilauroyl PE	3-D crystal	*PNAS* **71**:3036
PC, PC + cholesterol	Multibilayer	*NNB* **230**:69
Mycoplasma	Packed dispersion	*NNB* **230**:72
RBC ghost	Packed dispersion	*NNB* **230**:72
Myelin	Multibilayer	*NNB* **231**:46
Bacteriophage	Packed	*NNB* **229**:197
PC/GM$_1$ liposomes	Multibilayer	*Biophys. J.* **49**:94
Lipid phase transitions		
PC + cholesterol	Packed dispersion	*JBC* **247**:3694
Mycoplasma	Packed fragment	*JMB* **58**:153
Anacystis	Packed fragment	*BBA* **602**:673
Tetrahymena	Multibilayer	*BBA* **730**:17
Oligomers		
Alamethicin	3-D crystal	*Nature* **300**:325
Gramicidin A	3-D crystal	*Science* **241**:182
Nonactin	3-D crystal	*HCA* **55**:137
A23187	3-D crystal	*J. Antibiot.* **29**:4
Macrocyclic tetracarboxamide	3-D crystal	*Nature* **295**:526
Proteins		
Gap junction	Fragment pellet	*JCB* **74**:629
Cytochrome b$_5$	Pellet	*Biophys. J.* **49**:829
Photosynthetic reaction center	3-D crystal	*JMB* **180**:385
Hemagglutinin	3-D crystal	*Nature* **289**:366
Bacteriorhodopsin	2-D crystal[b]	*Nature* **257**:28
	2-D crystal	*JMB* **93**:123
	3-D crystal	*PNAS* **77**:1283
Porin	3-D crystal[c]	*Biophys. J.* **49**:96
	3-D crystal	*Science* **254**:1627
	2-D crystal	*FEBS Lett.* **205**:29
Aspartate receptor	3-D crystal	*Science* **254**:1342
Acetylcholine receptor	Pellet	*JMB* **116**:635
HLA class I	3-D crystal	*Nature* **329**:506
ras protein	3-D crystal	*Nature* **345**:309
Peripheral membrane structures		
Peptidoglycan	Pellet	*EJB* **95**:147
	Cold drying	*JMB* **117**:927
Cytoskeleton	Fixed oriented retina	*JMB* **158**:435

[a]Journal abbreviations: *BBA, Biochim. Biophys. Acta*; *EJB, Eur. J. Biochem.*; *HCA, Helv. Chim. Acta*; *JBC, J. Biol. Chem.*; *JCB, J. Cell Biol.*; *JMB, J. Mol. Biol.*; *NNB, Nature New Biol.*; *PNAS, Proc. Natl. Acad. Sci. USA.*
[b]Electron diffraction with medium resolution.
[c]Neutron diffraction.

plore fresh data obtained with emerging techniques such as scanning tunneling microscopy and atomic force microscopy.

4.1.1. X-ray Diffraction from Membranes

X-ray diffraction experiments yield structural information regarding a sample. This
information is obtained by analyzing one or more diffraction patterns which record
how a sample's molecular structures bend x rays. Figure 4.1 shows an x-ray diffraction
pattern created by an oriented multilamellar stack of PC membranes. A series of closely
spaced arcs cross the diffraction pattern's vertical axis. These arcs correspond to
multiple reflections from successive bilayers within the stack. Using Braggs' law, this
series of arcs indicate that the width of the bilayers is 4.97 nm. Large diffuse arcs cross
the horizontal axis in Figure 4.1. These reflections, corresponding to a distance of 0.46
nm, arise from neighboring fatty acyl chains and correspond to the distance between the
chains. Disorder among the fatty acyl chains smears out the arcs.

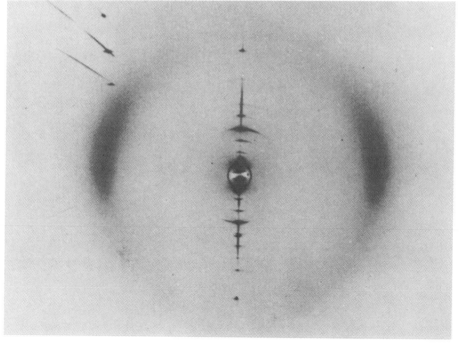

Figure 4.1. An x-ray diffraction pattern of oriented egg PC multibilayers. A hydrated PC sample (14% water)
was examined by x-ray diffraction. The vertical axis extends from the top to the bottom of the page; the
horizontal axis runs from side to side. (Reprinted with permission from *Nature New Biology*, Vol. 230, p. 69,
copyright © 1971 by Macmillan Magazines Ltd.)

The physical properties of membranes are reflected in their diffraction patterns. For example, as the temperature of DPPC multibilayers is decreased, the diffuse 0.46-nm band becomes a sharp 0.42-nm band. As a membrane becomes less fluid, its disorder decreases. The 0.42-nm distance corresponds to a reduction in the physical separation between adjacent fatty acyl chains. As the temperature of a DPPC membrane decreases, the bilayer's width increases from 5.0 to 5.5 nm.

The structure of bilayers composed of PC and II^3-NANA-Gg_4Cer (GM_1) has been studied by x-ray diffraction (McDaniel and McIntosh, 1986). Multibilayers were examined as described above. The negative charge of NANA was labeled with the electron-dense atom europium. The distance between the europium and the phospholipid head groups was found to be 1.5 nm. This distance indicates that the carbohydrate chain must extend perpendicular to the bilayer's plane.

This subsection has described how two-dimensional structural information can be gleaned from x-ray diffraction patterns, and three-dimensional structural information from crystals. We will now discuss structural information gleaned from x-ray diffraction studies and its implications.

4.1.2. Crystal Structures of Membrane Ionophores

The crystal structures of several membrane ionophores have been determined. These include high-resolution structures of valinomycin, nonactin, dibenzo 30-crown-10, macrocyclic polyethers, X537A, A23187, alamethicin, and gramicidin (Table 4.1). We will discuss the high-resolution structures of nonactin, a carrier ionophore, and alamethicin, a channel-forming ionophore.

The structures of nonactin (Dobler, 1972) and its K^+ complex (Kilbourn et al., 1967) have been reported. Figure 4.2 shows a high-resolution structure of native crystalline nonactin. Uncomplexed nonactin is a relatively flat doughnut-shaped molecule with dimensions of $1.7 \times 1.7 \times 0.85$ nm. The central cavity of nonactin is large enough to allow a hydrated K^+ ion to approach its oxygen atoms. X-ray crystallography of the K^+–nonactin complex has shown that nonactin assumes a roughly spherical shape about a K^+ ion. This complex has dimensions of $1.5 \times 1.5 \times 1.2$ nm. The entrapped K^+ ion has lost its waters of hydration. The ion adopts an approximately cubic coordination with 8 of nonactin's 12 oxygen atoms. A K^+ ion fits tightly into the complex; the distance between the centers of the potassium and oxygen atoms (0.27 nm) is the sum of their crystal radii. A K^+ ion is held in the complex by ion–dipole interactions with oxygen atoms. The K^+ ion's waters of hydration are believed to be replaced in a stepwise fashion by interactions with nonactin's oxygen atoms.

Fox and Richards (1982) have determined the crystal structure of the voltage-gated channel-forming antibiotic alamethicin to a resolution of 0.15 nm. Of course, these alamethicin crystals were not membrane-bound. Nonetheless, the structure of the monomer in combination with reasonable intermolecular interactions allow the construction of molecular membrane models of alamethicin channels. The secondary structure

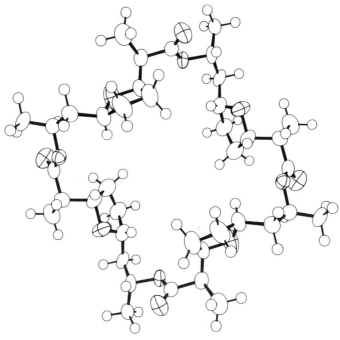

Figure 4.2. Crystal structure of nonactin. The nonactin molecule is viewed along its central axis. (From Dobler, 1972, *Helv. Chim. Acta* **55**:1371.)

and a portion of a proposed tertiary structure of membrane-bound alamethicin are shown in Figure 4.3. The primary structural features of the alamethicin monomer are two α-helical segments separated by a bend at Pro-14. The α-helical regions of each monomer are represented as cylinders in Figure 4.3. Polar amino acid side chains line the center of the cavity. A negatively charged glutamic acid residue is near the COOH-terminus as the opening of a pore. Interchain hydrogen bonding and van der Waals forces are believed to stabilize the overall structure. The mode of insertion is not certain. The monomers could oligomerize and then insert, or insert and then oligomerize. The end result is the formation of a funnel-like channel.

4.1.3. X-ray and Electron Diffraction Studies of Membrane Proteins

Three-dimensional crystals of several membrane proteins have been prepared. The hydrophobic character of membrane proteins is the principal difficulty encountered in preparing crystals. It has recently been shown that some detergents both solubilize membrane proteins and aid in their crystallization. Lower-molecular-weight detergents, such as octylglycoside, are particularly useful. They solubilize proteins and are small

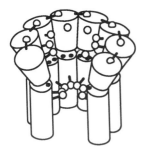

Figure 4.3. A schematic illustration of an alamethicin channel. The structure of an individual alamethicin molecule was determined by x-ray diffraction to be composed of two α-helical domains. These domains are drawn as cylinders in the illustration. The interactions among subunits leading to pore formation are presumed to occur based on reasonable physicochemical criteria. (Redrawn with permission from *Nature* Vol. 300, p. 325, copyright © 1982 by Macmillan Magazines Ltd.)

enough to avoid disruption of crystal packing. Some crystals prepared in this fashion are of very high quality and suitable for x-ray diffraction analysis. This approach has been especially useful in studies of bacterial photosynthetic reaction centers (Section 5.2.2.3).

One approach to avoid the inherent problems in crystallizing membrane proteins is to proteolytically remove the hydrophobic membrane-bound segment. The extra-membrane portion can then be crystallized from aqueous buffers. This approach is particularly useful when the membrane-bound domain is of little or no interest.

The crystal structure of an influenza virus hemagglutinin has been determined (Wilson *et al.*, 1981). This virus caused the "Hong Kong" flu epidemic of 1968. Wilson *et al.* first cleaved the influenza virus hemagglutinin glycoprotein near its membrane attachment site. Over 90% of this protein was released from the membranes. This large fragment was crystallized and then studied by x-ray diffraction. It contains the physiologically interesting antigenic sites, carbohydrates, and host receptor binding sites. The hemagglutinin's primary structure has been determined (Table 2.3). The mature protein's mass (224,640 Da) is distributed between two chains: HA1 (328 amino acids) and HA2 (221 amino acids). The native protein exists as a trimer in viral envelopes.

Figure 4.4 shows a schematic diagram of the hemagglutinin glycoprotein derived from a 0.3-nm structural determination. This protein contains three major extracellular structural domains: a distal globular region, a fibrous stem region, and a proximal globular region. An enhanced view of each of these three regions is shown on the right-hand side of Figure 4.4. Each domain participates in influenza virus infections.

The distal globular region is composed of eight strands of β-sheet structure and two loop structures between the strands (Figure 4.4). The loops possess the primary antigenic sites. The host receptor binding site is located in a groove at the distal end of this domain. This groove is composed of highly conserved amino acid residues. Virions attach to target cells by binding cell surface sialic acid residues within this pocket. X-ray diffraction has been used to study hemagglutinin proteins possessing bound sialic acid molecules (Weis *et al.*, 1988). Sialic acid slides laterally into the groove; one side of sialic acid's ring faces the bottom of the groove while the other side is exposed to the aqueous environment. The groove's conserved amino acids interact with sialic acid. Several of these side chains form hydrogen bonds with sialic acid's acetamido nitrogen atom, a carboxylate oxygen, and two hydroxyl groups. Van der Waals contact between

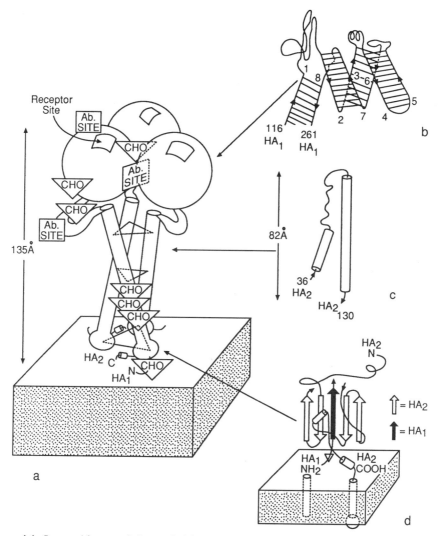

Figure 4.4. Structural features of a hemagglutinin trimer. A schematic illustration of a hemagglutinin trimer is shown. (a) A summary diagram is given. The antigenic and carbohydrate attachment sites are indicated as Ab and CHO, respectively. Enhanced views of the distal globular domain (b), stem region (c), and proximal globular region (d) are shown. (Adapted with permission from *Nature* Vol. 289, p. 366, copyright © 1981 by Macmillan Magazines Ltd.)

the protein and carbohydrate is especially important for the acetamido's methyl moiety. Sialic acid's negative charge likely interacts with nearby positive dipoles (Weis *et al.*, 1988). These interactions account for the tight binding of influenza virus envelopes to cell membranes.

The fibrous stem region is composed of two antiparallel α-helices. The hemagglutinin trimer is stabilized on viral envelopes by three α-helices of three hemagglutinin molecules forming a coiled coil structure. The globular domain near the membrane surface contains five strands of β-sheet. These strands are contributed by both the HA1 and HA2 subunits. This region undergoes a dramatic pH-dependent conformational change that initiates the fusion of an envelope with a target cell's membrane. We shall return to this process in Section 8.1. Many of hemagglutinin's carbohydrate sites are located on the proximal globular domain. This domain is conserved among similar viruses. Although its functional role is not well understood, the proximal domain is required for infectivity.

Wiley and colleagues have also studied the structure of HLA class I antigens by x-ray diffraction (Bjorkman *et al.*, 1987a). HLA molecules are expressed on the surfaces of human leukocytes where they bind to short processed forms of foreign antigens. When processed antigens are bound to HLA molecules, they stimulate an immune reaction against that antigen. The x-ray structural data provide a means for understanding HLA-mediated immune responses at a molecular level.

Figure 2.17 shows a line drawing of an HLA molecule. A second protein known as β_2-microglobulin ($\beta_2 m$) is very tightly, but noncovalently, bound to HLA. HLA's chain is composed of three domains: α_1, α_2, and α_3. Figure 4.5, panel b shows a side view of HLA's three-dimensional structure. $\beta_2 m$ and α_3 are structurally homologous to one another and are found near the lipid bilayer. The α_1 and α_2 domains are farthest from a membrane. The homologous α_1 and α_2 domains fit together to form an antigen-binding groove across the protein's distal end. Figure 4.5 shows top and side views of HLA's groove. The groove is 1 nm wide and 2.5 nm long, sufficient to hold a processed antigen. The sides of the antigen-binding site are composed of long α-helices contributed by the α_1 and α_2 domains. The groove's floor is built from a series of β-strands. The amino acid side chains surrounding the groove determine its antigen binding affinity and contribute to T cell recognition of the antigen–HLA complex. Variations in the groove's primary and tertiary structure likely contribute to some people's predisposition for certain diseases and their ability to respond to specific antigens.

In principle, electron diffraction could be used to study any well-ordered two-dimensional membrane crystal. Unfortunately, just two examples, bacteriorhodopsin and porin, have been reported. As described in Section 2.3.7, bacteriorhodopsin exists as a two-dimensional crystal in living bacteria. Unstained specimens of bacteriorhodopsin studied by electron diffraction have provided structural details to a resolution of 0.7 nm (Henderson and Unwin, 1975). Improvements in data analysis and the use of low-temperature microscope stages which minimize specimen damage have improved resolution to 0.37 nm (Hayward and Stroud, 1981) and 0.35 nm (Henderson *et al.*, 1990). Bacteriorhodopsin has seven transmembrane α-helices whose pathways are roughly perpendicular to a membrane's plane. A model of bacteriorhodopsin derived from these studies is shown in Figure 4.6. Only the transmembrane domains of the protein are shown. The polypeptide loops between the helices at the *cis* and *trans* membrane faces are not shown since their greater degree of flexibility "smears out" their image. This model of bacteriorhodopsin's three-dimensional structure provides insight into the structure of integral membrane proteins.

a

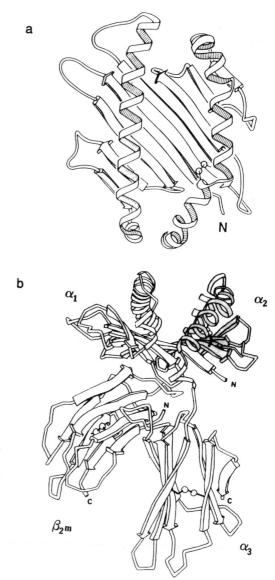

b

Figure 4.5. Structure of an HLA antigen. Two views of the three-dimensional structure of an HLA antigen are shown. The structure was determined by x-ray diffraction. Panel a shows the antigen-binding groove as viewed perpendicular to a membrane. Panel b shows a side view of this membrane protein that includes β_2-microglobulin and HLA's α_1, α_2 (gray), and α_3 subunits. (Reprinted with permission from *Nature* Vol. 329, p. 506, copyright © 1987 by Macmillan Magazines Ltd.)

4.1.4. Image Processing

To this point we have been concerned with information obtained by x-ray and electron diffraction. These data are collected in the form of diffraction patterns. We shall now discuss similar experiments in which data from two-dimensional crystals are

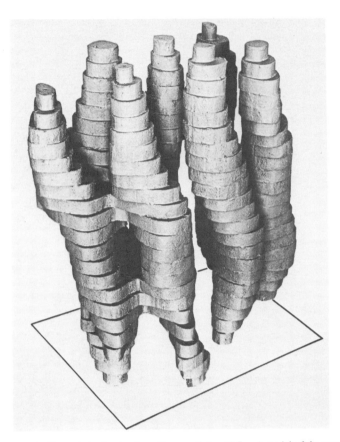

Figure 4.6. A model of bacteriorhodopsin. A medium-resolution (0.7 nm) model of the membrane-bound domain of bacteriorhodopsin is shown. It is composed of seven transmembrane α-helices. Data were collected by electron diffraction. (Reprinted with permission from *Nature* Vol. 257, p. 28, copyright © 1975 by Macmillan Magazines Ltd.)

collected as images instead of diffraction patterns. Diffraction patterns and crystalline images are closely related to one another. Using mathematical equations that relate an image with its diffraction pattern, scientists can filter noise from an image and reinforce or average many individual images.

The three major steps of image processing are graphically illustrated in Figure 4.7. In these experiments Caspar *et al.* (1977) studied the supramolecular structure of gap junctions by negative-stain electron microscopy. Panels a and b of Figure 4.7 show negative-stain electron micrographs of a gap junction. A well-ordered two-dimensional crystalline array is seen. Because of this order, an optical diffraction pattern (panel c) can be formed from the image in panel b. The arrangement and intensity of spots on the diffraction pattern are related to a sample's structural details. The diffraction pattern contains information from many points of the micrograph; it therefore represents an "average" structure. Random features or noise in the micrograph contribute to the

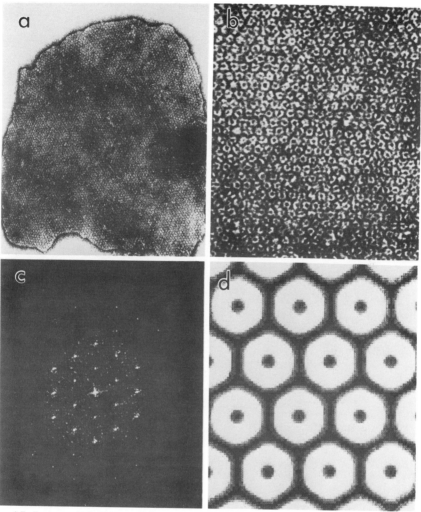

Figure 4.7. Image analysis of an isolated gap junction. Samples were negatively stained followed by electron microscopy. (a) A survey electron micrograph of a gap junction is shown (×90,000). (b) A high-magnification electron micrograph is shown (×310,000). (c) This panel shows an optical diffraction pattern of the image in panel B. (d) An optically filtered image of this gap junction is shown. (Reproduced from the *Journal of Cell Biology*, 1977, Vol. 74, p. 605, by copyright permission of the Rockefeller University Press.)

diffuse background of the optical diffraction pattern (panel c). This noise can be removed from the pattern. The inverse procedure called a Fourier synthesis is then performed to create an image from the diffraction pattern (panel d). Although the steps of image processing are particularly well illustrated by this figure, they are now performed by computers. Gap junctions are found to be hexagonal structures 8.7 nm center-to-center with a 2.0-nm central cavity. We shall return to the topic of gap junctions after considering two simple membrane systems.

4.1.5. Dark-Field Electron Microscopy of Membrane Ionophores

Most biological samples are visualized with electron microscopes by direct staining, indirect staining, or shadowing with heavy metal atoms. Examples of these approaches are: thin-section, negative-stain, and freeze-fracture electron microscopy. However, these visualization strategies have an inherent resolution of about 2.0 nm. In contrast, electron diffraction and dark-field electron microscopy do not rely on staining or shadowing a sample; therefore, they can produce higher-resolution images. Although a specimen's contrast is dramatically reduced, image processing methods are used to enhance dark-field electron micrographs.

Figure 4.8 shows images of the ionophore valinomycin. Panel A shows an averaged

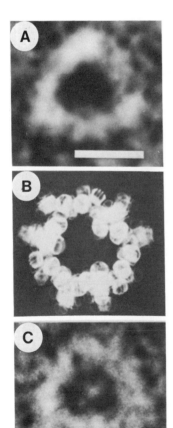

Figure 4.8. Dark-field electron microscopy of valinomycin. (A) An enhanced image of valinomycin is shown. (B) A space-filling model is shown for comparison. (C) The addition of potassium ions creates an additional electron density at the center of the image. Bar = 10 Å. (Reproduced with permission from the *Annual Review of Biophysics and Bioengineering*, Vol. 8, copyright © 1979 by Annual Reviews, Inc.)

dark-field electron micrograph of native valinomycin. A doughnut-shaped structure is observed. A space-filling model of valinomycin is given in panel B for comparison. In the presence of K^+ ions an additional central prominence can be found (panel C). Dark-field electron microscopy provides another method to investigate biological structures. It is most applicable to smaller membrane structures beyond the resolution of conventional electron microscopic techniques. Since samples do not require crystallization, it can provide data complementary to those obtained by other techniques.

4.1.6. Negative-Stain Electron Microscopy of Membrane Proteins

Two-dimensional crystals of at least 18 different membrane proteins or protein complexes have been obtained. Although these samples are not suitable for electron diffraction studies, structural information can be gleaned from negative-stain electron microscopy and image processing. However, these images must be carefully interpreted since different negative stains can yield different staining patterns of the same sample (Baker *et al.*, 1985).

We have previously introduced the supramolecular structure of gap junctions during our discussion of image processing (Section 4.1.4). Gap junctions allow small molecules to cross from one cell into a neighboring cell without entering the extracellular environment. They are composed of hexameric units called connexons. Each connexon has a central channel 2.0 nm in diameter. Connexons of one cell pair with connexons from another cell to form a channel connecting the two cells. An *en face* view of a gap junction is given in Figure 4.7.

To obtain information about the entire connexon, instead of just its membrane face, data are acquired from tilted samples. Information from several two-dimensional views at different angles are combined to yield a three-dimensional picture of a sample. This procedure was employed to obtain most of the three-dimensional protein models shown in this subsection.

Figure 4.9 shows two models of a connexon. Reconstituted samples were prepared in the presence or absence of calcium (Unwin and Ennis, 1984). Gap junctions exist in closed and open states *in vivo*. Calcium stimulates closing of a connexon's channel. This closing is accomplished by the change in protein conformation shown in Figure 4.9.

The nicotinic acetylcholine receptor is a pentamer composed of four similar transmembrane glycoprotein subunits in the stoichiometry of $\alpha_2\beta\gamma\delta$. All of these subunits have been sequenced by cDNA methodology (Table 2.3). This receptor participates in cell-to-cell signaling via the neurotransmitter acetylcholine. When acetylcholine binds to a nicotinic acetylcholine receptor, a large change in the target membrane's ionic permeability is induced. The permeability change is mediated by a membrane channel that allows the passage of Na^+ and K^+ ions. The membrane channel is formed by the receptor's five subunits.

The nicotinic acetylcholine receptor's supramolecular structure has been studied by electron microscopy and image processing (Brisson and Unwin, 1985; Toyoshima and Unwin, 1988). The receptor's overall structure resembles a cylinder passing through a membrane. Figure 4.10 shows a cylindrically averaged cross section of the receptor. It

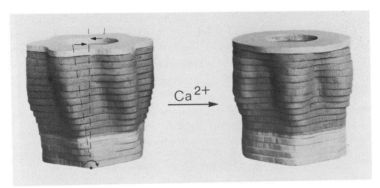

Figure 4.9. Three-dimensional models of gap junction connexons. Data were obtained by negative-stain electron microscopy. The shaded portions represent membrane-bound domains. In the presence of calcium, gap junctions undergo a conformation change that closes the channel (the *trans* face is at the bottom of the figure). (Reprinted with permission from *Nature* Vol. 307, p. 609, copyright © 1984 by Macmillan Magazines Ltd.)

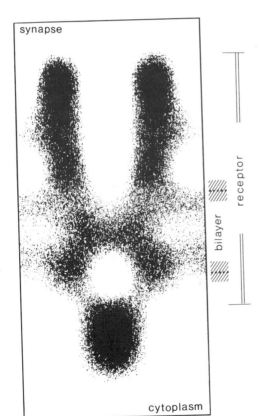

Figure 4.10. A model of the acetylcholine receptor. Data were obtained by negative-stain electron microscopy. The bilayer's location is shown. The long funnel-like region faces the exterior of a cell. (Reprinted with permission from *Nature* Vol. 336, p. 247, copyright © 1988 by Macmillan Magazines Ltd.)

extends 6.0 nm into the extracellular space. The channel's outer diameter is 2.5 nm at the tip of the receptor's *trans* face; its inner diameter of 1.0 to 1.3 nm narrows to roughly 0.7 nm as it nears the membrane bilayer. The circular structure shown at the cytoplasmic face is not an intrinsic component of the receptor. It is a peripheral 43-kDa protein that may provide a link with the membrane skeleton and participate in regulating the receptor's channel activity.

4.1.7. Negative-Stain Electron Microscopy of Cytoskeletal Structures

Negative-stain electron microscopy has been utilized to study the supramolecular structures of microfilaments and microtubules. F-actin has been studied as both individual filaments and two-dimensional crystalline sheets (Pollard and Cooper, 1986; Amos, 1985).

The helical properties of actin filaments were first noted in the 1950s using x-ray diffraction. However, it has not been possible to obtain a high-resolution structure of G- or F-actin. Negative-stain electron microscopy has provided the most detailed information regarding the structure of actin filaments. Figure 4.11 shows an electron micrograph and a model of an actin filament. The actin filament is a helix with two strands with many intermolecular contacts. The arrangement of actin monomers within a filament is uncertain. However, this should be resolved when the high-resolution x-ray diffraction structure becomes available. It should also be possible to map where actin-binding proteins bind to filaments using both electron microscopy and x-ray diffraction.

Several structural features of microtubules were discussed in Section 3.2.2.5.

Figure 4.11. A three-dimensional structural model of an actin filament. Actin filaments were studied by negative-stain electron microscopy. Images were analyzed by computer. Two views of the model with subunits numbered 1 through 4 are shown. The bar along the right-hand side is 5.5 nm. (Reproduced with permission from the *Annual Review of Biochemistry*, Vol. 55, copyright © 1986 by Annual Reviews, Inc.)

Microtubules have also been examined by negative-stain electron microscopy. Micro-tubules in the form of sheets have usually been examined (Schultheiss and Mandelkow, 1983). A model based on this work is shown in Figure 3.19.

4.1.8. Negative-Stain Electron Microscopy of Complex Bacterial Membranes

The most complex crystalline membrane domains studied by negative-stain elec-tron microscopy are bacterial cell envelopes and photosynthetic membranes. Although the molecular order of these systems has been appreciated for many years, recent technological advances have spurred interest in extracting their structural details. In some of these complex membrane crystals the subunit composition and/or stoichiometry are not well characterized. The bacterial systems offer insight into the molecular logic of membrane pores. Studies of photosynthetic membranes provide information regarding the electron conduction pathways during photosynthesis (Section 5.2.2.2).

Two-dimensional crystalline structures have been observed in the cell envelopes of several bacteria (Glauert and Thornley, 1969; Baumeister *et al.*, 1989). Gram-negative bacteria such as *Spirillum serpens* have hexagonal arrays of surface macromolecules. A few species with linear and tetragonal supramolecular structures have been found. Both hexagonal and tetragonal structures are frequently found in gram-positive bacteria. For example, *Bacillus anthracis* and *B. polymyxa* have hexagonal and tetragonal arrays, respectively. These membrane structures are amenable to further study by image processing.

Figure 4.12 shows averaged and filtered images of the surface of the gram-positive bacterium *Micrococcus radiodurans*. This is a rather atypical gram-positive bacterium since it contains an outer membrane. The outer membrane layer is characterized by a hexagonally packed array of proteins. Figure 4.12, panel a shows a negatively stained region of the outer envelope of *M. radiodurans*. Each circular element within the

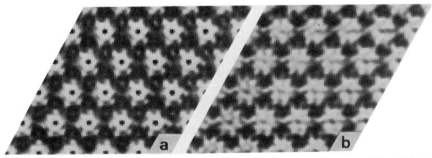

Figure 4.12. The crystalline outer membrane layer (HPI) of *Micrococcus radiodurans*. Filtered electron microscope images of negatively stained (panel a) and positively stained (panel b) samples are shown. Central pores of about 3 nm are visible by negative-staining. Lipid-rich domains are observed by positive-staining. (From Baumeister and Kubler, 1978, *Proc. Natl. Acad. Sci. USA* **75**:5525.)

processed image contains a central pore 3 nm in diameter. Small projections can be found around the perimeter of each element. Panel b shows a similar sample positively stained with osmium tetroxide. Since osmium tetroxide preferentially reacts with lipids, the lipid-containing regions appear dark in panel b. The channels seen in panel a are separated by lipid-rich domains. The positive-staining approach provides data complementary to the negative-staining experiments. Higher-resolution information has been obtained for *Deinococcus radiodurans* by preparing samples and making observations at very low temperatures (Rachel *et al.*, 1986).

4.1.9. Emerging Methods in Molecular Structural Analysis

Recently, several new techniques have been introduced and applied to the structural analysis of membranes. These include spot-scanning transmission electron microscopy, scanning tunneling microscopy, and atomic force microscopy (e.g., Downing, 1991; Hoh *et al.*, 1991; Butt *et al.*, 1990; Wang *et al.*, 1990; Zasadzinski *et al.*, 1988; Smith *et al.*; 1987). Under certain conditions these methods may allow higher resolution, less perturbative sample preparation, observations in aqueous media, and/or faster data acquisition. It seems likely that these new imaging methods will complement or replace some existing technology. These emerging methods will likely give us a deeper understanding of molecular structures in membranes.

Scanning Tunneling Microscopy of Bilayers

During scanning tunneling microscopy a finely sharpened tungsten probe is brought to within several tenths of a nanometer from a surface. Although there is no physical contact between the probe and a surface, they are so very close that electrons can "quantum mechanically" tunnel across the gap. The electrons' movements set up a measurable current. This current is very sensitive to the molecular topography of a surface; height changes of 0.01 nm are detected. Therefore, individual surface atoms and their spatial arrangements are visualized. If the current is measured as the probe is scanned across a surface, a two-dimensional map of the surface is created.

Scanning tunneling microscopy has recently been extended to the study of fatty acid bilayer surfaces (Smith *et al.*, 1987). These investigators examined inverted graphite-supported planar arachidic bilayers. The methyl groups of the fatty acyl chains pointed toward the graphite or the air. The carboxylic acid groups were near one another; they were likely stabilized by the presence of cadmium ions. The scanning tunneling microscope visualized the structure as viewed from the air. Figure 4.13 shows this unconventional bilayer structure. Each light oval represents an end-on view of one fatty acyl chain. The spacing along the *a* axis is 0.06 nm whereas that along the *b* axis is 0.04 nm. A resolution of 0.02 nm is obtained. Since scanning tunneling microscopy is nondestructive under certain conditions, multiple images or "movies" of a sample should become possible. It may become possible to "film" phase transitions and observe ionophores and integral membrane proteins in the not too distant future.

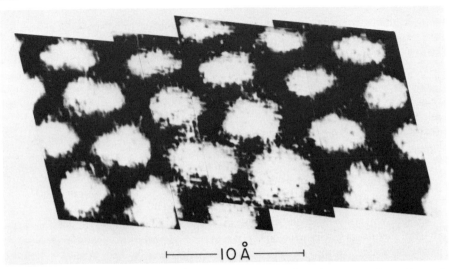

Figure 4.13. Scanning tunneling microscopy of supported arachidic bilayer. A computer-enhanced image of a bilayer is shown. Since the bilayer is inverted, the nearest part of the fatty acids are the terminal methyl groups. Bar = 10 Å. (From Smith *et al.*, 1987, *Proc. Natl. Acad. Sci. USA* **84**:969.)

4.2. TRANSMISSION ELECTRON MICROSCOPY

The preceding section described the three-dimensional molecular structures of several membrane components. These structures were obtained using diffraction or image processing techniques which require relatively simple and highly ordered structures. Complex living systems do not meet these criteria. However, the organization of complex biological systems can be studied by preparing intact organelles or cells for examination with transmission electron microscopes.

Transmission electron microscopes provide a resolution of 0.2 nm with magnifications of up to 200,000×. It easily surpasses the resolution of optical and scanning electron microscopes, but not that of scanning tunneling or atomic force microscopes. However, in cellular studies the resolution is established by a sample's preparatory methods, not an instrument. In transmission electron microscopy, electrons are passed through a specimen *in vacuo* to form an image. Fragile biological materials cannot withstand electron bombardment or a vacuum. Therefore, they must be converted to derivative structures that are both stable and nonvolatile. Since electrons have a low penetrating power, samples must also be very thin. In this section we will discuss thin-section and freeze-fracture electron microscopy.

4.2.1. Thin-Section

During the "golden age" of cell biology, electron microscopes revealed the presence of endoplasmic reticula and cytoskeletons in cells. In addition, after many years of uncertainty, the existence of the Golgi apparatus was confirmed using thin-section electron microscopy. The steps involved in thin-section electron microscopy are listed in Table 4.2. Fixation steps are employed to stabilize biological structures. The bifunctional reagent glutaraldehyde fixes samples by cross-linking various chemical groups. Cell structures are effectively polymerized to yield a gelatinous material. In some cases samples are then treated with cytochemical stains to visualize specific enzyme activities of endogenous or exogenous origin (see below). The object of these experiments is to precipitate an electron-dense material at the site of enzyme activity.

At this stage a specimen is generally postfixed with osmium tetroxide. Osmium tetroxide reacts with lipids and thereby reduces their loss during dehydration. Figure 4.14 shows a possible reaction product when bilayers are treated with osmium tetroxide. In this example two unsaturated fatty acids became cross-linked. Membranes are electron-dense after postfixation due to the presence of the heavy metal osmium. Cross-links form at the *cis* and *trans* faces and between the two faces of a membrane.

A fixed and postfixed sample is dehydrated by washing it with an organic solvent such as ethanol or acetone. A plastic monomer is added to a sample followed by its polymerization. The plastic supports a tissue and provides the hard resin needed for thin sectioning. Thin sections of roughly 50 nm are sliced from samples using a glass or diamond knife.

4.2.2. Applications of Thin-Section Electron Microscopy

Early thin-section electron microscopy studies revealed that all membrane boundaries of cells and organelles possess a characteristic trilamellar structure. The trilamellar structure is composed of two parallel electron-dense lines separated by an electron-lucent zone. This is the so-called "railroad track" or "sandwich" appearance of membranes. The width of this boundary is roughly 10 nm. On the basis of electron microscopic evidence available in the late 1950s, Robertson revised the Davson–Danielli model of membrane structure (Chapter 1). The study of the trilamellar properties of cell

Table 4.2.
Comparison of Sample Preparatory
Methodologies for Thin-Section
and Freeze-Fracture Electron Microscopy

Thin-section	Freeze-fracture
Prepare sample	Prepare sample
Glutaraldehyde fixation	Freeze
Cytochemical stain (optional)	Fracture
OsO$_4$ fixation	Etch (for freeze-etching)
Dehydration	Metal shadowing
Suspend in plastic monomer	Carbon support coating
Polymerize plastic	Clean replica
Cut thin sections	

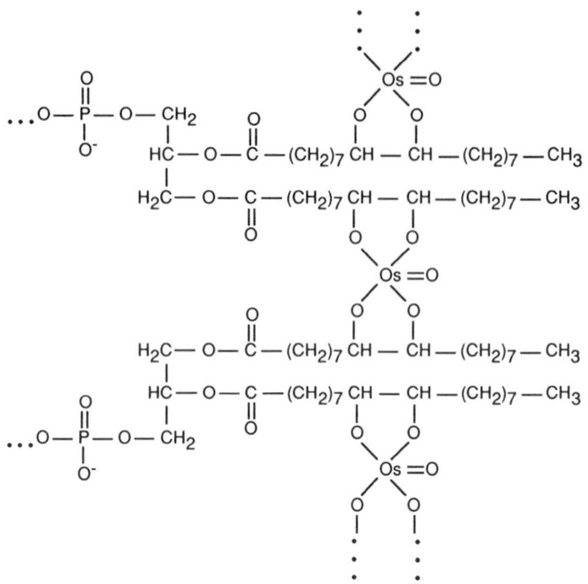

Figure 4.14. A possible reaction product of osmium tetroxide and lipid. Oleic acid is a major component of the phospholipids of this bilayer. Osmium tetroxide reacts with unsaturated moieties, thereby immobilizing the lipid and inserting an electron-dense osmium atom into the bilayer.

boundaries was instrumental in the early development of membrane biology. Many applications of thin-section electron microscopy such as cytochemistry, immuno-cytochemistry, and autoradiography remain useful in membrane studies.

The objective of cytochemical investigations is to render a specific cell component electron-dense. Several strategies are employed to create a specific electron-dense labeling of a membrane. For example, membrane components can be localized with: (1) a chemical reaction, such as ruthenium red (Figure 4.15), (2) an antibody or lectin conjugated to ferritin, and (3) an endogenous or exogenous enzyme activity that creates an electron-dense reaction product (Figure 4.16). Numerous membrane-bound enzymes have been examined by cytochemical techniques. For example, the membrane-bound phosphatases alkaline phosphatase, 5'-nucleotidase, ATPase, glucose-6-phosphatase, and adenylyl cyclase have been studied. In this type of experiment a substrate specific for a particular enzyme activity is incubated with glutaraldehyde-fixed cells. The enzyme activity cleaves its substrate, releasing a phosphate molecule. In the presence of lead ions, the phosphate is precipitated as electron-dense lead phosphate. Figure 4.16 shows that the 5'-nucleotidase activity of lymphocytes is located at the *trans* face of plasma membranes. The lead phosphate reaction product encircles the cell. In this fashion the chemical and biological properties of membrane-bound enzymes can be determined *in situ*.

Singer and Schick (1961) were the first to combine antibody labeling with thin-section electron microscopy. This technique, immunocytochemistry, provides a highly

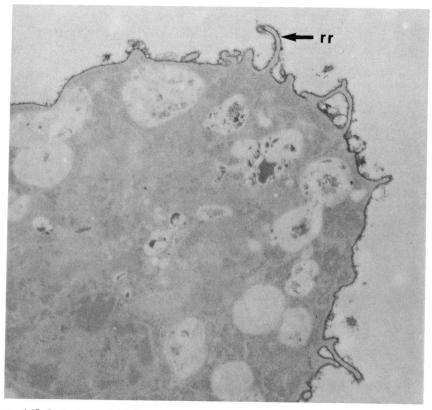

Figure 4.15. Ruthenium red labeling of a plasma membrane. Thin-section electron micrograph of a macrophage after antibody-dependent phagocytosis of ferritin-containing liposomes. The ruthenium red (rr) reaction product labels carbohydrates at the cell surface. Electron-dense clumps of ferritin can be seen within phagosomes. (This portion of the cell is roughly 3 μm wide.) (From Petty and McConnell, 1983, *Biochim. Biophys. Acta* **735**:77.)

sensitive and specific method for localizing antigens within cells. An antibody is tagged with an electron-dense marker such as ferritin or an enzyme that catalyzes the formation of an electron-dense precipitate. Since some enzymes have no known cytochemical staining procedure, they must be localized with immunocytochemistry. For example, antibodies specific for the enzyme carbon monoxide dehydrogenase have been prepared (Rohde *et al.*, 1984). This enzyme plays a central role in carbon monoxide metabolism of *Pseudomonas carboxydovorans*. It passes electrons derived from carbon monoxide oxidation into the carbon monoxide-insensitive respiratory chain. This enzyme is unique in that it is found in both membrane-bound and cytosolic forms. Immunocytochemistry has shown that during log-phase growth, most of the enzyme is translocated to the cytoplasmic membrane's *cis* face (Rohde *et al.*, 1985). It has been suggested that the rate of electron flow from carbon monoxide to oxygen is controlled by the amount of membrane-bound carbon monoxide dehydrogenase.

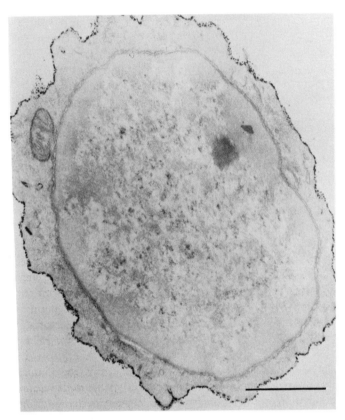

Figure 4.16. Cytochemical labeling of a plasma membrane-bound enzyme. The 5'-nucleotidase activity is visualized at the *trans* face of the lymphocyte plasma membrane. Bar = 1 μm. (Reproduced with permission from Uusitalo and Karnovsky, 1977, Surface localization of 5'-nucleotidase on the mouse lymphocyte, *J. Histochem. Cytochem.* **25**:87.)

4.2.3. Freeze-Fracture

Sample preparation for thin-section electron microscopy involves chemical fixation, dehydration, and embedding within plastic. Specimens treated in this fashion become rather far removed from their biological setting. Fixation is intended to preserve the native properties and positions of cell components. An alternative to chemical fixation is cryofixation. During cryofixation samples are rapidly frozen in a hydrated state. Indeed, most cryofixed cells will live if thawed.

Table 4.2 outlines the techniques of freeze-fracture and freeze-etch microscopy. These two terms are sometimes confused. Freeze-etch is freeze-fracture with an additional etching step (Table 4.2). In both approaches a sample must first be frozen. The conventional freezing methods utilize cryoprotectants (antifreeze) to prevent cell or membrane damage due to ice formation. Cryoprotectants such as polyvinylpyrrolidone, dextran, glycerol, dimethyl sulfoxide, and ethylene glycol have been used. All of these cryoprotectants have the ability to form hydrogen bonds with water and thereby interfere with ice formation. During freezing a sample is immersed in liquid Freon where it is frozen within 0.1 sec. More sophisticated techniques such as spray-freezing and quick-freezing provide faster cooling rates and, therefore, better structural preservation and greater temporal resolution of biological events.

Figure 4.17 illustrates the overall process of conventional freeze-fracture studies. A frozen sample is fractured *in vacuo* by a passing knife blade. If a sample is held at $-100°C$, water sublimes from its surface at a rate of 100 nm/min; this is known as etching the surface. At $-110°C$ the sample is in equilibrium with its surroundings and no etching occurs. A metallic copy of the fractured surface is now prepared by shadowing or "replication." A platinum gun held at an angle of 45° is used to coat a sample. Since platinum is electron-dense, it appears dark in electron micrographs. Because of a sample's surface topography, some areas receive more platinum than others. This is analogous to the shadows cast by objects when the sun is low in the sky. Shadowing creates a sample's three-dimensional appearance when viewed in an electron microscope. A platinum replica is given a uniform coat of carbon. Carbon is invisible in the microscope and provides a stable backing which holds a replica together. A platinum–carbon replica is ready for viewing after cell debris has been washed away.

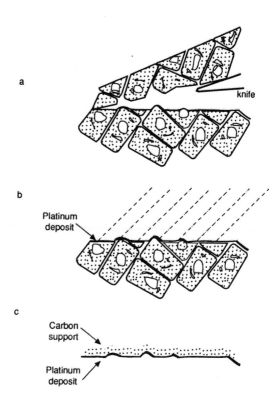

Figure 4.17. Methods of specimen preparation during freeze-fracture. (a) At $-110°C$ a sample is fractured by the impact of a knife. (b) platinum exposure creates shadows at the surface of the specimen. (c) The cleaned carbon-coated replica is ready for viewing in an electron microscope.

4.2.4. Interpretation of the Fracture Plane

During a freeze-fracture experiment, a fracture plane is propagated through a sample including its membranes. Two critical questions regarding freeze-fracture experiments are: (1) how is a fracture plane propagated at a membrane and (2) what is observed when replicas are viewed in a microscope?

Two models were proposed to account for fracture plane propagation in membranes. Figure 4.18 shows two models proposed by groups from Switzerland (panel a) and California (panel b). The first group thought that a fracture plane passes along the interface between a medium and a cell surface. Since cleavage occurs outside a bilayer, the particles observed by electron microscopy must be outside a cell. Alternatively, the California group led by Daniel Branton (1966) held that a fracture plane moves through a bilayer's center. The particles, therefore, could originate from a membrane's core. The reasons for this interpretation were based upon physicochemical principles. Covalent, hydrogen, and ionic bonds are strong at both room temperature and $-110°C$. However, hydrophobic interactions are strong at room temperature but weak at $-110°C$ because water is in the form of solid ice. It was reasoned, therefore, that membranes were split down the middle during freeze-fracture.

Several lines of evidence have shown Branton's interpretation to be the correct one. Branton and Park (1967) showed that aldehyde fixation has no effect on the appearance of freeze-fracture replicas. However, lipid extraction of aldehyde-fixed cells abolished membrane splitting in freeze-fracture although no effect was observed in thin-section studies. Furthermore, postfixation with osmium tetroxide, which would be expected to cross-link the two bilayer faces, destroyed the fracture plane. These findings suggested that lipids or lipid-rich regions were required to freeze-fracture membranes.

Deamer and Branton (1967) employed a model membrane system to show that membranes are split along the hydrophobic core during freeze-fracture. Bilayers were formed on a glass insert using the monolayer trough technique (Section 2.1.8). Radio-labeled fatty acids were incorporated into either or both sides of a bilayer. One leaflet of the bilayer was apposed to the glass while the other was adjacent to the aqueous buffer. The sample was then frozen and fractured. In this experiment there could be four possible results: (1) no preferred fracture plane, (2) fracture at the glass/bilayer interface, (3) fracture at the ice/bilayer interface, and (4) fracture along the bilayer's center. When both membrane leaflets were labeled, the radioactivity was equally distributed between the ice and the glass surfaces. Similarly, if only the outer monolayer was labeled, all of the radioactivity was associated with the ice. These results indicate that model membranes must be split along the hydrocarbon tail region during freeze-fracture.

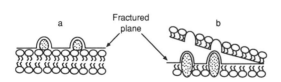

Figure 4.18. Two hypotheses regarding the propagation of the fracture plane at membranes. (a) The fracture plane was proposed to pass along the outside of a membrane. (b) The fracture plane was proposed to pass through the center of a bilayer. The latter model is correct.

Although the data discussed above were quite convincing, a few skeptics believed that results with model membranes were somehow inapplicable to cell membranes. Any lingering doubts regarding the location of the fracture plane were removed by the experiments of Pinto da Silva and Branton (1970). Ferritin was linked to both the *cis* and *trans* faces of erythrocyte ghosts using a bifunctional covalent cross-linking reagent. Freeze-fracture electron micrographs of this sample are shown in Figure 4.19. Intramembrane particles are the predominant feature of both images. The specimen shown in panel B was etched whereas that in panel A was not. Large ferritin molecules are only seen in panel B after etching. They are visible around the perimeter of the cell's image which is the cell's outer surface. Therefore, the fracture plane must have been propagated through the bilayer's center.

4.2.5. Intramembrane Particles

As Figure 4.19 reveals, numerous intramembrane particles are visualized during freeze-fracture experiments. But what are intramembrane particles? Intramembrane particles are largely or exclusively composed of integral membrane proteins. Several lines of evidence, both direct and indirect, support this conclusion. Membranes rich in proteins are rich in intramembrane particles whereas membranes that contain few proteins have few intramembrane particles. This correlation was extended to an association by Pinto da Silva *et al.* (1971) and Pinto da Silva and Nicolson (1974). These freeze-etch studies showed that clusters of intramembrane particles were contiguous with clusters of antigens or Con A binding sites in erythrocyte ghost membranes. Furthermore, aggregation of red blood cell membrane proteins was accompanied by aggregation of intramembrane particles. Similarly, treatment with antispectrin antibodies led to aggregation of intramembrane particles. This suggests that intramembrane particles extend to both membrane surfaces. Unambiguous evidence demonstrating that membrane proteins were intramembrane particles was obtained by reconstitution studies (Hong and Hubbell, 1972; Grant and McConnell, 1974; Segrest *et al.*, 1974; Yu and Branton, 1976). When reconstituted into phospholipid liposomes, large integral membrane proteins such as rhodopsin, cytochrome c oxidase, and band 3 form intramembrane particles. However, some small membrane proteins which traverse a bilayer only once, such as glycophorin, do not form intramembrane particles. Therefore, intramembrane particles are large integral membrane proteins or aggregates of small integral membrane proteins. Further evidence comes from a study by Sowers and Hackenbrock (1981). They showed that intramembrane particles of inner mitochondrial membranes exhibit a lateral diffusion coefficient similar to that of membrane proteins. Therefore, most, if not all, intramembrane particles are integral membrane proteins.

4.2.6. Applications of Conventional Freeze-Fracture

Several properties of intramembrane particles can be quantitated. These include: (1) the density of particles, (2) the particles' sizes, (3) their distribution (random versus

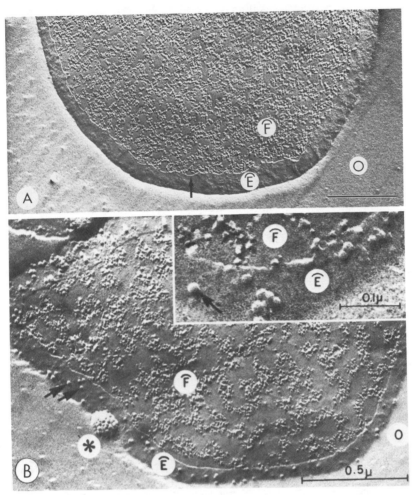

Figure 4.19. Membrane splitting during freeze-fracture of erythrocyte membranes. Erythrocyte ghosts were freeze-fractured and then etched. Panel A shows a freeze-etch micrograph of a control erythrocyte ghost. A ridge (arrow) separates the etched surface from the fractured surface. The external surface of the cell is smooth. Panel B shows a similar experiment with the inclusion of a ferritin label at the external surface. Intramembrane particles can be seen in both panels. Ice is found along the perimeter of the cell in panel A. After etching the water level is lowered. The etched sample in panel B shows ferritin molecules along the outer perimeter of the cell. This indicates that the exterior or *trans* face is exposed by etching. Therefore, a fracture plane generally travels through the center of a membrane. Bars = 0.5 μm. (Panel B is reproduced from the *Journal of Cell Biology*, 1970, Vol. 45, p. 598, by copyright permission of the Rockefeller University Press.)

aggregates), (4) their pattern (e.g., rosettes), and (5) lateral diffusion. The distribution of intramembrane particles is sensitive to various processes including oncogenic transformation and aging. Chloroplast membranes contain four distinct classes of intramembrane particles which are nonrandomly distributed between grana and stroma membranes (Staehelin, 1976). In a few cases intramembrane particles form organized patterns such as the two parallel rows found at motor nerve terminals (Figure 4.20).

Freeze-fracture is also used to study the lateral distribution of cholesterol within the fracture plane and between the two leaflets of a bilayer. Robinson and Karnovksy (1980) employed filipin, which aggregates with cholesterol (Section 2.2.2.1), to study the distribution of cholesterol in cell membranes. Figure 4.21 shows an example of a filipin-labeled LM cell. Cells were labeled after glutaraldehyde fixation to avoid both cell damage and artifactual redistribution of membrane components. The intramembrane particles and filipin–cholesterol complexes are randomly distributed in the fracture plane. Although the asymmetry of membrane glycerophosphatides and glycolipids has been supported by a variety of experiments, the distribution of cholesterol between membrane faces has received little attention. Fisher (1976) devised a method to freeze-fracture a monolayer of cells. Since a membrane is split along the hydrocarbon tail region, the *cis* and *trans* faces can be separated from one another in the frozen state. The split membranes were confirmed by electron microscopic observations of intramembrane particles. These studies indicate that cholesterol is enriched at the *trans* face of erythrocyte membranes.

4.2.7. Quick-Freezing

A seminal paper by Heuser *et al.* (1979) introduced the quick-freezing technique. With this technique they provided the first *definitive* structural evidence that neurotransmitters are released at nerve terminals via exocytosis. The success of this study rested on three important factors. (1) Since quick-freezing is complete in less than 2 msec, very rapid events could be frozen in time. Moreover, since specimens are not converted into derivative structures by chemical fixation, they are likely to be more accurate representations of living cells. (2) 4-Aminopyridine augments neurotransmitter release by about 100-fold, thereby making exocytotic events easy to find by electron microscopy. This strategy allowed many exocytotic events to be frozen in time for observation. Figure 4.20 shows two freeze-fracture electron micrographs of 4-amino-pyridine-treated motor nerve terminals. (3) The quick-freezing apparatus was electronically rigged to stimulate neuromuscular junctions at various times before freezing. The two parallel rows of intramembrane particles seen in panel A are called active zones because they participate in neurotransmitter release. In panel A the cells were stimulated 3 msec prior to freezing. Since it is indistinguishable from unstimulated cells, this must represent a point in time prior to neurotransmitter discharge. Panel B shows an identical sample in which the nerve was stimulated 5 msec before freezing. Numerous exocytotic vesicle openings are seen along the active zone. Each opening corresponds to the fusion of one synaptic vesicle with the plasma membrane and to the release of one burst of

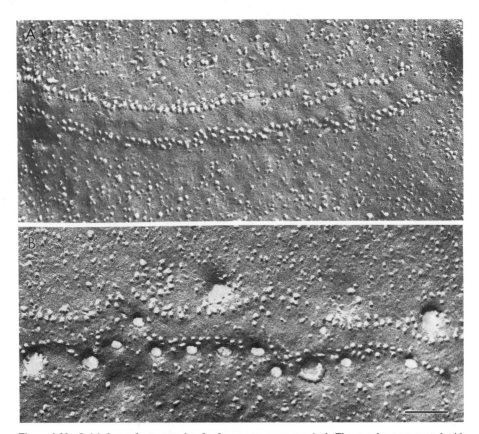

Figure 4.20. Quick-freeze-fracture study of a frog motor nerve terminal. The samples were treated with 4-aminopyridine. Panel A shows an active zone stimulated 3 msec before freezing. Two parallel rows of large intramembrane particles identify an active zone. Panel B shows a similar experiment except that electrical stimulation was provided 5 msec prior to freezing. Many membrane perturbations appear along the active zone. These perturbations correspond to synaptic vesicle exocytotic events. Bar = 0.1 μm. (Micrographs produced by Dr. John E. Heuser of Washington University School of Medicine, St. Louis.)

neurotransmitter. This represents the first case in which exocytotic phenomena have been visualized at various points in time.

Quick-freeze fracture has been combined with deep-etching and rotary replication. In this type of experiment a fractured surface is etched for 2.5 min, allowing 0.1 to 0.24 μm of water to sublime. To avoid formation of large shadows due to the amount of water removed, samples were rotated while being coated with platinum. Figure 4.22 shows the membrane morphology of fibroblasts during the endocytosis of LDL. Panel A shows a view from outside a cell. The spherical LDL particles are seen in the invagination. Panel B shows endocytosis as viewed from inside a cell. A honeycomb of sub-membranous filaments surrounds the coated pit. The hexagon network is formed by the

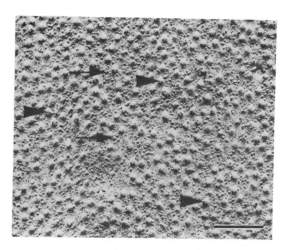

Figure 4.21. Filipin labeling of sterols. Fibroblasts were treated simultaneously with filipin and glutaraldehyde. Intramembrane particles (arrows) and filipin–sterol complexes (arrowheads) are uniformly distributed in the membrane. Bar = 0.2 μm. (Reproduced with permission from Robinson and Karnovsky, 1980, Evaluation of the polyene antibiotic filipin as a cytochemical probe for membrane cholesterol, *J. Histochem. Cytochem.* **28:**161.)

assembly of clathrin trimers or "triskelians." Figure 3.24 shows a few of the building blocks which contribute to the coated membrane seen in Figure 4.22, panel B.

4.3. SCANNING ELECTRON MICROSCOPY

Scanning electron microscopes provide dramatic quasi-three-dimensional views of cells and membranes. The surfaces of organs, cells, and organelles are studied by this technique. The gains in aesthetic quality of an image are surely lost in resolution. A typical scanning electron microscope has at best a resolution of 7 nm in comparison with 0.2 nm for transmission electron microscopes. Due to limitations in biological specimen preparation and image interpretation, a resolution of 7 nm is rarely achieved. Nonetheless, scanning electron microscopy is valuable in assessing overall membrane features.

4.3.1. Background

The preparation of a biological sample for scanning electron microscopy generally involves: fixation, dehydration, critical point drying, and heavy metal coating. Fixation is performed with glutaraldehyde and, frequently, osmium tetroxide. These derivative structures are then dehydrated by washing with ethanol or acetone. Since a scanning electron microscope's sample chamber operates *in vacuo*, all volatile solvents (e.g., ethanol) must be removed. Drying could be performed by allowing a solvent to evaporate in air. As a solvent evaporates, its surface moves closer to a sample. A solvent's surface tension pulls and distorts a sample as it moves downward. Critical point drying avoids this experimental artifact by not passing a solvent surface across a sample. The critical point is defined as the temperature and pressure at which a solvent's liquid and

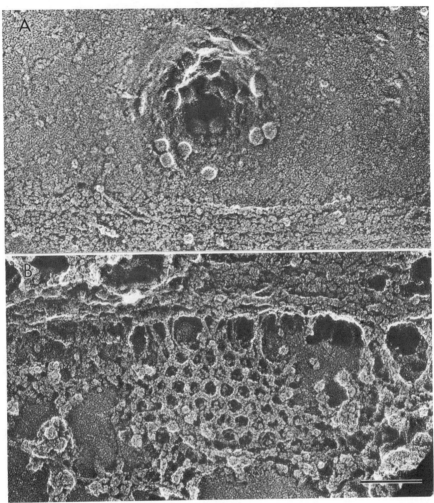

Figure 4.22. Three-dimensional views of LDL endocytosis. Fibroblasts were prepared for electron micro-scopic studies by quick-freeze deep-etch methodology. Panel A shows a pit containing several LDL droplets as viewed from outside the cell. Panel B shows a clathrin network of a coated pit as viewed from inside the cell. Bar = 0.1 μm. (Micrographs produced by Dr. John E. Heuser of Washington University School of Medicine, St. Louis.)

gas phases have the same specific gravity. As a consequence, the liquid and gas phases are completely miscible in one another. Surface tension forces disappear because there is no liquid–gas interface. After critical point drying, samples are coated with a metallic substance such as gold–palladium. The process is similar to the metallic coating used in freeze-fracture electron microscopy. The coated specimen is now ready to be placed in the vacuum chamber of a scanning electron microscope.

4.3.2. Applications

The principal applications of scanning electron microscopy are: (1) the morphological evaluation of cell surfaces and intracellular organelle membranes and (2) the localization of specific cell surface components.

Cell surface features such as microvilli and folds are studied by scanning electron microscopy. Several cell surface events such as exocytosis and phagocytosis may alter cell surface area. Cell surface folds may increase or decrease in number to accommodate changes in surface area while encompassing a constant volume. Burwen and Satir (1977) developed a method to quantitate cell surface folds. Mast cell exocytosis leads to an increase in cell surface folds. Macrophage endocytosis is accompanied by a decrease in surface folds (Figure 4.23). A dramatic reduction in cell surface folds is evident (Petty *et al.*, 1981). The reduction in surface area accounts for a portion of the plasma membrane which became wrapped about the phagocytosed particles during internalization.

Scanning electron microscopy is also used to visualize the surface distribution of membrane components. Receptors, antigens, and carbohydrates are observed using ligands, antibodies, or lectins, respectively, coupled to large markers. These large markers can be any recognizable structure. For example, viruses, hemocyanin, polystyrene latex spheres, silica particles, and gold granules have been used as markers. Figure 4.24 shows carbohydrate-specific labeling of *D. discoideum* using wheat germ agglutinin conjugated to latex microspheres. Cells fixed with glutaraldehyde prior to

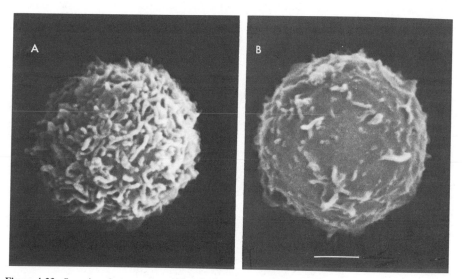

Figure 4.23. Scanning electron micrographs of membrane folds. The effect of phagocytosis on macrophage membrane folds is illustrated. Panel A shows a typical micrograph of a control cell. Panel B shows a cell after phagocytosis. A considerable decrease in the number of cell surface folds is observed. Bar = 2 μm. (Reproduced from the *Journal of Cell Biology*, 1981, Vol. 89, p. 223, by copyright permission of the Rockefeller University Press.)

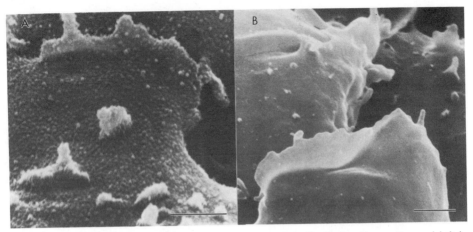

Figure 4.24. Scanning electron microscopy of microsphere-labeled cells. *D. discoideum* cells were labeled with wheat germ agglutinin-conjugated microspheres. Panel A shows a uniform distribution of microspheres at the cell surface. Panel B shows a control utilizing 0.01 M *N*-acetylchitobiose as a hapten sugar. The binding of microspheres is carbohydrate-specific. Bars = 1 μm. (Reproduced with permission from the *Journal of Cell Biology*, 1976, Vol. 71, p. 314, by copyright permission of the Rockefeller University Press.)

labeling exhibit a uniform distribution of microspheres and, by inference, wheat germ agglutinin binding sites.

Conventional scanning electron microscopy cannot be used to visualize intracellular membranes. However, the development of instrumentation and preparatory techniques of increased sophistication have permitted the observation of intracellular structures (Tanaka, 1980). Special microscopes with a resolution of 3 nm have been

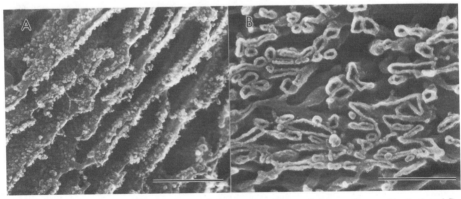

Figure 4.25. High-resolution scanning electron micrographs of intracellular membranes. Panels A and B show rough and smooth endoplasmic reticula, respectively. Bars = 0.5 μm. (From Tanaka, 1980, *Int. Rev. Cytol.* **68:**97.)

constructed. To reveal intracellular structures, cells must be broken open. Cryofracturing techniques in combination with fixation and critical point drying methods are used to prepare specimens. Figure 4.25 shows scanning electron micrographs of rough (panel A) and smooth (panel B) endoplasmic reticula. The *cis* face of the rough endoplasmic reticulum is studded with ribosomes whereas the *trans* face is smooth (data not shown). The branching tubules and cavities of the smooth endoplasmic reticulum are evident. These vivid micrographs provide rich three-dimensional detail of intracellular membranes.

REFERENCES AND FURTHER READING

X-ray and Electron Diffraction and Image Processing

Amos, L. A. 1985. Structure of muscle filaments studied by electron microscopy. *Annu. Rev. Biophys. Biophys. Chem.* **14**:291–313.

Baker, T. S., *et al.* 1985. Gap junction structures. VII. Analysis of connexon images obtained with cationic and anionic negative stains. *J. Mol. Biol.* **184**:81–98.

Baumeister, W., and Kubler, O. 1978. Topographic study of the cell surface of *Micrococcus radiodurans*. *Proc. Natl. Acad. Sci. USA* **75**:5525–5528.

Baumeister, W., *et al.* 1989. Principles of organization in eubacterial and archaebacterial surface proteins. *Can. J. Microbiol.* **35**:215–227.

Bjorkman, P. J., *et al.* 1987a. Structure of the human class I histocompatibility antigen, HLA-A2. *Nature* **329**:506–512.

Bjorkman, P. J., *et al.* 1987b. The foreign antigen binding site and T cell recognition regions of class I histocompatibility antigens. *Nature* **329**:512–518.

Brisson, A., and Unwin, P. T. N. 1985. Quaternary structure of the acetylcholine receptor. *Nature* **315**:474–477.

Butt, H.-J., *et al.* 1990. Imaging the membrane protein bacteriorhodopsin with the atomic force microscope. *Biophys. J.* **58**:1473–1480.

Caspar, D. L. D., and Kirschner, D. A. 1971. Myelin membrane structure at 10Å resolution. *Nature New Biol.* **231**:46–52.

Caspar, D. L. D., *et al.* 1977. Gap junction structures. I. Correlated electron microscopy and x-ray diffraction. *J. Cell Biol.* **74**:605–628.

Deisenhofer, J., *et al.* 1984. X-ray structure analysis of a membrane protein complex. Electron density map at 3Å resolution and a model of the chromophores of the photosynthetic reaction center from *Rhodopseudomonas viridis*. *J. Mol. Biol.* **180**:385–398.

Dobler, M. 1972. The crystal structure of nonactin. *Helv. Chim. Acta* **55**:1371–1384.

Dorset, D. L., *et al.* 1983. Two-dimensional crystal packing of matrix porin: A channel forming protein in *Escherichia coli* outer membranes. *J. Mol. Biol.* **165**:701–710.

Downing, K. H. 1991. Spot-scan imaging in transmission electron microscopy. *Science* **251**:53–59.

Eisenberg, D., *et al.* 1984. Three-dimensional structure of membrane and surface proteins. *Annu. Rev. Biochem.* **53**:595–623.

Engelhardt, H., *et al.* 1986. Stoichiometric model of the photosynthetic unit of *Ectothiorhodospira halochloris*. *Proc. Natl. Acad. Sci. USA* **83**:8972–8976.

Fox, R. O., and Richards, F. M. 1982. A voltage-gated ion channel model inferred from the crystal structure of alamethicin at 1.5-Å resolution. *Nature* **300**:325–330.

Franks, N. P., *et al.* 1982. Structure of myelin lipid bilayers changes during maturation. *J. Mol. Biol.* **155**:133–153.

Freymann, D. M., *et al.* 1984. 6Å-resolution x-ray structure of a variable surface glycoprotein from *Trypanosoma brucei. Nature* **311**:167–169.

Garavito, R. M., *et al.* 1983. X-ray diffraction analysis of matrix porin, an integral membrane protein from *Escherichia coli* outer membranes. *J. Mol. Biol.* **164**:313–327.

Glauert, A. M., and Thornley, M. J. 1969. The topography of the bacterial cell wall. *Annu. Rev. Microbiol.* **23**:159–198.

Harrison, S. C., *et al.* 1971. Lipid and protein arrangement in bacteriophage PM2. *Nature New Biol.* **299**:197–201.

Hayward, S. B., and Stroud, R. M. 1981. Projected structure of purple membrane determined to 3.7Å resolution by low temperature electron microscopy. *J. Mol. Biol.* **151**:491–517.

Hebert, H., *et al.* 1985. Structure of two-dimensional crystals of membrane-bound Na^+, K^+-ATPase as analyzed by correlation averaging. *J. Ultrastruct. Res.* **92**:28–35.

Henderson, R. 1975. The structure of the purple membrane from *Halobacterium halobium*: Analysis of the x-ray diffraction pattern. *J. Mol. Biol.* **93**:123–138.

Henderson, R., and Unwin, P. N. T. 1975. Three-dimensional model of purple membrane obtained by electron microscopy. *Nature* **257**:28–31.

Henderson, R., *et al.* 1990. Model for the structure of bacteriorhodopsin based on high-resolution electron cryo-microscopy. *J. Mol. Biol.* **213**:899–929.

Hitchock, P. B., *et al.* 1974. Structural chemistry of 1,2 dilauroyl DL-phosphatidylethanolamine: Molecular conformation and intermolecular packing of phospholipids. *Proc. Natl. Acad. Sci. USA* **71**:3036–3040.

Hoh, J. H., *et al.* 1991. Atomic force microscopy and dissection of gap junctions. *Science* **253**:1405–1408.

Jap, B. K., *et al.* 1991. Structural architecture of an outer membrane channel as determined by electron crystallography. *Nature* **350**:167–170.

Kilbourn, B. T., *et al.* 1967. Structure of the K^+ complex with nonactin, a macrolide antibiotic possessing highly specific K^+ transport properties. *J. Mol. Biol.* **30**:559–563.

Kuhlbrandt, W. 1984. Three-dimensional structure of the light harvesting chlorophyll a/b-protein complex. *Nature* **307**:478–480.

Levine, Y. K., and Wilkins, M. H. F. 1971. Structure of oriented lipid bilayers. *Nature New Biol.* **230**:69–72.

McDaniel, R. V., and McIntosh, T. J. 1986. X-ray diffraction studies of cholera toxin receptor GM_1. *Biophys. J.* **49**:96–98.

Mannella, C. A. 1986. Mitochondrial outer membrane channel (VDAC, porin) two-dimensional crystals from *Neurospora. Methods Enzymol.* **125**:595–611.

Michel, H., and Oesterhelt, D. 1980. Three-dimensional crystals of membrane proteins: Bacterio-rhodopsin. *Proc. Natl. Acad. Sci. USA* **77**:1283–1285.

Michel, H., *et al.* 1982. Three-dimensional crystals of membrane protein complexes. The photosynthetic reaction centre from *Rhodopseudomonas viridis. J. Mol. Biol.* **158**:567–572.

Milburn, M. V., *et al.* 1991. Three-dimensional structures of the ligand-binding domain of the bacterial aspartate receptor with and without a ligand. *Science* **254**:1342–1347.

Ottensmeyer, F. P. 1979. Molecular structure determination by high-resolution electron micros-copy. *Annu. Rev. Biophys. Bioeng.* **8**:129–144.

Pascher, I., and Sundell, S. 1977. Molecular arrangements in sphingolipids. The crystal structure of cerebroside. *Chem. Phys. Lipids* **20**:175–191.

Pollard, T. D., and Cooper, J. A. 1986. Actin and actin-binding proteins. A critical evaluation of mechanisms and functions. *Annu. Rev. Biochem.* **55**:987–1035.

Rachel, R. *et al.* 1986. Projected structure of the surface protein of *Deinococcus radiodurans* determined to 8 Å resolution by cryomicroscopy. *Ultramicroscopy* **20**:305–316.

Rossmann, M. G., and Henderson, R. 1982. Phasing electron diffraction amplitudes with the molecular replacement method. *Acta Crystallogr. Sect. A* **38**:13–20.

Sass, H. J., *et al.* 1989. Densely packed β-structure at the protein–lipid interface of porin is revealed by high-resolution cryo-electron microscopy. *J. Mol. Biol.* **209**:171–175.

Schultheiss, R., and Mandelkow, E. 1983. Three-dimensional reconstruction of tubulin sheets and re-investigation of microtubule surface lattice. *J. Mol. Biol.* **170**:471–496.

Smith, D. P. E., *et al.* 1987. Images of a lipid bilayer at molecular resolution by scanning tunneling microscopy. *Proc. Natl. Acad. Sci. USA* **84**:969–972.

Toyoshima, C., and Unwin, P. N. T. 1988. Ion channel of acetylcholine receptor reconstructed from images of postsynaptic membranes. *Nature* **336**:247–250.

Unwin, P. N. T., and Ennis, P. D. 1984. Two configurations of a channel-forming membrane protein. *Nature* **307**:609–613.

Wallace, B. A., and Ravikumar, K. 1988. The gramicidin pore: Crystal structure of a cesium complex. *Science* **241**:182–187.

Wang, H., *et al.* 1990. Thickness determination of biological samples with z-calibrated scanning tunneling microscope. *Proc. Natl. Acad. Sci. USA* **87**:9343–9347.

Weis, W., *et al.* 1988. Structure of the influenza virus haemagglutinin complexed with its receptor, sialic acid. *Nature* **333**:426–431.

Weiss, M. S., *et al.* 1991. Molecular architecture and electrostatic properties of a bacterial porin. *Science* **254**:1627–1630.

Wilkins, M. H. F., *et al.* 1971. Bilayer structure of membranes. *Nature New Biol.* **230**:72–76.

Wilson, I. A., *et al.* 1981. Structure of the haemagglutinin membrane glycoprotein of influenza virus at 3Å resolution. *Nature* **289**:366–373.

Zampighi, G. *et al.* 1984. The structural organization of (Na$^+$ + K$^+$)-ATPase in purified membranes. *J. Cell Biol.* **98**:1851–1864.

Zasadzinski, J. A. N., *et al.* 1988. Scanning tunneling microscopy of freeze-fracture replicas of biomembranes. *Science* **239**:1013–1015.

Transmission Electron Microscopy

Andrews, L. D., and Cohen, A. J. 1983. Freeze-fracture studies of photoreceptor membranes: New observations bearing upon the distribution of cholesterol. *J. Cell Biol.* **97**:749–755.

Armond, P. A., and Staehelin, L. A. 1979. Lateral and verticle displacement of integral membrane proteins during lipid phase transition in *Anacystis nidulans*. *Proc. Natl. Acad. Sci. USA* **76**: 1901–1905.

Branton, D. 1966. Fracture faces of frozen membranes. *Proc. Natl. Acad. Sci. USA* **55**:1048–1056.

Branton, D., and Park, R. B. 1967. Subunits in chloroplast lamellae. *J. Ultrastruct. Res.* **19**:283–303.

Deamer, D. W., and Branton, D. 1967. Fracture planes in an ice-bilayer model membrane system. *Science* **158**:655–657.

Fisher, K. A. 1976. Analysis of membrane halves: Cholesterol. *Proc. Natl. Acad. Sci. USA* **73**: 173–177.

Fisher, K., and Branton, D. 1974. Application of the freeze fracture technique to natural membranes. *Methods Enzymol.* **32B**:35–44.

Forsman, C. A., and Pinto da Silva, P. 1988. Fracture-flip: New high-resolution images of cell surfaces after carbon stabilization of freeze-fractured membranes. *J. Cell Sci.* **90**:531–541.

Grant, C. W. M., and McConnell, H. M. 1974. Glycophorin in lipid bilayers. *Proc. Natl. Acad. Sci. USA* **71**:4653–4657.

Heuser, J. 1980. Three-dimensional visualization of coated vesicle formation in fibroblasts. *J. Cell Biol.* **84**:560–583.

Heuser, J., and Kirchhausen, T. 1985. Deep etch views of clathrin assemblies. *J. Ultrastruct. Res.* **92**:21–27.

Heuser, J., and Salpeter, S. R. 1979. Organization of acetylcholine receptors in quick-frozen, deep-etched, and rotary-replicated *Torpedo* postsynaptic membrane. *J. Cell Biol.* **82**:150–173.

Heuser, J., *et al.* 1979. Synaptic vesicle exocytosis captured by quick freezing and correlated with quantal transmitter release. *J. Cell Biol.* **81**:275–300.

Hirokawa, N., and Heuser, J. 1982. The inside and outside of gap junction membranes visualized by deep etching. *Cell* **30**:395–406.

Hong, K., and Hubbell, W. L. 1972. Preparation and properties of phospholipid bilayers containing rhodopsin. *Proc. Natl. Acad. Sci. USA* **69**:2617–2621.

Napolitano, C. A., *et al.* 1983. Organization of calcium pump protein dimers in the isolated sarcoplasmic reticulum membranes. *Biophys. J.* **42**:119–125.

Ossmer, R. *et al.* 1986. Immunocytochemical localization of component c of the methyl reductase system in *Methanococcus voltae* and *Methanobacterium thermoautotrophicum*. *Proc. Natl. Acad. Sci. USA* **83**:5789–5792.

Pinto da Silva, P., and Branton, D. 1970. Membrane splitting in freeze-etching. *J. Cell Biol.* **45**:598–605.

Pinto da Silva, P., and Nicolson, G. L. 1974. Freeze-etch localization of concanavalin A receptors to the membrane intercalated particles of human erythrocyte ghost membranes. *Biochim. Biophys. Acta* **363**:311–319.

Pinto da Silva, P. *et al.* 1971. Localization of A antigen sites on human erythrocyte ghosts. *Nature* **232**:194–196.

Rohde, M, *et al.* 1984. Immunocytochemical localization of carbon monoxidase in *Pseudomonas carboxydovorans*. *J. Biol. Chem.* **259**:14788–14792.

Rohde, M., *et al.* 1985. Attachment of CO dehydrogenase to the cytoplasmic membrane is limiting the respiratory rate of *Pseudomonas carboxydovorans*. *FEMS Microbiol. Lett.* **28**:141–148.

Segrest, J. P., *et al.* 1974. Association of the membrane penetrating polypeptide segment of the human erythrocyte MN-glycoprotein with phospholipid bilayers. I. Formation of freeze-etch intra-membrane particles. *Proc. Natl. Acad. Sci. USA* **71**:3294–3298.

Shnitka, T. K., and Seligman, A. M. 1971. Ultrastructural localization of enzymes. *Annu. Rev. Biochem.* **40**:375–396.

Singer, S. J., and Schick, A. F. 1961. The properties of specific stains for electron microscopy prepared by the conjugation of antibody molecules with ferritin. *J. Biophys. Biochem. Cytol.* **9**:519–537.

Sowers, A. E., and Hackenbrock, C. R. 1981. Rate of lateral diffusion of intramembrane particles: Measurement by electrophoretic displacement and rerandomization. *Proc. Natl. Acad. Sci. USA* **78**:6246–6250.

Staehelin, L. A. 1976. Reversible particle movements associated with the unstacking and restacking of chloroplast membranes *in vitro*. *J. Cell Biol.* **71**:136–158.

Tillack, T. W, and Marchesi, V. T. 1970. Demonstration of the outer surface of freeze-etched red blood cell membranes. *J. Cell Biol.* **45**:649–653.

Uusitalo, R. J., and Karnovsky, M. J. 1977. Surface localization of 5'-nucleotidase on the mouse lymphocyte. *J. Histochem. Cytochem.* **25**:87–96.

Yu, J., and Branton, D. 1976. Reconstitution of intramembrane particles in recombinants of erythrocyte protein band 3 and lipid: Effects of spectrin–actin association. *Proc. Natl. Acad. Sci. USA* **73**:3891–3895.

Scanning Electron Microscopy

Burwen, S. J., and Satir, B. H. 1977. Plasma membrane folds on the mast cell surface: Their relationship to secretory activity. *J. Cell Biol.* **74**:690–697.

Cammisuli, S., and Wofsy, L. 1976. Hapten-sandwich labeling. III. Bifunctional reagents for immunospecific labeling of cell surface antigens. *J. Immunol.* **117**:1695–1704.

Carter, D. P., and Wofsy, L. 1976. Immunospecific labeling of mouse lymphocytes in the scanning electron microscope. *J. Supramol. Struct.* **5**:139–153.

Crusberg, T. C., *et al.* 1979. Spreading behavior and surface characteristics of young and senescent WI38 fibroblasts revealed by scanning electron microscopy. *Exp. Cell Res.* **118**: 39–46.

Hayat, M. A. 1978. *Introduction to Biological Scanning Electron Microscopy*. University Park Press, Baltimore.

Jan, L. Y., and Revel, J.-P. 1975. Hemocyanin-antibody labeling of rhodopsin in mouse retina for a scanning electron microscope study. *J. Supramol. Struct.* **3**:61–66.

Linthicum, D. S., and Sell, S. 1975. Topography of lymphocyte surface immunoglobulin using scanning immunoelectron microscopy. *J. Ultrastruct. Res.* **51**:55–68.

Molday, R. S. 1976a. A scanning electron microscope study of concanavalin A receptors on retinal rod cells labeled with latex microspheres. *J. Supramol. Struct.* **4**:549–557.

Molday, R. 1976b. Concanavalin A and wheat germ agglutinin receptors on *Dictyostelium discoideum*. Their visualization by scanning electron microscopy with microspheres. *J. Cell Biol.* **71**:314–322.

Petty, H. R., *et al.* 1981. Disappearance of macrophage surface folds after antibody-dependent phagocytosis. *J. Cell Biol.* **98**:223–229.

Porter, K. R., *et al.* 1973. A scanning electron microscope study of surface features of viral and spontaneous transformants of mouse BALB/3T3 cells. *J. Cell Biol.* **59**:633–642.

Tanaka, K. 1980. Scanning electron microscopy of intracellular structures. *Int. Rev. Cytol.* **68**: 97–125.

Weller, N. K. 1974. Visualization of concanavalin A-binding sites with scanning electron microscopy. *J. Cell Biol.* **63**:699–707.

Chapter 5

Bioenergetics

Putting Membranes to Work

The genesis and maintenance of life requires an energy source. Radiant energy from our sun directly or indirectly provides energy for all microbes, plants, and animals. Light energy is harvested during photosynthesis to produce carbohydrates. Respiration, on the other hand, is used to break down complex compounds and thereby release their stored chemical energy to perform useful work. On a biochemical level, one of life's fundamental properties is the interconversion of energy.

The most important form of energy transduction, which is common to both photosynthesis and respiration, relies on the formation of transmembrane electrochemical potential gradients. During nutrient oxidation or light absorption, an electrochemical proton potential is generated across a membrane. Proton potentials are then transduced by membrane-bound proteins into the synthesis of ATP, the currency of life. In this chapter we will explore the role of membranes in energy transduction. Since proton potentials are the *essential* feature of membrane bioenergetic pathways, we will describe their formation and utilization. We will begin by considering the importance of proton pumping.

5.1. THE IMPORTANCE OF PROTON PUMPING

The electrochemical potential of animal plasma membranes is largely due to the ATP-driven distribution of Na^+ and K^+ ions whereas that of plant and fungal plasma membranes is due to protons. The transmembrane potentials of prokaryotic membranes, inner mitochondrial membranes, and thylakoid membranes are also established by protons. The primary function of these specialized membranes is to produce ATP, although electrochemical potentials also participate in secondary active transport (Section 6.1). Transmembrane proton pumping is driven by light energy or a respiratory chain. This migration of protons is very much like storing energy in a battery. The stored energy is released to perform work, such as ATP synthesis. Most of a cell's energy requirements are met through the utilization of transmembrane proton potentials. In a very real sense, transmembrane proton pumping is the only thing standing between the vital you and rigor mortis. In this section we will explore the evolution of membrane energy transduction systems, the chemiosmotic hypothesis and proton potentials, and the contributions of redox reactions to proton pumping.

5.1.1. Origins of Membrane Bioenergetic Pathways

Life emerged on our planet roughly 4 billion years ago. At that time Earth possessed a reducing atmosphere devoid of oxygen. Haldane and Oparin were the first to hypothesize that the earliest organisms must have had constituents such as membranes, DNA, RNA, NADH, and enzymes. These early cells were anaerobic chemotrophic prokaryotes, much like modern clostridia. These primitive cells used fermentation to metabolize energy-rich carbon and nitrogen compounds from their environment. Although ATP and metabolic precursors were synthesized, potentially toxic acids such as lactic acid were also formed. To maintain cytoplasmic neutrality, cells necessarily developed a capacity to remove protons. Raven and Smith (1976) were the first to propose that the original function of H^+-ATPases was to pump protons out of cells. Organic acids were both pumped out of cells and repelled by their inside-negative membrane potentials. As time passed, environmental nutrients diminished while the pH dropped. Only cells possessing efficient means of proton removal survived because of their need to maintain a neutral cytoplasmic environment in the presence of fermentative metabolism and an acidic extracellular environment. Electron transport systems evolved as a more efficient means of pumping protons; primitive electron transport systems likely complemented the activity of membrane H^+-ATPases. As electron transport systems improved, they became more than sufficient to maintain a neutral cytoplasm. They eventually became so efficient that the proton ATPases could run "backwards" to generate ATP. Therefore, electron transport became useful in both the removal of protons and ATP formation. As atmospheric oxygen levels rose, cells developed the capacity to use molecular oxygen as a terminal electron acceptor. The capacity to use oxygen increased the efficiency of electron transport and removed the potentially toxic oxygen molecules from the cytoplasm, which needs to be maintained as a reducing environment. Thus, the components of modern bioenergetic systems are the result of millions of years of evolution (Wilson and Lin, 1980).

5.1.2. The Chemiosmotic Hypothesis and Its Verification

In the early 1960s, scientists believed that electron transfer reactions and ATP formation were linked by some activated chemical intermediate, generally thought to reside within bioenergetic membranes. This mechanism of substrate-level phosphorylation was an attractive idea to biochemists at that time because it resembled glycolysis. In a radical departure from conventional thinking, Mitchell (1961) proposed the chemiosmotic hypothesis, which was uniformly ignored for several years. The chemiosmotic model suggests that electron transfer and phosphorylation reactions are not chemically linked, as proposed by substrate-level phosphorylation models. A transmembrane electrochemical proton potential was proposed to link the respiratory chain with ATP production (Figure 5.1). The free energy released by the electron transport chain is used to form an electrochemical proton potential; the potential is then used to power a H^+-ATPase to produce ATP.

Many lines of evidence support the chemiosmotic model of energy transduction (Mitchell, 1979; Ferguson and Sorgato, 1982). As stipulated by the chemiosmotic model, a proton potential is a central feature of biological energy transduction. Proton gradients of 1 to 2 pH units are often formed across biological membranes when they become energized. Furthermore, proton extrusion is tightly coupled to metabolic reactions. For example, a transmembrane proton gradient and ATP formation are only observed in thylakoid preparations during or after their illumination. Transmembrane ion gradients can only be formed across intact and sealed membranes. Similarly, oxidative phosphorylation only occurs in the presence of closed membrane-bound compartments such as organelle or vesicle membranes. Mutations of electron transport chain components lead to diminished electron transport and energy production.

The first direct evidence supporting the chemiosmotic hypothesis came from pH-jump experiments (Jagendorf and Uribe, 1966). In these experiments an artificial pH gradient was imposed on isolated spinach chloroplasts to determine if a transmembrane pH difference is able to promote ATP synthesis. Chloroplasts in a pH 7 medium were immersed in a pH 4 buffer. After 1 min, which allowed sufficient time for the acidification of thylakoids, the sample was rapidly injected into a pH 8 broth that contained the necessary precursors for ATP synthesis. ATP was produced during these conditions, thus showing that a pH difference can support ATP synthesis in the *absence* of electron transport. Furthermore, it did not matter which acid was chosen for the acidification step, suggesting that enzymes were not directly involved in converting the acid molecules *per se* into ATP. Similar results have been observed in other

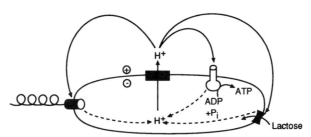

Figure 5.1. The chemiosmotic hypothesis and a few of its consequences. Protons are actively pumped out of this model bacterium. This pumping creates an electrochemical potential gradient of protons. According to the chemiosmotic hypothesis, oxidative reactions are linked to ATP production by an electrochemical proton potential. The outward pumping of protons creates a pH gradient (inside alkaline) and an electrical gradient (inside negative). This gradient is used to make ATP and to power some forms of membrane transport and cell motility.

experimental systems, including mitochondria and bacteria. These results provide important support for Mitchell's hypothesis.

Evidence supporting the chemiosmotic hypothesis is also found in studies of uncouplers of oxidative phosphorylation. Hydrophobic weak acids such as 2,4-dinitrophenol (1)

$$O_2N-\langle\bigcirc\rangle-OH$$
$$\overset{|}{NO_2}$$

(1)

uncouple oxidation and ATP formation, i.e., oxidation continues without phosphorylation of ADP. These uncouplers bind protons at one membrane face and then release them at the other. This short-circuits the proton current, collapses the proton gradient, and stops ATP production. This observation supports the idea that a proton gradient is required for oxidative phosphorylation.

One of the most dramatic confirmations of the chemiosmotic hypothesis comes from reconstitution studies (Racker and Stoeckenius, 1974; Drachev *et al.*, 1974). Racker and Stoeckenius (1974) reconstituted bacteriorhodopsin and/or the bovine mitochondrial H^+-ATPase in phospholipid vesicles. When bacteriorhodopsin-containing vesicles are illuminated, they translocate protons. When H^+-ATPase is included in the vesicles' membranes, they produce ATP when illuminated. These very different membrane proteins cooperate in the formation of ATP through a transmembrane proton potential. These simple and elegant experiments provide further direct evidence for the chemiosmotic hypothesis. The many different lines of investigation discussed above have firmly established the chemiosmotic model of energy transduction. The chemiosmotic hypothesis unified the field of membrane bioenergetics just as the double-helical model of DNA structure crystallized studies on the molecular basis of inheritance.

5.1.3. Proton Potentials

When a proton is pumped across a membrane, two things change. First, the difference in H^+ concentration creates a transmembrane chemical potential gradient. Since protons are charged, their transport also contributes to an electrical potential gradient. When protons are pumped out of bacteria or mitochondria, they develop an inside-negative electrical potential that attracts protons. A proton potential or proton motive force (pmf) is the electrochemical force pulling protons back into the matrix. It is quantitatively expressed according to the equation:

$$\frac{\Delta\mu_{H^+}}{F} = \Psi - 2.3\frac{RT}{F}\Delta pH \qquad (5.1)$$

where $\Delta\mu_{H^+}$ is the proton's electrochemical potential, F is Faraday's constant, Ψ is the transmembrane electrical potential, R is the gas constant, and T is absolute temperature. Thus, a proton potential is determined by both electrical and pH components. The importance of these two components varies among cell types.

5.1.4. Oxidation–Reduction Reactions Participate in Most Forms of Proton Pumping

During oxidation–reduction reactions electrons are moved from place to place but not created or destroyed. Chemists define oxidation as the loss of electrons and reduction as the gain of electrons. Atoms or molecules that donate electrons are called reducing agents because they reduce other substances. Oxidizing agents, such as O_2, accept electrons and thereby oxidize other atoms or molecules. For example, during the oxidation of methane:

$$CH_4 + 2O_2 \rightarrow 2H_2O + CO_2 \qquad (5.2a)$$

the valence of carbon changes from -4 to $+4$; it is said to be oxidized because its valence or oxidation state is raised.

In some of the cases described below, electrons are directly moved from place to place, such as the redox reactions occurring within light-harvesting complexes. However, in many biological redox reactions electron transfer is mediated by a hydrogen atom (a proton plus an electron) or a hydride ion (a proton and two electrons). Dehydrogenation reactions are equivalent to oxidation whereas hydrogenation is equivalent to reduction. For example, the oxidation (or dehydration) of NADH:

$$2\ NADH + O_2 + 2H^+ \rightarrow 2H_2O + 2\ NAD^+ \qquad (5.2b)$$

involves the transfer of one proton and two electrons per NADH molecule. These sorts of redox reactions play a central role in proton pumping.

5.1.4.1. Free Energy Yield of Redox Reactions

The ability of an atom or molecule to donate or accept an electron is described by its redox potential. The redox potential of an atom is measured in the following way. Let us consider an atom X which can be in a reduced (X) or oxidized (X^+) state. A solution containing 1 M X and 1 M X^+ is placed in a sample container. A reference cell contains 1 M H^+ and 1 atm H_2. When these two cells are electrically linked, a flow of electrons is established. If X has a lower affinity for electrons than H^+, electrons will flow to the reference cell, thus reducing H^+. On the other hand, if X^+ has a higher affinity for electrons than H_2, electrons will move to the sample cell, thus reducing X^+. The standard redox potential, E_0, is determined by the relative ability of electrons to oxidize or reduce a sample. To make E_0 values more useful in a biochemical setting, they are corrected to pH 7 where they are referred to as E_0' values.

Table 5.1 lists several conjugate redox pairs and their standard redox potentials. Strong oxidizing agents such as molecular oxygen have large positive E_0' values whereas reducing agents have smaller or negative redox potentials. A molecule or protein listed in

Table 5.1.

Standard Oxidation–Reduction Potentials for Several Conjugate Pairs

Oxidant	Reductant	Number of electrons	E_0' (mV)
$2H^+$	H_2	2	-421
$NADP^+$	$NADPH + H^+$	2	-324
NAD^+	$NADH + H^+$	2	-320
FAD	$FADH_2$	2	-220
FMN	$FMNH_2$	2	-220
Pyruvate	Lactate	2	-190
Fumarate	Succinate	2	-30
Cytochrome b (Fe^{3+}) (oxid)	Cytochrome b (Fe^{2+}) (red)	1	$+60$
Q	QH_2	2	$+100$
Cytochrome c_1 (oxid)	Cytochrome c_1 (red)	1	$+220$
Cytochrome c (oxid)	Cytochrome c (red)	1	$+250$
Cytochrome a (oxid)	Cytochrome a (red)	1	$+250$
Cytochrome a_3 (oxid)	Cytochrome a_3 (red)	1	$+380$
$\frac{1}{2}O_2 + 2H^+$	H_2O	2	$+820$

Table 5.1 can accept electrons from materials listed higher in the table and donate electrons to compounds listed lower in the table.

The standard free energy change ($\Delta G°$) is related to the difference in redox potential between two half-reactions ($\Delta E_0'$). This is calculated according to the equation:

$$\Delta G° = -nF\Delta E_0' \tag{5.3}$$

where n is the number of electrons transferred and F is Faraday's constant (23.06 kcal/mole). However, biological reactions generally do not take place at standard conditions (e.g., 1 M concentrations). To correct for nonstandard conditions, the initial concentrations of the reactants and products must be known. For the equation $xA \rightarrow yB$, the correction is:

$$\Delta G = \Delta G° + RT \ln \frac{[B]^y}{[A]^x} \tag{5.4}$$

where ΔG is the actual free energy change.

5.1.4.2. Free Energy Released by Redox Reactions Is Captured in Proton Potentials

Redox reactions participate in forming proton potentials across membranes. Since electron transport necessarily involves the simultaneous oxidation and reduction of two molecules, the electron transport chains of photosynthetic and respiratory membranes are a series of redox reactions. Some of the free energy released by electron transport is captured as a transmembrane electrochemical proton potential. In this chapter we will discuss the structural and functional properties of these membrane-associated redox sites, the molecular nature of electron transport, and integrate some of the molecular properties of these membranes into the physiology of bioenergetics.

5.2. PHOTOSYNTHETIC MEMBRANES

Since the Earth's food web begins with photosynthesis, it seems appropriate to begin our discussion of biological energy conversion mechanisms with photosynthesis. All plants and many microorganisms rely on light energy to generate a transmembrane proton potential. Photosynthetic organisms fall into two broad categories: those that evolve molecular oxygen and those that do not. The more primitive photosynthetic bacteria are anoxygenic. Anoxygenic mechanisms are characterized by a cyclic flow of electrons within a photosynthetic apparatus. In Section 5.2.2 we will discuss cyclic electron flow during light harvesting by the purple photosynthetic bacterium *Rhodopseudomonas*. Cyanobacteria and the eukaryotic green plants and algae produce oxygen during noncyclic electron transport (Section 5.2.3). The first key steps in photosynthesis are similar among all of these organisms. Light energy collects at protein–pigment complexes called photosynthetic reaction centers. Photosynthetic reaction centers are found in cytoplasmic membranes or in cytoplasmic membrane-associated "thylakoid-like" membranes of prokaryotes and in the chloroplast's thylakoid membranes of eukaryotes. Photosynthetic reaction centers use excited state energy to transport electrons across membranes; this eventually leads to a proton electrochemical potential across these membranes. In contrast to plants and most photosynthetic bacteria, *Halobacteria* generate a transmembrane proton potential in the absence of electron transport. We will now consider the properties of photosynthetic membranes beginning with the simple and unique features of bacteriorhodopsin.

5.2.1. Direct Proton Pumping by Bacteriorhodopsin

Halobacteria, a genus of archaebacteria, live in salt beds and concentrated brine. They maintain a 1 M intracellular sodium concentration in the presence of environmental levels of 4 to 5 M NaCl, which creates a transmembrane sodium gradient. A 3 M intracellular potassium concentration helps to minimize the transmembrane osmotic gradient. Since *Halobacteria* are aerobes, they generate a proton potential in the presence of oxygen via a respiratory chain. In the absence of oxygen, they capture light energy by a unique mechanism to form an electrochemical proton potential gradient. This proton potential is used by membrane-bound ATP synthases to make ATP. In contrast to other photosynthetic organisms, *Halobacteria* do not use electron transport, chlorophylls, or cytochromes to capture light energy. Oesterhelt and Stoeckenius (1973) discovered that a single membrane protein, bacteriorhodopsin, was responsible for transmembrane proton pumping. Since this protein promises to be the first completely understood bioenergetic system, it has attracted the attention of scientists worldwide.

Bacteriorhodopsin is one of the most thoroughly studied membrane proteins. As such, it is frequently used as a standard when comparing the properties of membrane proteins. We have previously discussed bacteriorhodopsin's amino acid sequence and three-dimensional structure (Sections 2.3.7 and 4.1.3). We will now focus on bacteriorhodopsin's physiological role: proton pumping. Bacteriorhodopsin's chromophore retinal (**2**)

$$\text{(structure)} \quad \text{N-Lys}$$

(2)

plays a central role in proton pumping. Retinal is attached to the ε-amino of Lys-216 via a Schiff base. Its delocalized electrons result in a broad absorption of green light, resulting in a membrane's purple color. The absorption peak of the all-*trans* isomer (2) is 568 nm. In the dark both the all-*trans* and 13-*cis* forms of retinal are present in roughly equal amounts. Exposure to light shifts retinal to the all-*trans* conformation. Bacteriorhodopsin's photocycle is triggered by light absorption by all-*trans* retinal (Figure 5.2). Raman spectroscopy has shown that the Schiff base is protonated. The chromophore then cycles through a series of intermediates resulting in the extracellular release of protons.

Figure 5.2 shows the two resting states and the four primary ground-state intermediates of bacteriorhodopsin's photocycle. These are: the 13-*cis* bacteriorhodopsin (BR548), all-*trans* bacteriorhodopsin (BR568), the K, L, M, and O states. The four intermediates absorb at 630, 550, 410, and 640 nm, respectively. The absorption of light energy by BR568 affects the resonance of its double bonds leading to isomerization to 13-*cis* BR590 (this intermediate state is called K). This isomerization alters the spatial position of the protonated Schiff base and thereby changes its ionic interactions. The next intermediate, L, is formed in 1 μsec. This intermediate absorbs at 550 nm. Bacteriorhodopsin then undergoes a conformational change that involves at least Tyr-26 and -64, Trp-182 and -189, and a Pro residue. This structural change alters retinal's environment, causing the proton on its Schiff base to be released to a neighboring dissociable group such as Tyr, Ser, or Asp. Several dissociable groups are found on the sixth transmembrane domain, near Lys-216 of the seventh helix. This forms the M state which

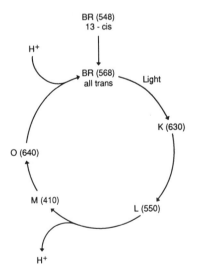

Figure 5.2. Bacteriorhodopsin's photocycle. Six well-defined states of bacteriorhodopsin's photocycle and their relationships are illustrated. The absorption maximum of each state is given in parentheses.

absorbs at 410 nm. Resonance Raman spectroscopy and Fourier transform infrared spectroscopy have shown that this intermediate is deprotonated. Resonance energy transfer studies suggest that retinal swings toward a membrane's *trans* face by the end of the M state (Hasselbacher and Dewey, 1986). The photocycle's next step is the M→O transition which requires about 7 msec. During this step the retinal reverts from the 13-*cis* to the all-*trans* isomer. The proton is released from a membrane during this transition. Current models suggest that during an M→O transition a proton "hops" about 3 nm along functional groups of bacteriorhodopsin such as Asp and Tyr (Dollinger *et al.*, 1986; Butt *et al.*, 1989; Marinetti *et al.*, 1989). The absorption maximum of state O is 640 nm. When the O state obtains a proton from a membrane's internal side, it is ready to begin the photocycle again. One complete cycle requires about 12 msec.

The seven transmembrane α-helices of bacteriorhodopsin are thought to form a channel-like structure. Dissociable groups at either side of retinal may form so-called "proton wires." However, the alteration of certain Ser or Tyr residues by site-directed mutagenesis of helix six did not significantly decrease proton pumping (Hackett *et al.*, 1987). Recent studies have indicated that Asp-96 is directly involved in shuttling protons *to* the Schiff base whereas Asp-84 participates in moving protons *from* the Schiff base (Butt *et al.*, 1989; Martino *et al.*, 1989). These observations have led to improved proton pumping models (e.g., Henderson *et al.*, 1990).

Additional intermediate states have been suggested to occur within the photocycle given above. The J625, X, and N states have been proposed to follow the BR568, L550, and M410 states, respectively (Stoeckenius and Bogomolni, 1982). However, these states have been defined using purified bacteriorhodopsin, not cell membranes. Furthermore, some of these studies required very low temperatures to observe a transient state. Therefore, we cannot be sure that these intermediates ever occur in living cells.

Each bacteriorhodopsin molecule pumps about two dozen protons per second. Although the photocycle is complete in only 12 msec, inefficiencies limit its measured overall rate. The optimal level of proton pumping is two protons per cycle, although this stoichiometry is sensitive to solvent conditions. The models of Schiff base protonation/deprotonation cycles do not easily account for this stoichiometry (Stoeckenius and Bogomolni, 1982). It will be important to identify all of the physiological intermediates of proton pumping and reconcile these intermediate states with the stoichiometry of pumping.

5.2.2. Bacterial Photosynthetic Reaction Centers: Cyclic Electron Flow

The best understood photosynthetic light-harvesting pathway is found in the gram-negative nonsulfur purple photosynthetic bacteria. *Rhodopseudomonas viridis*, *Rhodobacter sphaeroides*, *Rhodobacter capsulatus*, and *Rhodospirillum rubrum* are included among these bacteria. Although we will generally focus on *Rps. viridis* in our discussions, most of these observations apply to all of these photosynthetic bacteria.

In the preceding section we described how bacteriorhodopsin pumps protons across membranes. In contrast, photosynthetic reaction centers transport photoexcited electrons across membranes, but do not *directly* move protons across a membrane. Figure 5.3 illustrates the light-harvesting strategy of organisms such as *Rps. viridis*. As this figure shows, photosynthetic membranes contain four essential features: light-harvesting proteins, photosynthetic reaction center proteins, electron transport proteins, and quinones. Light is absorbed by bacteriochlorophyll molecules of light-harvesting complexes or by a special pair of bacteriochlorophyll molecules in photosynthetic reaction centers. Light energy absorbed by a light-harvesting complex is rapidly transferred to a photosynthetic reaction center by resonance energy transfer. When captured light energy reaches a photosynthetic reaction center, it causes an electron of bacteriochlorophyll to become promoted to a higher energy level. An excited electron has a much lower affinity for bacteriochlorophyll than its ground-state counterpart. Consequently, electrons are transferred via a series of oxidation–reduction reactions through redox sites until they reach a mobile quinone (Q) pool. Quinones carry high-energy electrons to cytochrome b–c_1 complexes. Cytochrome b–c_1 complexes in concert with quinones catalyze the transmembrane movement of protons, thus forming an electrochemical proton potential. Electrons are transferred from the cytochrome b–c_1 complex to a cytochrome c. Cytochrome c_2 then shuttles electrons to the photochemical

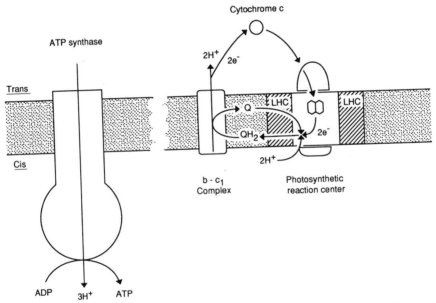

Figure 5.3. A highly schematic view of cyclic electron flow in purple photosynthetic bacteria. The photosynthetic strategy of purple photosynthetic bacteria is shown. Light absorption excites electrons within the photosynthetic reaction center. The electrons are transferred to ubiquinone (Q) to form dihydroubiquinone (QH_2). Dihydroubiquinone is reduced by the b–c_1 cytochrome. During this process Q is formed, protons are released at the *trans* membrane face, and the lower-energy electrons return to the photosynthetic reaction center. The proton gradient formed by this cyclic flow of electrons is used to produce ATP by the ATP synthase.

reaction center's periplasmic or *trans* surface. Each electron then migrates to the special pair of bacteriochlorophyll molecules, thus completing cyclic electron flow. Now that we have sketched cyclic electron flow in broad terms, we will examine bacterial photosynthetic light-harvesting in greater detail.

5.2.2.1. Supramolecular Organization of Bacterial Photosynthetic Membranes

Photosynthetic bacteria possess specialized membrane systems known as thylakoids or chromatophores than contain the photosynthetic apparatus. These stacked (lamellar), vesicular, or tubule-shaped membranes are derived from, and generally confluent with, the cytoplasmic membrane. In contrast to plants and true algae, most bacteria have only one photosystem. Photosystems are composed of a light-harvesting complex, which greatly increases the efficiency of light collection, and a photosynthetic reaction center which is directly involved in electron transport.

The stacked thylakoid membranes of *Rps. viridis* are composed of large roughly cylindrical structures on a two-dimensional lattice. These regular arrays are well-suited to structural analysis using image processing techniques. Figure 5.4 shows processed images of one subunit within the photosynthetic membrane of *Rps. viridis* as viewed from the *cis* (or exoplasmic) face (Stark *et al.*, 1984; Jay and Wildhaber, 1988). These processed images were derived from electron microscopic images of negatively stained membranes, which has a resolution of about 2 nm. A side view of one photosystem subunit is shown in panel c. The entire structure has a diameter of 12 nm. The central core region is the photosynthetic reaction center (Section 5.2.2.3). This core is surrounded by a ring of 12 subunits that comprise the light-harvesting complex. These 12 subunits are distributed about the periphery as six dimers, thus creating a sixfold symmetry (Figure 5.4). In turn, these light-harvesting complexes are composed of three polypeptides. This sort of supramolecular organization may apply to many types of photosynthetic bacteria (Engelhardt *et al.*, 1983).

5.2.2.2. Molecular and Macromolecular Structures of the *Rps. viridis* Light-Harvesting Complex

The light-harvesting complexes of *Rps. viridis* are believed to act like antennae for their photosynthetic reaction centers. In analogy with PSI and PSII of eukaryotes (Section 5.2.3), bacterial photosystems are believed to absorb and then transfer radiant energy to the photosynthetic reaction center in the middle of the photosystem.

Each of the 12 light-harvesting subunits of *Rps. viridis* is composed of three polypeptides and two bacteriochlorophyll b molecules. These three polypeptides are known as B1015-α (M_r = 6800), B1015-β (M_r = 6100), and B1015-γ (M_r = 4000). The indicated molecular weights are all based on the amino acid sequences; the apparent molecular weights found by SDS-PAGE are all greater. Figure 5.5. shows a silver-stained SDS-PAGE gel of solubilized proteins from *Rps. viridis* thylakoid membranes. The proteins of the B1015 light-harvesting complex and photosynthetic reaction center are

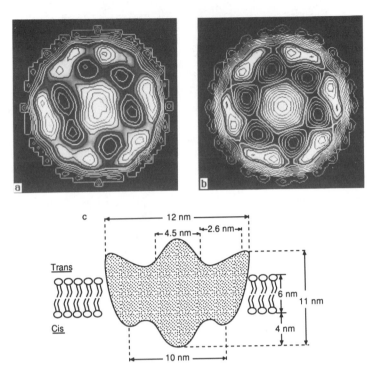

Figure 5.4. Supramolecular structure of the photosynthetic membrane of *Rhodopseudomonas viridis*. The photosynthetic membrane of *Rps. viridis* was studied by negative-stain electron microscopy. Panel a shows a contour map of computer-averaged *en face* images of the photosystem. The light central area is the photosynthetic reaction center. The peripheral ring is the light-harvesting complex. Panel b shows a computer-enhanced image of the photosystem based on its sixfold symmetry. A side view of a single photosystem is given in panel c. (From Jay and Wildhaber, 1988, *Eur. J. Cell Biol.* **46**:227.)

indicated. (The B1015-γ subunit stains anomalously at a low level; other stains such as ^{125}I label B1015-γ much better.) The proteins are called B1015 because their associated bacteriochlorophyll molecules absorb at 1015 nm. These three proteins are present at a 1:1:1 stoichiometry resulting in an apparent molecular mass of 20 to 25 kDa for each light-harvesting structure.

Figure 5.6 shows the covalent structure of bacteriochlorophyll b (Bchl b) which is found associated with both the light-harvesting antennae and photosynthetic reaction centers. The Mg atom at the center of the tetrapyrrole ring is coordinated to His residues of certain transmembrane helices. Bchl b molecules of the light-harvesting complex and photosynthetic reaction center absorb at 1015 nm and at 830 and 970 nm, respectively. The extensive conjugated ring system gives rise to absorption in the visible and near-infrared regions of the electromagnetic spectrum. The variations in the absorption spectrum of Bchl when attached to different proteins or protein locations

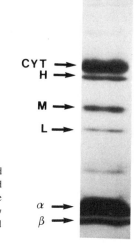

Figure 5.5. Membrane proteins of the *Rps. viridis* photosystem. Thylakoid membranes of *Rps. viridis* were isolated, dissolved in SDS, and then subjected to PAGE. The separated proteins were visualized by silver staining. The identities of the proteins are shown along the left-hand side. The B1015-γ subunit is present but stains very poorly during this procedure. (From Jay and Wildhaber, 1988, *Eur. J. Cell Biol.* **46:**227.)

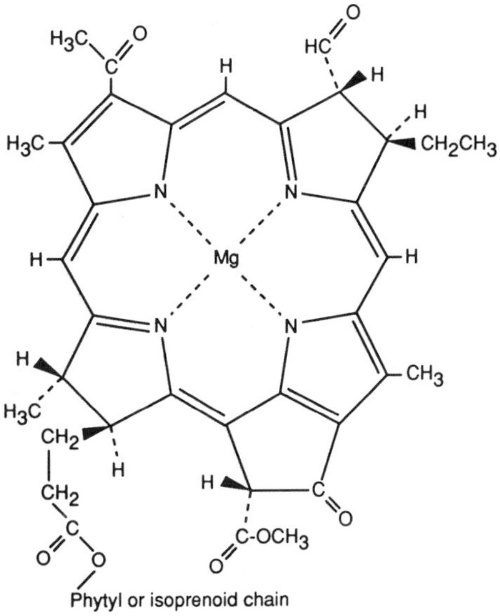

Figure 5.6. Molecular structure of bacteriochlorophyll b. The covalent structure of bacteriochlorophyll b is shown. Other prokaryotic and eukaryotic chlorophylls are slight variations on this structure.

are due to its local environment. When exposed to light, the system undergoes $\pi \rightarrow \pi^*$ or $n \rightarrow \pi^*$ electronic transitions.

The amino acid sequences of B1015-α, β, and γ have been deduced (Brunisholz *et al.*, 1985); these proteins determine the position and spectral characteristics of their associated Bchl molecules. These light-harvesting polypeptides map to the photosynthetic gene cluster. All of these proteins have stretches of 20 hydrophobic amino acids which are thought to span a bilayer. Labeling experiments have shown that these proteins are definitely exposed at the *trans* face, although their sequences suggest that they may also possess sequences at the *cis* face. The B1015-α and -β proteins have relatively hydrophilic NH_2- and COOH-termini. The locations of the NH_2- and COOH-termini are uncertain. However, if the NH_2-termini are *cis*, as in the light harvesting proteins of *Rs. rubrum* (Brunisholz *et al.*, 1985), this would place the B1015-α His-36 and the B1015-β His-37 residues and their associated Bchl b molecules near the *trans* membrane face. As we will discuss later, the photosynthetic reaction center's special pair of Bchl b molecules is at this same relative position.

The B1015-α and β proteins are homologous to each other and to light-harvesting proteins of other photosynthetic bacteria. B1015-α has one His residue whereas B1015-β has two His residues. The conserved His residues, which are located within their transmembrane helices, are thought to be axial ligands for the Mg atom of Bchl. B1015-α and β are unusually rich in aromatic amino acids, especially Trp. The strongly redshifted absorption band of Bchl may be due both to the hydrophobicity of its environment and to coupled interactions between Bchl b and nearby aromatic side chains. The B1015-α and β chains provide the structural framework supporting the light-absorbing cofactors and alter their spectral properties.

B1015-γ is a very hydrophobic polypeptide that is rich in aromatic amino acids. However, since B1015-γ possesses no His residues, it is not likely to be directly associated with Bchl b. B1015-γ is believed to be involved in linking the α and β subunits together and/or promoting the formation of the pseudocrystalline structures of the lightharvesting complex (Brunisholz *et al.*, 1985).

5.2.2.3. Structure of a Bacterial Photosynthetic Reaction Center

The structural information regarding photosynthetic reaction centers of nonsulfur purple bacteria is nearly complete. The primary structures of its constituent proteins have been determined by gene sequencing experiments (Michel *et al.*, 1985, 1986). More importantly, high-resolution x-ray crystal structures have been obtained for the reaction centers of *Rps. viridis* and *Rb. sphaeroides* (Deisenhofer and Michel, 1989). The photosynthetic reaction center is composed of three proteins and six cofactors. The three constituent integral membrane proteins are H (heavy), M (medium), and L (light). These unimaginative names were derived from the proteins' relative positions after SDS-PAGE. In fact, gene sequencing experiments have shown that the H protein is lightest whereas the M protein is heaviest. These three proteins possess six associated cofactors including: a special pair of Bchl b molecules, two additional Bchl b molecules, two bacteriopheophytin (BPh) molecules, two quinones (Q), one carotenoid, and one nonheme iron. These cofactors collaborate with the membrane proteins to transport electrons across membranes.

Gene cloning experiments have provided the deduced amino acid sequences of the *Rps. viridis* cytochrome, H, M, and L proteins (Michel *et al.*, 1985, 1986; Weyer *et al.*, 1987). The L (M_r = 30,571) and M (M_r = 35,902) proteins form the photosynthetic reaction center's core. These proteins are rich in His residues; this amino acid is generally coordinated with cofactors, as previously described for B1015-α and β. Each of these core proteins possesses five transmembrane domains, as illustrated in Figure 5.7. The NH$_2$-termini of both the M and L proteins are exposed at a membrane's *cis* face. The H subunit (M_r = 28,345) possesses a single transmembrane domain. Its N$_2$-terminus is *trans* but most of its mass is at the photosynthetic reaction center's *cis* face. The genes encoding these three proteins of the photosynthetic gene cluster are transcribed from a polycistronic operon that includes at least B1015-β, B1015-α, L, M, cytochrome, and H, in order of appearance (5'→3').

Figure 5.7 shows several important sites on the M and L proteins. For example, several mutations in the L chain's sequence between its fourth and fifth transmembrane helices confer herbicide resistance. As mentioned above, the photosynthetic reaction center contains four Bchl molecules that are associated with His residues. Monomeric Bchl b molecules are associated with His-180 and His-153 of the M and L proteins, respectively. The special pair of Bchl b molecules, which are known as P$_{960}$ because of their change in absorbance at 960 nm when illuminated, are linked to the M chain's His-200 and the L chain's His-173. The nonheme iron is coordinated to His-199 and His-225 of the L subunit and His-259, His-225, and Glu-232 of the M chain.

The final protein component of the *Rps. viridis* photochemical reaction center is a "c-type" cytochrome which is tightly associated with the reaction center's *trans* face. This protein delivers electrons to the photooxidized special pair of Bchl b molecules. The cytochrome's primary structure has been determined by nucleic acid and amino acid sequencing (Weyer *et al.*, 1987). In addition to its constituent amino acids, the

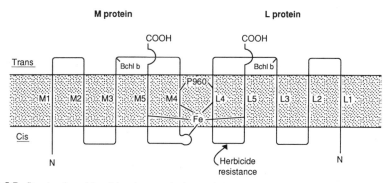

Figure 5.7. Structural models of the photosynthetic reaction center's L and M proteins. The L and M proteins of the *Rps. viridis* photosynthetic reaction center each possess five transmembrane α-helices labeled L1 to L5 and M1 to M5, respectively. These two proteins cooperate in binding the special pair of Bchl b molecules (P$_{960}$) and the Fe atom. The attachment sites for two additional Bchl b molecules are shown.

cytochrome has four covalently linked heme groups and two covalently linked fatty acids, yielding an aggregate molecular weight of 40,500. The cytochrome's primary structure is not homologous to other cytochromes. Although there is little internal homology in primary structure, there is considerable homology in tertiary structure (see below). Fatty acids are linked to the NH_2-terminus via an s-glycerocysteine, similar to that of lipoprotein (Section 2.3.4.1).

The cytochrome's four heme groups are attached at -Cys-X-X-Cys-His- sequences. Each heme is attached to four Cys residues via thioether linkages. A His residue acts as the heme groups' fifth ligands. The heme groups' sixth ligands are contributed by either Met or His. The four hemes are classified into two groups: a low-potential C_{553} and a high-potential C_{558}. In the following section we will explore functional aspects of electron transport.

In an important series of experiments, Deisenhofer and Michel (1989) determined the crystal structure of the *Rps. viridis* photosynthetic reaction center. As we mentioned in Section 4.1, integral membrane proteins are notoriously difficult to crystallize. An analysis of solvent conditions, especially the detergent's structure, led to high-quality reaction center crystals. X-ray diffraction studies have produced crystal structures of the photosynthetic reaction centers from *Rps. viridis* and *Rb. sphaeroides* at resolutions of 0.23 and 0.3 nm, respectively (Deisenhofer *et al.*, 1985; Deisenhofer and Michel, 1989; Allen *et al.*, 1987).

The photosynthetic reaction center sits at the center of the *Rps. viridis* photosystem as illustrated in Figure 5.4. The overall structure of the photosynthetic reaction center is shown in Figure 5.8. The resolution of this crystal structure is roughly ten times greater than that of the negative-stain electron microscopic images. The crystalline reaction center has a height of 13 nm, which is in good agreement with the size estimated by electron microscopy. By comparing the "side" and "front" views of the photochemical reaction center (Figure 5.8, top and bottom panels, respectively), it is apparent that the reaction center's core is elliptical with major and minor axes of 7 and 3 nm, respectively. As shown in Figure 5.8, the α-helices of the L and M subunits form walls that surround the chromophores. The L and M proteins are related by a twofold axis of symmetry. The photochemical reaction center appears as a 4.5-nm circle in Figure 5.4 because: (1) the size difference between the major and minor axes is not much greater than the stain's resolution, (2) the core is covered by the fairly symmetric cytochrome, and (3) the enhancement of the peripheral light-harvesting complexes of P6 symmetry tends to obscure the reaction center's P2 symmetry. Therefore, the photochemical reaction center's crystal structure is consistent with its proposed localization within photosynthetic membranes.

As Figures 5.7 and 5.8 show, the L and M subunits each have five transmembrane domains. Helices 4 and 5 of both proteins are associated with cofactors whereas 1 and 2 are peripheral. The H subunit has a single transmembrane α-helix and extracellular β-sheet structure. Less than half of the reaction center's mass is within the membrane. The comparatively large amount of extracellular mass may promote crystallization.

The three-dimensional structure of the photosynthetic reaction center's cytochrome subunit is shown at the top of Figure 5.8. This large protein rests at the reaction center's *trans* face. The locations of the heme groups are also shown. The heme groups display a

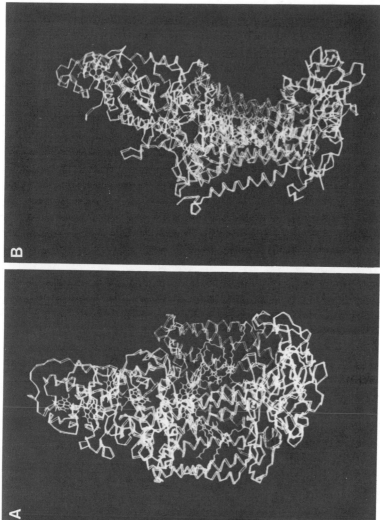

Figure 5.8. Tertiary structure of the *Rps. viridis* photosynthetic reaction center. The crystal structure of the photosynthetic reaction center is shown. Panels A and B show views from the side and front of the complex, respectively. The cytochrome and H protein are located at the top and bottom of these two panels. (Reprinted with permission from *Nature* Vol. 318, p. 618, copyright © 1985 by Macmillan Magazines Ltd. Illustrations courtesy of J. Deisenhofer and H. Michel.)

twofold symmetry. Although there is little internal homology in primary amino acid sequence, the tertiary structures in the vicinity of the heme groups are quite similar. For example, the hemes are always connected to an α-helical stretch of 17 amino acids.

The protein subunits described in the preceding paragraphs provide the necessary framework and environment for electron transfer reactions. The primary photochemical events are mediated by cofactors associated with the subunits. Figure 5.9 shows the cofactors associated with the L and M chains in their native protein environment. Two cofactor branches are seen. The "A" cofactor branch is linked with the L protein whereas branch "B" is attached to the M protein. Early spectroscopic data (see below) indicated that the electron transfer reaction occurred in the sequence: special pair→BPh→Q→Fe. These cofactors appear in the same order from the special pair at the *trans* face to the nonheme Fe at the *cis*. Each cofactor's ring is separated from the previous and/or next ring by about 1 nm. These active sites are close enough to allow electron transfer.

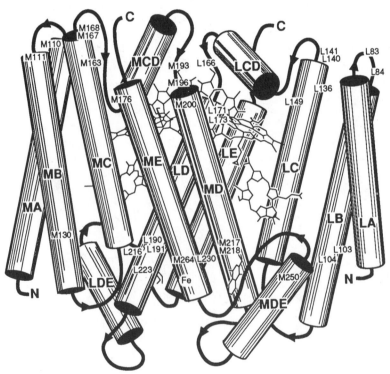

Figure 5.9. Spatial organization of the photosynthetic reaction center's cofactors. Three-dimensional line drawings of the cofactors associated with the M and L proteins are presented. This shows the cofactors in their native protein environment. (Adapted with permission from *Biochemistry* **27**:1, copyright © 1988 by American Chemical Society. Illustration courtesy of J. Deisenhofer and H. Michel.)

Previous ESR studies indicated the presence of a special pair of Bchl b molecules (Norris *et al.*, 1971; Feher *et al.*, 1975). However, the discovery of two cofactor branches in the crystal structure was unexpected. Earlier experiments suggested that only a single branch was present. Recent experiments (Knapp *et al.*, 1985) have shown that the A branch associated with the L chain is functional. The nonfunctional B branch may represent an evolutionary remnant.

5.2.2.4. Primary Structural Features of the *Rb. capsulatus* Cytochrome b–c_1 Complex

In the preceding paragraphs we discussed the structural components of photosynthetic reaction centers which cooperate to move electrons across membranes. As Figure 5.3 details, the cytochrome b–c_1 complex (or ubiquinone–cytochrome c reductase) releases protons at a membrane's *trans* face. The proton electrochemical potential is used to drive the ATPase while the electrons are recycled to reaction centers.

The cytochrome b–c_1 complex of bacterial photosynthetic membranes is composed of three integral membrane proteins: cytochrome b, cytochrome c_1, and the Fe–S protein. The genes of *Rb. capsulatus* encoding these three proteins have been studied by Gabellini and Sebald (1986) (for a review see Gabellini, 1988). The Fe–S protein, cytochrome b, and cytochrome c_1 genes are called *fbcF*, *fbcB*, and *fbcC*, respectively. These genes, in the order listed, are transcribed as a single polycistronic mRNA from the *fbc* operon. These prokaryotic proteins are highly homologous to their mitochondrial counterparts, but less so with those of chloroplasts.

The first gene of the *fbc* operon is *fbcF* which encodes the Fe–S protein (M_r = 21,000). Its deduced amino acid sequence revealed a single transmembrane domain near the NH_2-terminus. Most of the protein's mass is found in a hydrophilic domain at a membrane's *trans* face. The Fe–S protein has a cluster of two Fe and two S atoms that act as a prosthetic group. Four Cys residues (positions 133, 138, 153, and 155) may serve as ligands for the Fe–S cluster. The Fe–S cluster is near cytochrome b's low-potential heme. These two prosthetic groups form the catalytic site that oxidizes ubiquinone.

Cytochrome b of the *Rb. capsulatus fbc* operon is a hydrophobic 48.1-kDa membrane protein. Hydropathy analysis suggests that 9 or 10 transmembrane domains are present. The orientation of this protein in membranes is unknown. Cytochrome b has two associated heme groups with potentials of -90 and 50 mV. Each of these hemes is believed to link the second transmembrane domain (M2) to the fifth transmembrane domain (M5) (Figure 5.10). The heme groups are coordinated to four His residues 97 and 212 and 111 and 198. Four conserved positive charges are found in the vicinity of the heme binding pockets. These residues likely modify the environment and, hence, the hemes' redox potentials.

The *fbcC* gene encodes cytochrome c_1 (M_r = 30,100). Most of this protein is located at the *trans* face. It possesses a single transmembrane domain (positions 249–269) that anchors it to a membrane. The "c-type" heme moiety binds near the NH_2-terminus. It is linked to Cys-55 and -58 and His-59. A sixth ligand may be Met-205. A series of residues including Glu-84, Asp-90, Glu-94, Glu-217, and Asp-218 may form a nega-

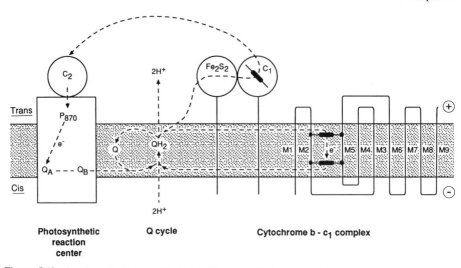

Figure 5.10. A schematic diagram illustrating the flow of electrons and protons across a prokaryotic photosynthetic membrane. The cytochrome b–c_1, Fe–S protein, photosynthetic reaction center, and Q cycle of *Rps. viridis* are shown. The vectorial movement of charge is illustrated: negatively charged electrons (e^-) move in one direction on the right- and left-hand sides of the illustration while protons move in the opposite direction. Cytochrome b possesses nine apparent transmembrane domains labeled M1 to M9. Two heme groups, which participate in electron migration, are attached as shown. The Q cycle couples electron and proton trafficking across membranes.

tively charged domain that binds cytochrome c_2 to permit electron transfer. We will return to this step of electron transport during our discussion of mitochondrial cytochrome c.

5.2.2.5. Functional Aspects of Bacterial Photosynthetic Membranes: The Q Cycle

In the preceding paragraphs we have described the structural elements of the photosynthetic light-harvesting pathway of purple bacteria. We have also mentioned the roles of these proteins and their associated cofactors in electron trafficking through membranes. Of course, the *whole point* of these structural elements and redox sites is ultimately to pump protons across membranes. Proton pumping requires membrane-associated quinones that participate in both proton and electron trafficking in bioenergetic membranes. The biochemical pathway of quinone-mediated proton translocation is known as the Q cycle, which was originally proposed by Mitchell.

As Figure 5.10 depicts, the photosynthetic reaction center and cytochrome b move electrons from one side of a membrane to the other. Charged species, including electrons, cannot freely dissolve in a membrane's hydrophobic core. Electrons must be escorted across a hydrophobic environment by the redox sites of reaction centers and cytochromes b. Two electrons and two protons at the *cis* face combine with Q to yield QH_2. QH_2 is oxidized near the *trans* face through a semiquinone intermediate. The reaction liberates two protons at a membrane's *trans* face and two electrons which ultimately return to a photochemical reaction center via cytochrome c_2. We shall return to the Q cycle later in this chapter.

5.2.2.6. Functional Aspects of Bacterial Photosynthetic Membranes: Molecular Energy Flow to Photosynthetic Reaction Centers

Photosynthetic membranes absorb most visible wavelengths of light and some near-infrared radiation. The most poorly absorbed visible wavelengths are reflected and thereby give a bacterium or plant its color. Pigment molecules of light-harvesting complexes generally absorb light at shorter wavelengths than photochemical reaction centers. We will now consider how energy migrates within photosynthetic membranes.

Light is rapidly absorbed by a sample and thereby promotes an electron to a higher energy level. To maintain efficient photosynthesis, most of this energy should be transmitted to a reaction center, not lost as heat or fluorescence. This is accomplished by transmitting the excitation energy to another pigment molecule in a time shorter than its fluorescence lifetime.

Energy transmission takes place by two different mechanisms within or between the light-harvesting complexes and between a light-harvesting complex and a photosynthetic reaction center. The pigments of a light-harvesting complex are all less than 2 nm from the pigments of adjoining subunits (e.g., Figure 5.4). Therefore, the molecular orbitals of two separate pigment molecules overlap significantly. This allows energy migration via excitons (*excit*ed elect*rons*) (Zuber, 1986). During exciton migration, an excited-state electron in one pigment molecule (P_1) trades places with a resting electron in a neighboring pigment (P_2):

$$P_1(e_1) + h\nu \rightarrow P_1(e_1{}^*) \tag{5.5}$$

$$P_1(e_1{}^*) + P_2(e_2) \rightarrow P_1(e_2) + P_2(e_1{}^*) \tag{5.6}$$

Exciton migration is very rapid (10^{-15} sec). A photoexcited electron could change partners many times before its energy is donated to a reaction center.

As indicated in Section 5.2.2.1, many bacterial photosystems such as that of *Rps. viridis* are assembled on a hexagonal lattice. This places a light-harvesting subunit of one photosystem immediately adjacent to a light-harvesting subunit of a neighboring photosystem. This allows excitation energy to be shared among many photosystems. Detailed spectroscopic studies of *Rs. rubrum* have shown that a single photosynthetic reaction center can use almost 1000 Bchl molecules as antennae (Glazer and Melis, 1987). Clearly, energy migrates within and between light-harvesting complexes of bacterial photosynthetic membranes.

A second mechanism of energy transmission is responsible for moving excited-state energy from a light-harvesting complex to a photosynthetic reaction center. Since the light-harvesting complex is more than 2 nm from a reaction center, exciton coupling is not possible. However, since the distance is about 4 nm, resonance energy transfer is very efficient. This provides a unidirectional flow of energy to the reaction center. An electron's excited-state energy is transferred to a reaction center in about 10^{-13} sec. The rapidity of exciton migration and resonance energy transfer minimizes energy loss via fluorescence. Energy transfer occurs between a light-harvesting pigment's excited electron and the reduced form of the reaction center's special pair. When the special pair is oxidized (Bchl $b^+)_2$, its absorption spectrum shifts out of resonance with the light-harvesting pigments. The oxidized state of the special pair is a "wait" state that allows light energy to be gained and stored by the light-harvesting complex. When the special pair is reduced, the energy contained in its storage ring is transferred to the special pair, thus maximizing the reaction center's efficiency.

5.2.2.7. Functional Aspects of Bacterial Photosynthetic Membranes: Charge Separation along the Redox Chain and Proton Pumping

In Sections 5.2.2.3 and 5.2.2.4 we discussed the structural features of the photosynthetic reaction center and the cytochrome $b-c_1$ complex. Each of these protein complexes contains several redox sites that participate in electron transport. Figure 5.11 shows the redox potentials (or "energy") of functional sites known to participate in cyclic electron flow. The excited special pair of Bchl b molecules (P^*_{870}) is a good electron donor. The electron then sequentially reduces the series of redox sites along the electron transport chain; a portion of the free energy released by electron transport is conserved in the form of a transmembrane electrochemical proton potential. We will now examine some of these processes in more detail.

The function of photosynthetic light absorption is ultimately to separate charges. The primary charge separation event occurs at the special pair of Bchl molecules after it receives energy from light or the light-harvesting complex according to:

$$(Bchl)_2 + acceptor \xrightarrow{h\nu} (Bchl)_2{}^+ + acceptor^- \qquad (5.7)$$

The formation of a photooxidized special pair has been directly observed by ESR spectroscopy. In its oxidized state, the special pair has one unpaired electron. If $(Bchl)_2{}^+$ is continuously produced by constant illumination, a strong ESR signal is produced. Figure 5.12 shows ESR spectra of $Bchl^+$ and the oxidized reaction center. A single broad line is observed for these samples indicating that the free electron of Bchl b is extensively delocalized over its π electron system (Figure 5.6). Consequently, many different interactions take place with nearby nuclei thereby broadening the spectrum (Figure 5.12).

A careful inspection of Figure 5.12 shows that the spectrum of $Bchl^+$ is broader than the spectrum of the oxidized reaction center. This shows that the free radical of the photosynthetic reaction center's special pair is distributed over both Bchl b molecules (Norris et al., 1971; Feher, et al., 1975). The linewidth of the special pair decreases because the electron is spread over twice the number of nuclei. The ratio of the monomer's to dimer's linewidth is $\sqrt{2}$ (Feher et al., 1975) which is experimentally observed (Figure 5.12). This provided the first evidence that a special pair of Bchl molecules was present.

ESR experiments have also shown that free radical formation is related directly to the primary photochemical reaction. Figure 5.13 shows the relationships between

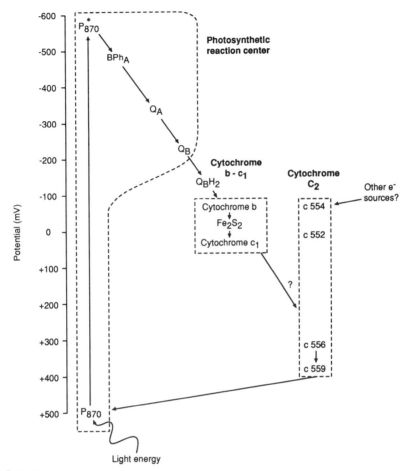

Figure 5.11. Redox potentials associated with cyclic electron flow. The redox sites linked to electron flow are shown. The ordinate gives the reduction potential of each of these sites. The electron path is indicated by solid arrows. The primary membrane protein complexes participating in electron transport and proton pumping are labeled and surrounded by dashed lines.

illumination, ESR detection of free radical formation, and absorbance changes of *Rb. sphaeroides*. When these cells or their membranes are illuminated, the ESR peak height and optical absorbance at 795 nm rise simultaneously. Similarly, these parameters fall together when the light is extinguished.

A great deal of effort has been expended to identify the intermediate acceptor of Eq. 5.7. As Figure 5.9 shows, the nearest cofactor is the monomeric Bchl b molecule. However, despite repeated attempts, it has not been possible to observe Bchl b$^-$ (Brenton *et al.*, 1986; Kirmaier and Holten, 1987). The first unambiguously identified electron carrier is BPh. The initial charge separation complex [(Bchl)$_2^+$ BPh$_L^-$] is formed 2.8 psec after the special pair's excitation. It is remotely possible that Bchl b$^-$ is the

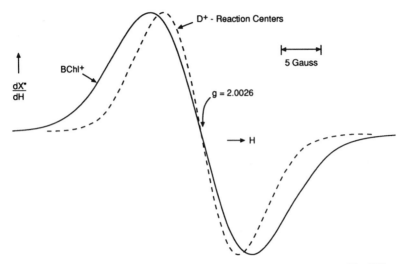

Figure 5.12. ESR spectra of bacteriochlorophyll b and the photosynthetic reaction center. The ESR spectra of Bchl b⁺ (----) and the oxidized reaction centers of *Rb. sphaeroides* (D⁺) (–––) are shown. (Redrawn from Feher *et al.*, 1975, *Ann. N.Y. Acad. Sci.* **244**:239.)

initial electron carrier, but its lifetime is immeasurably short. Although the function of the reaction center's monomeric Bchl b is unknown, its proximity with BPh suggests that their molecular orbitals overlap. It has been suggested that these overlapping orbitals allow Bchl b to shepherd excited-state electrons from the special pair to BPh.

The electron is shuttled from BPh_L to Q_A with a time constant of 0.2 nsec. The excited electron is then transferred from Q_A to Q_B in about 100 μsec. The nonheme iron separating these quinones does not play a role in electron transfer. The Q_B molecule acquires two electrons and two protons, thus forming Q_BH_2 (Figure 5.14). The Q_BH_2 molecule then dissociates from the photosynthetic reaction center. Due to its hydro-

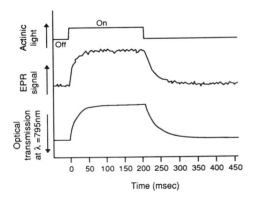

Figure 5.13. Relationships between illumination, free radical formation, and optical transmission. The kinetics of ESR and optical changes in the *Rb. sphaeroides* photochemical reaction center are plotted. Panel A shows the presence or absence of exciting light. Panel B shows the intensity of the ESR signal. Panel C shows the kinetics of optical transmission changes at 795 nm. (Redrawn from McElroy *et al.*, 1969, *Biochim. Biophys. Acta* **172**:180.)

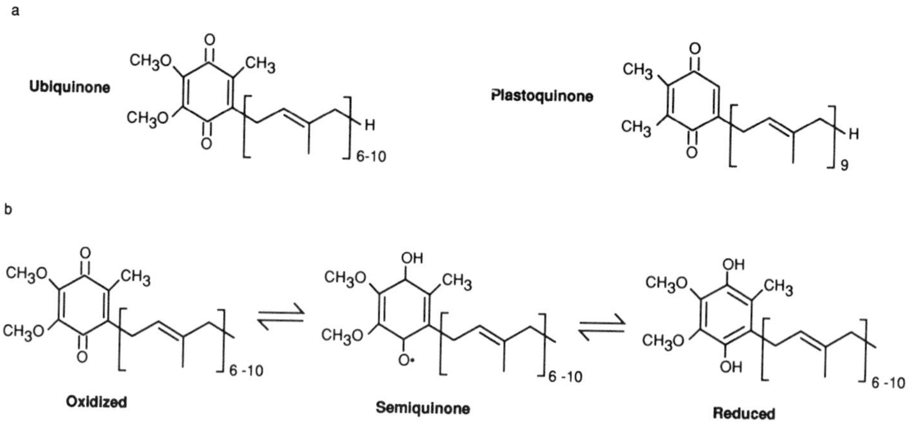

a

b

Figure 5.14. Quinone structures. Panel a shows the covalent structures of ubiquinone and plastoquinone. Panel b shows the oxidized, semiquinone, and reduced forms of ubiquinone.

phobicity, it diffuses within the photosynthetic membrane. Free Q in a membrane rapidly binds to the vacant Q_B binding site.

As indicated in Figures 5.3 and 5.10, Q_BH_2 interacts with cytochrome $b–c_1$ within a photosynthetic membrane. As mentioned in the previous section, the three integral membrane proteins of the cytochrome $b–c_1$ complex oxidize Q_BH_2 to Q_B (Figure 5.14), thus releasing protons and electrons at the *trans* face. The electrons are transferred back to the photosynthetic reaction center's cytochrome via cytochrome c_2. The electrons are first transferred to heme c-556, a high-potential heme of the reaction center's cytochrome (Dracheva *et al.*, 1988). Heme c-556 then reduces heme c-559 in 2.5 μsec. Heme c-559 then donates the electron to (Bchl b)$_2^+$ in about 3 μsec. When all of the individual steps are considered, the oxidized special pair is found to be reduced with a time constant of 270 μsec. The rate-limiting step is likely the diffusion of cytochrome c_2 to the reaction center's cytochrome. This completes the cyclic molecular flow of electrons in photosynthetic bacterial membranes.

5.2.2.8. Functional Aspects of Bacterial Photosynthetic Membranes: Reconstitution of Cyclic Electron Flow

As we learned above, the mechanism of electron transport and transmembrane proton pumping in purple photosynthetic bacteria are becoming rather well characterized. The *in vitro* reconstitution of a membrane system is the best way to test our understanding of its mechanism of action. If it is possible to disassemble a bioenergetic pathway, purify its components, and functionally reconstitute the system in lipid vesicles, then we can be reasonably confident about the roles of these components in living cells. Recently, Gabellini *et al.* (1989) have successfully reconstituted a bacterial photosynthetic system.

The photosynthetic reaction center with its associated light-harvesting polypeptides, cytochrome $b-c_1$ complex, ATP synthase, and cytochrome c_2 were purified from *Rb. capsulatus* (Gabellini *et al.*, 1989). The photosynthetic reaction center and cytochrome $b-c_1$ complex were reconstituted in lipid vesicles using detergent dialysis. Cytochrome c_2 in solution was simultaneously entrapped within the vesicles. ATP synthase solubilized in detergent was incorporated into the vesicles' membranes by the detergent dilution technique. Due to its hydrophobicity, ubiquinone spontaneously inserts into vesicle membranes. The structural elements of the photosynthetic phosphorylation machinery were thereby reassembled in artificial membranes.

The membrane topologies of the reassembled components were analyzed. The location of cytochrome c_1 was established by adding the impermeable reagent ascorbate. Roughly one-third of the cytochrome c_1 proteins could not be reduced by ascorbate. This indicates that one-third of the cytochrome c_1 molecules and, by inference, one-third of the $b-c_1$ complexes are exposed at the inner face of the vesicles. The photosynthetic reaction center is randomly oriented in these reconstituted membranes. This was demonstrated by the fact that only one-half of the reaction centers could be rapidly reduced by cytochrome c_2. Consequently, only a fraction of the total number of reaction centers and $b-c_1$ complexes could interact productively with cytochrome c_2. Nevertheless, when these vesicles are illuminated they generate 36 nmole ATP/ng ATP synthase per min. Furthermore, these vesicles demonstrate the same sensitivities to a broad range of inhibitors as are shown by living bacteria. Therefore, it is possible to reassemble working bioenergetic membranes *in vitro*.

5.2.3. Photosynthetic Membranes of Higher Organisms: Noncyclic Electron Flow

As photosynthetic organisms evolved, they developed more complicated strategies for harvesting light energy. Eukaryotic plants, cyanobacteria, and prochlorophytes generate molecular oxygen during noncyclic electron flow. Just as we have seen in the preceding paragraphs, light energy is used to pump electrons through an electron transport chain. However, in oxygenic photosynthesis water is used as the source of reducing power (electrons). The excited electrons are ultimately transferred to $NADP^+$, thus forming NADPH. During electron transport a proton gradient is formed that drives ATP synthesis. Therefore, ATP and NADPH are generated by oxygenic photosynthesis. These compounds are used during dark reactions to reduce carbon dioxide to carbohydrates. We will now discuss the membrane processes participating in oxygenic photosynthesis.

5.2.3.1. Noncyclic Photosynthetic Electron Flow: The Big Picture

The most distinguishing feature of oxygenic photosynthesis is the cooperative interaction of two photosystems. As illustrated in Figure 5.15, these two photosystems are linked in series to transport electrons from water to $NADP^+$. The core of photosystem II (PSII) is homologous to the photosynthetic reaction center of nonsulfur purple

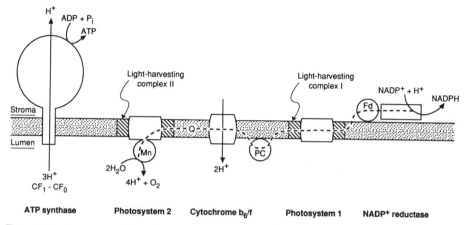

Figure 5.15. A summary of interactions during noncyclic electron flow during photosynthesis. The electron transfer and proton gradient-forming reactions during noncyclic photosynthesis are schematically illustrated.

bacteria. Consequently, higher oxygenic photosynthetic organisms are thought to have evolved from these purple bacteria. Similarly, photosystem I (PSI) may have evolved from green bacteria (e.g., *Chlorobiaceae*). Therefore, oxygenic organisms are thought to have evolved a hybrid photosynthetic apparatus.

Nearly 40 different membrane proteins distributed among five major complexes (Figure 5.15) participate in oxygenic photosynthesis. In the order of electron flow, these complexes are: (1) PSII with its peripheral water-splitting enzyme and the surrounding chlorophyll a/b proteins that constitute PSII's light-harvesting complex (LHCII), (2) the cytochrome b_6–f complex, (3) PSI, and (4) $NADP^+$ reductase. The fifth component is the CF_1–CF_0 ATP synthase that produces ATP. In addition to these major complexes, several additional proteins and cofactors such as ferredoxin (fd), plastocyanin (Pc), chlorophyll a, and quinones are used to shuttle electrons.

As Figure 5.15 shows, two water molecules bind to the Mn cluster of the water-splitting enzyme which catalyzes the reaction:

$$2H_2O \xrightarrow{4h\nu} 4H^+ + 4e^- + O_2 \tag{5.8}$$

The four protons released contribute to a transmembrane proton potential. The electrons removed from H_2O are used to sequentially oxidize a chlorophyll a molecule at PSII's reaction center. Four electrons from two water molecules are sequentially removed by four photons to liberate one oxygen molecule. The electrons then reduce quinones in the membrane. The reduced quinones shuttle electrons to the cytochrome b_6–f complex, which is structurally homologous to the cytochrome b–c_1 complex discussed above. In concert with quinones, cytochrome b_6–f pumps protons into the thylakoid space, further enhancing the proton electrochemical potential. The electrons are then passed to PSI via plastocyanin, a copper-containing protein. Each electron is used to reduce an oxidized chlorophyll molecule of PSI. Two photons are used by PSI to excite one of chlorophyll's resting electrons to an excited state. The electron is then shuttled to NADP⁺ reductase by the Fe–S complex of ferredoxin. The electron then participates in reducing NADP⁺ to NADPH. This reaction produces NADPH and removes protons from the stromal space, further enhancing a proton potential. We will begin our detailed discussion of this process by considering the spatial locations of these complexes in thylakoid membranes.

5.2.3.2. Organization of a Chloroplast's Membranes

Chloroplasts possess two membrane systems: the envelope and thylakoid membranes (Figure 5.16). Chloroplasts are surrounded by a double membrane or envelope that separates a plant's cytosol from the chloroplast's stroma. The envelope resembles those of gram-negative bacteria and contains about 75 proteins. Porins in the outer membrane allow metabolite passage. The inner membrane contains a variety of transport proteins. The stroma contains enzymes involved in lipid and pigment synthesis and the chloroplast's DNA and protein synthesis machinery.

Bioenergetic functions are associated with thylakoid membranes. Thylakoid membranes are stacked one upon another from 2 to over 20 times to form grana (Figure 5.16). Alternatively, these membranes can also be unstacked stromal thylakoids (Figure 5.16). Thylakoid membranes separate their internal compartment (lumen) from the stroma. A thylakoid's lumen is continuous among the stacks of membranes forming a granum. When illuminated, the thylakoid's lumen becomes acid. Thylakoid membranes are composed of 50% protein, 40% lipid, and 10% pigment by weight (Gounaris *et al.*, 1986). Although the dispositions of membrane components within the stacked and

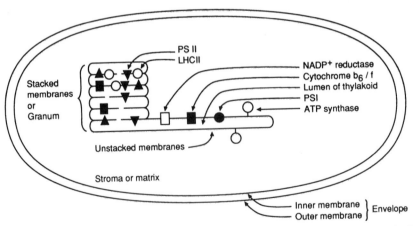

Figure 5.16. Chloroplast membranes. Chloroplasts are surrounded by an envelope composed of two membranes separated by an intermembrane space. Thylakoid membranes are found in the chloroplast's stroma. These membranes are either unstacked or stacked to form grana.

unstacked thylakoid membranes are still being sorted out, there is general agreement that considerable asymmetries exist. We shall begin by considering the disposition of lipids in thylakoid membranes.

The primary lipid components of thylakoid membranes are: 50% monogalactosyl diacylglycerol (MGDG), 30% digalactosyl diacylglycerol (DGDG), 10% sulfolipids, and 10% PG. The potential transverse and lateral asymmetries of these lipids have been studied (Gounaris *et al.*, 1986). Although some uncertainties remain, it appears that PG and MGDG are enriched on the stromal leaflet of thylakoid membranes. Studies of the lateral distribution of lipids have suggested that MGDG is enriched in granum membranes. The membrane distribution and functional roles of thylakoid sulfolipids are unknown. However, they are believed to be tightly bound to PSII and the ATP synthase.

The protein compositions of the stromal and granum thylakoid membranes differ qualitatively and quantitatively. Stacked thylakoid membranes are richer in protein than are unstacked membranes. The lipid protein mole ratio of stromal membranes is 4:1 whereas stacked membranes have a ratio of 2.5:1. Figure 5.16 illustrates some of the lateral protein heterogeneities of thylakoid membranes. Both biochemical and microscopic studies have provided evidence supporting the structural and functional differentiation of these two membrane domains. The ATP synthase, NADP$^+$ reductase, and PSI are located on unstacked membranes. The cytochrome b_6–f is found in both stacked and unstacked membranes. PSII and its light-harvesting complex are primarily associated with stacked membranes. A subpopulation of PSII complexes (PSII β) are found in stromal membranes. These PSII β complexes may play an important role in photosynthetic regulation (Section 5.2.3.9).

5.2.3.3. Thylakoid Membrane Stacking

Izawa and Good (1966a,b) were the first to show the reversibility of thylakoid stacking. When added to isoosmotic low-salt solutions, granum membranes unstack. This is accompanied by the intermixing of granum and stromal membrane components. However, the addition of low concentrations of divalent cations or higher salt concentrations led to spontaneous restacking. It has been suggested that granum unstacking is due to the increase in electrostatic repulsion between membranes at low ionic strength. Under physiological conditions, the stacking and unstacking of thylakoid membranes are thought to be triggered by specific molecular interactions.

The great enrichment of PSII and its associated light-harvesting polypeptides in granum membranes makes them ideal candidates for mediators of membrane stacking. Mullet and Arntzen (1980) and McDonnel and Staehelin (1980) tested the ability of LHCII polypeptides to trigger stacking. These polypeptides were reconstituted into lipid vesicles composed of PC. When divalent cations were added to these vesicles, they spontaneously aggregated. However, if the vesicles were first exposed to mild trypsinization, they were unable to aggregate. The 26- and 27-kDa proteins of LHCII are responsible for membrane stacking. The membrane stacking function has been localized to the NH$_2$-terminal regions of these proteins (Mullet, 1983). This short sequence is rich in basic amino acids and Thr. The positively charged residues may interact with acidic groups of appressed membranes or neutralize the negative charges of stacked membranes, such as those of sulfolipids. The Thr and neighboring residues match a consensus sequence for a cAMP-dependent protein kinase (PKA) phosphorylation site. The phosphorylation of this site may play a role in the regulation of photosynthesis including the distribution of excitation energy between PSII and PSI. We will return to regulatory interactions and the control of thylakoid membrane stacking later in this section.

Although these experiments help explain why thylakoid membranes stack, they do not explain the lateral heterogeneity discussed above. Hinshaw and Miller (1989) showed that LHCII is almost exclusively found on stacked membranes. The appressed membrane faces have very smooth surfaces, consistent with their short separation distances. Membrane protein complexes such as PSI and ATP synthase have large protruding domains. Consequently, they could be sterically excluded from stacked membranes. LHCII complexes have strong lateral (side-to-side) interactions. This is demonstrated by the fact that they spontaneously form two-dimensional crystals in solubilized or membrane-bound forms (e.g., Mullet and Arntzen, 1980). As LHCII hexagonal lattices form, only associated proteins (PSII) or proteins that do not disturb its packing interactions are included. This provides another possible mechanism of lateral heterogeneity. Therefore, interactions between two smooth membrane faces and lateral interactions within a membrane promote membrane stacking.

5.2.3.4. Photosystem II: Harvesting the Light

In the 1930s Emerson and Arnold first showed that many chlorophyll molecules of plants act cooperatively to harvest light energy which they then transmit to a few special chlorophyll molecules. Since chlorophyll absorbs in the red and blue, photosynthetic plant cells are typically green. As described above for *Rps. viridis*, the light-harvesting complexes of higher plants also surround their photosynthetic reaction centers. However, the photosystems of higher plant cells are more highly evolved. In this section we will describe some of the features of light-harvesting complexes of higher organisms.

Figure 5.17 shows freeze-fracture electron micrographs of normal and detergent-treated thylakoid membranes. Detergent treatment solubilizes PSI and ATP synthase but leaves PSII and its associated LHC within grana (Lyon and Miller, 1985). Panels A and B show two magnifications of untreated granum membranes. The randomly distributed tetrameric clumps are the photosynthetic reaction cores of PSII. The smaller particles in the background are LHCII molecules. LHCII proteins form two-dimensional crystals in thylakoids when exposed to a detergent. This is illustrated by the micrographs of panels C and D; separation into two well-defined regions is evident. The ability to form two domains rich in core or LHCII molecules indicates that they are not covalently attached to one another.

LHCII is a large oligomeric integral membrane protein complex. Typically, LHCIIs consist of one or two major proteins of 21 kDa to 29 kDa and two or three minor protein species. The LHCII of peas contains four proteins of 28, 26, 25.5, and 24.5 kDa. The 26-kDa protein is by far the most abundant under normal conditions as judged by SDS-

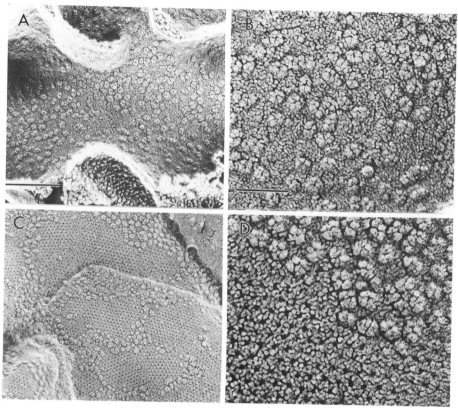

Figure 5.17. Freeze-fracture electron micrographs of thylakoid membranes. Electron micrographs of untreated (panels A and B) and detergent-treated (panels C and D) thylakoid membranes are shown. Intermediate (panels A and C; bar = 0.19 μm) and high (panels B and D; bar = 0.055 μm) magnification images are shown. The larger intramembrane particles represent PSII whereas the smaller particles are LHCII components. After detergent treatment the LHCII molecules form crystalline-like domains. (Reproduced from the *Journal of Cell Biology*, 1985, Vol. 100, p. 1139, by copyright permission of the Rockefeller University Press. Illustration courtesy of M. Lyon and K. Miller.)

PAGE gels and antibody binding experiments. Several antigenic relationships among these polypeptides have been mapped by Darr *et al.* (1986). The 26- and 28-kDa proteins possess unique antigenic determinants suggesting that they are not derived from a common precursor protein or mRNA. Quantitative binding studies indicate that there are 27 copies of the 26-kDa protein and 2 copies of the 28-kDa protein per electron transport chain. Eight monoclonal antibodies were found to react with the 26-, 25.5-, and 24.5-kDa proteins. No antibodies could be found which distinguished the 24.5- and/ or the 25.5-kDa from the 26-kDa protein. The smaller proteins likely arise from proteolytic processing of the 26-kDa polypeptide.

Deduced amino acid sequences of several LHCII polypeptides have been obtained (e.g., Murphy, 1986). These proteins are encoded by a nuclear multigene family. LHCII sequences from several plant species display similar structures and compositions. These transmembrane proteins are rich in hydrophobic amino acids and Pro residues but poor in His residues. Hydropathy analyses indicate the presence of three transmembrane domains. The presence of three transmembrane α-helices is consistent with the 44% α-helical content measured by circular dichroism. The NH_2-terminus is located at the stromal surface where it interacts to stack membranes.

Chlorophyll molecules of LHCII complexes account for over 50% of the pigments associated with thylakoid membranes. Chlorophyll is noncovalently but specifically bound to the transmembrane domains of LHCII polypeptides. Their tetrapyrrole rings are located within a membrane's hydrophobic core. Spectroscopic studies indicate that they make a 60° angle with the plane of the membrane. Three chlorophyll a and/or b molecules are associated with each polypeptide.

The LHCII proteins dimerize. Three dimers assemble into a triangular hexamer with threefold symmetry. When LHCII molecules reconstituted in lipid vesicles are viewed using freeze-fracture electron microscopy, they form 8-nm intramembrane particles (Mullet and Arntzen, 1980; McDonnel and Staehelin, 1980). The particles correspond to individual hexameric LHCII complexes. Figure 5.18 shows a model of LHCII based on analysis of negatively stained electron microscopic images. Protrusions, which are individual dimers, are 2 nm high and 2.6 nm in diameter (Li, 1985). Since there is on average one chlorophyll molecule per transmembrane domain, exciton coupling could occur along this two-dimensional lattice.

Chlorophyll's absorption spectrum is dependent on: its host protein, the neighboring molecules and environment, and its covalent structure (chlorophyll a versus b).

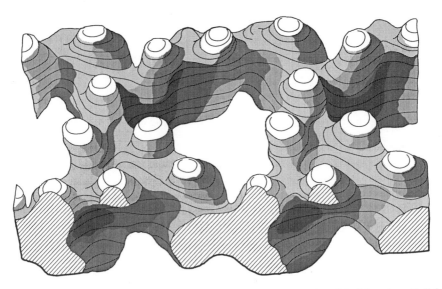

Figure 5.18. A model of the light-harvesting complex. A three-dimensional model of the eukaryotic light-harvesting complex is shown.

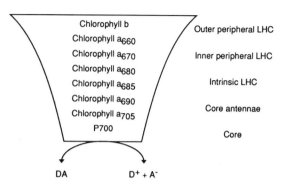

Figure 5.19. An illustration of the spatial relationships between components of higher eukaryotic photosystems. The left-hand side shows the spatial relationships among the spectroscopically distinct chlorophyll molecules. The right-hand side shows the spatial relationships among the physical components of a photosystem.

PSII's chlorophyll molecules are spectrally and spatially distinct. Figure 5.19 illustrates the spatial dependence of chlorophyll's absorbance. The periphery of LHCII absorbs higher-energy photons whereas the photochemical reaction center absorbs lower-energy photons. This allows light-harvesting pigments to operate with broad spectral sensitivity. It also promotes a unidirectional flow of energy from the periphery to a reaction center. At each step along the LHC a small amount of energy is lost. Therefore, excitation energy can never move toward the periphery; it is always funneled to a reaction center.

Each PSII contains about 250 chlorophyll molecules associated with its photochemical reaction core and LHCII. Chlorophyll molecules associated with the core and core antennae will be discussed in the following section. LHCII can be operationally divided into two regions: the intrinsic LHC and the peripheral LHC (Glazer and Melis, 1987). The intrinsic LHC is tightly bound to the core complex with its 80 chlorophyll molecules. This domain of LHCII may be enriched in the higher-molecular-mass (28 kDa) protein. The peripheral LHCII contains roughly 120 chlorophyll (a + b) molecules. The peripheral LHCII is likely identical to the mobile LHCII component. When the smaller LHCII polypeptides are phosphorylated (Glazer and Melis, 1987), grana loosen or unstack, the peripheral LHC uncouples from PSII, and lateral rearrangements of the thylakoid membrane redistribute excitation energy between PSII and PSI. We will return to photosynthetic regulation later in this chapter.

5.2.3.5. Photosystem II: Electrons on the Move

A few structural details regarding PSII are beginning to emerge. The luminal (inner or *trans*) faces of stacked thylakoid membranes are shown in Figure 5.17. The larger tetrameric complexes (14 nm) are PSII's core structure. The tetrameric appearance may arise, in part, from the water-splitting enzymes associated with the luminal face of PSII's core complex. Negative-stain electron microscopy has shown that the isolated PSII core complex is in the shape of an elliptical disk 6.5 nm high with major and minor axes of 15.5 and 10.5 nm, respectively (Rogner *et al.*, 1987). This size is consistent with that of the intramembrane particles derived from PSII's core.

Figure 5.20 illustrates PSII's core protein complex. It is composed of five major integral membrane proteins: D1, D2, 47 kDa, 43 kDa, and the cytochrome b559 dimer. The D1 and D2 proteins participate in electron transport. The 47- and 43-kDa proteins are "core" antennae. The physiological role of the cytochrome is uncertain. There are three additional peripheral proteins of 33, 23, and 17 kDa. These peripheral proteins are thought to participate in water-splitting reactions.

The 47- and 43-kDa proteins shuttle excitation energy from LHCII to a photochemical reaction center. Their deduced amino acid sequences have shown their true molecular masses to be 56 and 51.8 kDa, respectively. Each protein possesses seven apparent transmembrane domains. Both of these proteins contain about 25 chlorophyll a (Chl a) molecules. The 43-kDa protein is believed to accept energy from LHCII and then donate it to the 47-kDa protein. The 47-kDa protein, which is closely associated with P_{680}, directly donates energy to P_{680}. These two proteins link PSII's light-harvesting complex with its reaction center.

The photosynthetic reaction center of PSII is formed by the D1 and D2 proteins. They are homologous to the L and M proteins of *Rps. viridis*. In addition to similarities in amino acid sequences, these proteins are also similar in their cofactors and spectroscopic properties. In analogy with purple bacteria, these two proteins of higher plants possess four Chl a molecules, two phenophytin a molecules (Pheo), one β-carotene molecule, a nonheme iron, and two plastoquinones (PQ) that act as electron carriers.

The D1 and D2 proteins are encoded by the *psbA* and *psbD* genes of chloroplasts, respectively. The deduced amino acid sequences of these two proteins have been obtained from several species (e.g., Alt *et al.*, 1984; Murphy, 1986). Both proteins possess five transmembrane domains. Their membrane topology, as shown in Figure

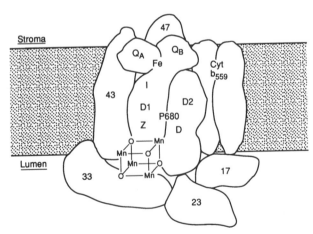

Figure 5.20. A structural model of photosystem II's core complex. The cofactors of the D1 and D2 proteins carry out the charge separation reactions. The 43 and 47 kDa are core antennae. The Mn cluster and peripheral proteins facing the lumen may function in water-splitting reactions. (Redrawn with permission from the *Annual Review of Biophysics and Bioengineering*, Vol. 18, copyright © 1989 by Annual Reviews, Inc.)

5.21, has been worked out by antibody binding and chemical modification studies. The similarities between Figures 5.21 and 5.7 should be noted.

The D1 protein binds PQ_B and herbicides, in analogy with the L protein of *Rps. viridis*. A special pair of Chl a molecules are bound to His residues 198 of the D1 and D2 proteins, thus forming the primary electron donor P_{680}. However, there are no compelling data indicating that the free radical formed upon excitation is distributed across both Chl a pigments. The nonheme iron is attached near the *cis* side of a bilayer similar to that of *Rps. viridis*. Trp-250 and Ala-261 of D2 contribute to the PQ_A binding pocket. The PQ_B binding region likely includes Phe-255 and Ser-264 of D1. When herbicides bind to the wild-type protein, they alter the conformation of the PQ_B pocket.

Hill and Bendall originally postulated the Z scheme of photosynthesis illustrated in Figure 5.22. This diagram shows the redox potential of several cofactors or redox sites along the ordinate. It describes how two photosystems cooperate in generating a proton potential and NADPH by converting light energy into chemical intermediates. The early intermediates are associated with the photosynthetic reaction center of PSII.

When light energy is absorbed by the special pair of Chl a molecules, an electron is promoted to a higher energy level. This creates a change in the molecule's absorbance at 680 nm. In less than a picosecond the charge separation complex P_{680}^+ $Pheo^-$ is formed. This intermediate has been detected by both ESR and optical spectroscopy. Within 100 psec the electron is shuttled to PQ_A, which is associated with the D2 protein. Although the neighboring nonheme iron does not participate in electron transport, it may help maintain the protein–pigment complex in an appropriate conformation; it likely stabilizes the semiquinone intermediate. The electron migrates to PQ_B, which is associated with the D1 protein. The reduced quinone PQ_BH_2, which retains much of the excitation energy, then dissociates from D1 and diffuses within thylakoid membranes.

The remaining integral membrane component of the photosynthetic reaction center is cytochrome b_{559}. Cytochrome b_{559} is a small heterodimer of 10- and 6-kDa proteins. The 10- and 6-kDa proteins are encoded by the chloroplast's *psbE* and *psbF* genes,

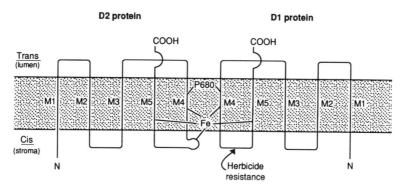

Figure 5.21. Structural models of PSII's D1 and D2 proteins. The D1 and D2 proteins each possess five transmembrane α-helices. The cofactors associated with these proteins participate in the primary charge separation events of photosynthesis. Sites of cofactor binding and herbicide resistance-conferring mutations are shown.

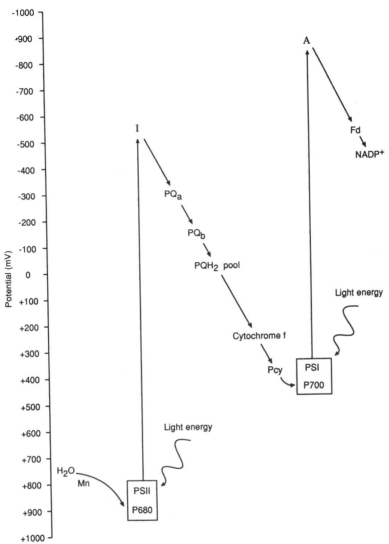

Figure 5.22. Z scheme of photosynthesis. Several photosynthetic states are shown in this diagram. The ordinate lists the states' redox potentials. PSII and PSI absorb light energy, as shown. NADP$^+$ and a proton potential (not shown) are produced by this cycle.

respectively. Their deduced amino acid sequences revealed the presence of a single transmembrane domain in each protein (e.g., Heinemeyer *et al.*, 1984). Both proteins have a His residue in the transmembrane domain near the luminal face. These two His residues coordinate a heme between them, in the middle of the heterodimer. However, the function of cytochrome b_{559} is incompletely understood. This cytochrome does not undergo photooxidation during normal physiological conditions. However, if the oxygen-evolving proteins are dissociated or inhibitors are added, cytochrome b_{559}

donates electrons to P_{680}^+. Electrons may follow this route during stressful conditions, such as high illumination levels, to protect the reaction center from damaging photo-oxidation reactions.

5.2.3.6. Water Splitting at Photosystem II Replenishes Electrons

After an electron leaves P_{680}^+, a positive charge or "hole" is left behind that must be filled to begin another photosynthetic cycle. Electrons are obtained by splitting water molecules into protons, electrons, and oxygen molecules [Eq. (5.8)]. The four electrons from water are sequentially transferred to P_{680}^+ where they enter the energy-conserving reactions of photosynthesis. The four protons contribute to a transmembrane proton potential.

The oxygen-evolving complex is composed of three peripheral membrane proteins associated with PSII's luminal face. They are encoded by three nuclear genes: *psbI*, *psbII*, and *psbIII*. The 33-kDa protein is associated with a manganese cluster, which participates in the catalytic breakdown of water. This protein has two conserved Cys residues. It is homologous to Mn-requiring superoxide dismutases, although the details of water oxidation remain obscure. The 23- and 17-kDa proteins participate in binding Ca^{2+} and Cl^-, two cofactors in oxygen evolution. These proteins are all hydrophilic and only bind to membranes containing PSII. They hold Mn, Ca^{2+}, and Cl^- in association with PSII. Both the 33- and 23-kDa proteins and their cofactors are known to be required for oxygen production.

A cluster of four Mn atoms found between the 33-kDa protein and the photosynthetic reaction center (Figure 5.20) play a central role in water oxidation. For example, Mn-deficient mutants of algae are unable to produce oxygen. In addition, displacement of Mn from the complexes also abolishes oxygen production. Mn and O atoms form a cluster, as indicated in Figure 5.20 (Burdvig *et al.*, 1989). Various physical techniques have detected the presence of these clusters. ESR spectroscopy has shown the presence of free radical intermediates associated with Mn clusters. Mn clusters are believed to act as a reservoir for oxidizing equivalents during water-splitting reactions.

Figure 5.23 shows a model of the catalytic cycle of PSII's oxygen-evolving complex developed by Kok *et al.* (1970). Each state (S_0 to S_4) represents a different oxidation

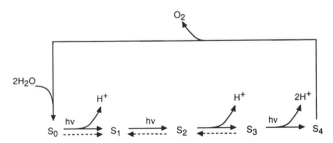

Figure 5.23. Catalytic cycle of the oxygen-evolving complex. The five oxidation states (S_0 to S_4) of the Mn cluster are indicated. Light absorption, proton release, and the production of molecular oxygen are indicated.

state of the Mn cluster. The resting dark-adapted state corresponds to S_1. Each time a quantum of light is absorbed, an electron is donated to the reaction center. Four quanta must be absorbed during illumination to produce one oxygen molecule. Protons are released from the oxygen-evolving complex as indicated in Figure 5.23. Electrons are sequentially removed from water by the oxidizing equivalents of the Mn cluster. The electrons are then shuttled to P_{680}^+ via a Tyr residue of the D1 protein. Although the chemical reactions participating in water oxidation are not established, models consistent with experimental results have been proposed (Burdvig *et al.*, 1989).

5.2.3.7. The Eukaryotic Cytochrome b_6–f Complex, Q Cycle, and Plastocyanin

A common theme among electron transport pathways is the utilization of a quinone to shuttle electrons to a cytochrome complex that pumps protons. The cytochrome b_6–f complex of plants resembles the b–c_1 complex of bacteria and mitochondria. All cytochrome b_6–f complexes contain four integral membrane proteins: cytochrome f, cytochrome b_6, an Fe–S protein, and a 17-kDa protein in the ratio 2:4:2:4. When reconstituted into vesicles composed of phospholipids and galactolipids, the complex forms intramembrane particles 8.8 nm in diameter. This transmembrane protein complex catalyzes electron transport from PSII and PSI and concomitantly translocates protons into the thylakoid's lumen.

Cytochrome f is encoded by the chloroplast's *petA* gene. It is a 31.8-kDa protein with a single transmembrane domain near its COOH-terminus (Alt and Herrmann, 1984). The deduced amino acid sequence also revealed five pairs of basic residues in the NH$_2$-terminal domain within the lumen. These positive charges may interact with negative charges on plastocyanin. The consensus heme binding sequence -His-Cys-X-X-Cys-His- is near its NH$_2$-terminus. The heme is covalently attached via thioether linkages. A large hydrophilic NH$_2$-terminus protrudes into the lumen.

Cytochrome b_6 is a hydrophobic heme-containing protein encoded by the chloroplast's *petB* gene. Five transmembrane domains are suggested by its deduced amino acid sequence (Heinemeyer *et al.*, 1984; Murphy, 1986). Five His residues of its sequence are highly conserved when compared with other b-type cytochromes. These residues participate in coordinating two heme groups between the M2 and M5 transmembrane domains, as we have previously described for purple photosynthetic bacteria (Figure 5.10). The heme's rings are oriented in the plane of the membrane to facilitate transmembrane electron movement.

The Fe–S protein is a 20-kDa transmembrane protein encoded by the nuclear *petC* gene. Its active site is exposed at the luminal face where it interacts with cytochrome f. The Fe–S protein and cytochrome b_6 both bind PQ as shown by the covalent binding of photoaffinity-labeled PQ.

A 17-kDa integral membrane protein is also associated with the cytochrome b_6–f complex. Its sequence indicates the presence of three transmembrane domains. The 17-kDa protein does not perform a redox function. However, it is believed to help attach a 37-kDa ferredoxin reductase to the stromal side of some cytochrome b_6–f complexes. We will soon return to the functional role of the 37-kDa reductase.

   A few key elements of proton translocation in this system are illustrated in the
Q cycle of Figure 5.24, which is generally true for all b–c_1-type cytochromes. PQH_2
interacts with the Fe–S protein to liberate a semiquinone, an electron, and a proton.
Since the proton is released at the luminal face, it contributes to a transmembrane proton
potential. The electron is transferred to the Fe_2–S_2 cluster, then to cytochrome f's heme.
The electron is picked up by plastocyanin (Pc) and then shuttled to PSI. The semiquinone
binds to cytochrome b_6 where it releases an electron, a proton (at the luminal face),
and PQ. The electron then hops between the cytochrome's two heme groups. It combines
with a proton from the stromal face to regenerate, after two cycles, PQH_2. The net result is the removal of two protons from the stroma and the addition of two protons to the
lumen.

   Plastocyanin receives an electron from cytochrome f. This luminal protein binds
the electron at its copper-containing prosthetic group. Pc is roughly cylindrical in
shape (3 nm in diameter and 4 nm in length). Since the distance between the apposing
luminal faces within a granum is about 5 nm, Pc's motion within a lumen is highly
restricted. Fluorescence recovery after photobleaching experiments have shown that
Pc's lateral diffusion coefficient between multilamellar phospholipid membranes is
5×10^{-8} cm²/sec (Fragata *et al.*, 1984). Since the hydration forces (or ordered water)
extend about 2 nm from each surface, the effective local viscosity within the lumen is
unusually high. A 30-fold increase in luminal viscosity over bulk-phase water has been
seen using ESR spectroscopy (Murphy, 1986). In addition to the bulky proteins of the
oxygen-evolving complex, cytochrome f also impedes luminal diffusion *in vivo*. Since
the ratio of Pc to PSI is about 1:1, it is very difficult to imagine how Pc in the center of a
stack migrates to unstacked PSI regions to effect electron transfer. Therefore, Pc may
principally act in comparatively short-range electron transfer events in unstacked
membranes. The small lipid PQH_2 is more likely the electron carrier from stacked to
unstacked membrane domains.

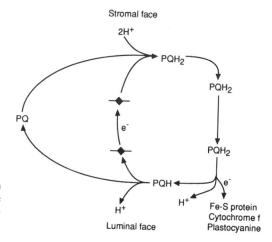

Figure 5.24. A possible model of electron
flow in the cytochrome b_6–f complex. The
electron flow and proton release of the cytochrome b_6–f complex are shown.

5.2.3.8. Photosystem I

Our molecular understanding of PSI is lagging behind that of PSII. Furthermore, there are no well-characterized PSI-like bacterial systems analogous to *Rps. viridis* to aid in the study of PSI. Consequently, our discussions of PSI will be a bit sketchy.

About one-third of a chloroplast's chlorophyll is associated with PSI. Forty percent of this chlorophyll is bound to LHCI whereas the remainder is linked to the core or its antennae. Each LHCI contains roughly 60 Chl a and 20 Chl b molecules. Three transmembrane proteins of 22, 23, and 20 kDa bind these chlorophyll molecules at a ratio of 1 chlorophyll per 2 kDa of protein mass. Since antibodies directed against LHCII demonstrate some cross-reactivity with LHCI, there may be some structural similarities. Unfortunately, there is little structural information regarding LHCI. The three known polypeptides of LHCI are separated by sucrose density gradient centrifugation. The light fraction containing the 22- and 23-kDa proteins have a fluorescence emission maximum of 730 nm whereas the heavy fraction (20-kDa protein) has an emission maximum of 680 nm. Reasoning by analogy with LHCII, it has been suggested that the 20-kDa protein is part of a loosely bound LHCIb complex whereas the 22- and 23-kDa proteins form a tightly bound LHCIa complex.

PSI has been imaged by freeze-fracture electron microscopy. PSI forms 12-nm intramembrane particles in both thylakoid and reconstituted membranes. When the core complex is studied separately, it forms 8-nm particles. The photochemical reaction center is believed to be at the center of the core surrounded by the core antennae, LHCIa, and LHCIb, in that order.

It is difficult to separate the structural and functional properties of the core antennae and photosynthetic reaction center because their structures and identities have not been completely worked out. The largest structure of the core complex is a tetramer of 70-kDa proteins. The entire tetramer contains the P_{700} reaction center, 130 Chl a molecules, and 16 carotenoids. One 70-kDa dimer is thought to contain the reaction center whereas a second may be the core antenna complex. The core antennae presumably funnel energy to the reaction center via Chl a molecules.

PSI's photosynthetic reaction center, P_{700}, undergoes a dramatic change in absorbance at 700 nm upon photooxidation. P_{700} is a chlorophyll dimer, although their electron clouds are apparently not delocalized over one another. Chemical cross-linking experiments have suggested the P_{700} is held between two heterodimeric 70-kDa proteins. These two homologous proteins are encoded by the chloroplast's *psaA* gene. The deduced amino acid sequences of these proteins exhibit molecular masses of 82.5 and 83.2 kDa (Fish *et al.*, 1985). Furthermore, these homologous proteins each possess 11 apparent transmembrane domains. The transmembrane domains contain conserved His residues that may participate in chlorophyll binding. The luminal faces of these proteins possess a region of negatively charged amino acids. This region may interact with the positive charges of Pc. Several Fe–S centers that participate in electron transport through the reaction center are covalently bound to the 70-kDa proteins.

In addition to the 70-kDa proteins, the core complex contains five additional proteins of 20, 16, 14, 10, and 8 kDa. The functions of these proteins are uncertain. One or two of these proteins contain Fe–S centers that shuttle electrons to peripheral ferredoxins (Gounaris *et al.*, 1986). Although the precise locations of these several Fe–S clusters are now known, they play a central role in electron transport.

Within 10 psec after excitation, P_{700} donates an electron to A0, the primary electron acceptor. A0 has been identified as 13-hydroxy-20-chloro-chlorophyll a. In about 100 psec the electron migrates to A1, which is believed to be a quinone. The photochemical pathway contains three Fe–S centers, X, B, and A, which all follow A1 in electron flow. Each of these Fe–S centers is composed of a cluster of four Fe and four S atoms. These clusters are generally linked to proteins via Cys residues. There is some uncertainty in the trafficking of electrons through the Fe–S centers to ferredoxin. The electrons could flow in series X→B→A→Fd or parallel X→B→Fd and X→A→Fd. Current evidence favors parallel electron flow (Murphy, 1986). Electrons are shuttled from either the A or B center to ferredoxin.

Ferredoxin is a small 11-kDa protein found in a thylakoid's stroma. It carries electrons from either the A or B sites of the PSI reaction center to the $NADP^+$ reductase. Ferredoxin carries electrons in an Fe_2S_2 cluster. The $NADP^+$ reductase is the final protein in the noncyclic electron flow pathway. It is a 34-kDa protein found in a thylakoid's lumen. It can be easily removed from membranes by salt solutions, indicating that it is a peripheral protein. It forms ternary complexes with Fd and $NADP^+$ in solution. Presumably, these ternary complexes form in the lumen, bind to PSI, and then accept an electron leading to the reduction of $NADP^+$.

In some instances electrons cycle around PSI instead of through PSI. During cyclic electron flow, electrons are shuttled from ferredoxin to cytochrome b_6–f instead of the $NADP^+$ reductase. The cycle produces a proton potential without NADPH formation. During this pathway a peripheral 37-kDa protein oxidizes ferredoxin (ferredoxin–quinone oxidoreductase). Electrons and protons from the stromal surface are returned to PQ to form PQH_2 which then liberates two protons and two electrons at the luminal face; the latter are then returned to PSI via Pc.

5.2.3.9. Regulation of Photosynthesis

Plants have evolved several strategies to maximize their efficiency of light collection and minimize potential photochemical damage. For example, shade-adapted plants have more thylakoids, grana, LHCI, and LHCII than sun plants. In addition to these long-term adaptive strategies, plants possess short-term mechanisms to regulate light collection and electron transfer reactions that rely on the spatial distribution of membrane complexes.

To maintain maximal photosynthetic efficiency during noncyclic electron flow, PSI and PSII must become excited at equivalent rates. Any difference in rate would seriously affect the PQ pool, the efficiency of photosynthesis, and lead to possible photochemical damage. Under typical conditions the PSII reaction cores outnumber the PSI reaction centers by a 1.7:1 ratio. PSII receives less light because it is shaded by neighboring stacks within grana, and light associated with PSII is more easily absorbed and scattered by other plant organelles, resulting in a slightly diminished intensity at P_{680}'s peak (Glazer and Melis, 1987). Plants can alter the ratio of PSII/PSI and their relative efficiencies to counterbalance the spectral characteristics (color) and intensity (position of sun) of light.

When PSII receives more excitation energy than PSI, nearly all of the PQ becomes reduced, thus activating a membrane-bound protein kinase. The kinase phosphorylates the NH_2-terminal region of LHCII proteins, as described above, thus weakening the interactions holding grana together. As grana unstack, the components of stromal and stacked thylakoid membranes intermix. This allows a redistribution of thylakoid components. The mobile component of LHCII dissociates from PSII and then collects in stromal membranes. The membrane reorganization of LHCII decreases the light-collecting ability of PSII, thus balancing electron flow between the two photosystems. Although it is controversial, some investigators believe that the mobile LHCII molecules couple with PSI and thereby increase its effective antenna size.

When the thylakoid membrane's PQ pool becomes too oxidized, the kinase is deactivated. A membrane phosphatase then removes phosphoryl groups from LHCII molecules, triggering their reassembly with PSII. Since these phosphorylation and dephosphorylation reactions are occurring on a fairly confined two-dimensional surface, they are rapid (~ 20 sec). When a cloud occludes the sun, a plant's grana rapidly alter the lateral distribution and functional properties of their membrane complexes using reversible phosphorylation reactions (Anderson and Andersson, 1988; Anderson, 1986).

5.3. MITOCHONDRIAL MEMBRANES: THE RESPIRATORY CHAIN

The appearance of oxygenic photosynthetic organisms on Earth led to the accumulation of atmospheric oxygen. This allowed cells to use molecular oxygen as a terminal electron acceptor during the oxidation of organic substrates to CO_2 and water. The respiratory chain converts much of a substrate's chemical energy into an electrochemical potential across inner mitochondrial membranes. In contrast to photosynthetic light absorption, the respiratory chain achieves charge separation by a series of vectorial chemical reactions.

5.3.1. Mitochondria and Their Membranes

Mitochondria are surrounded by a double membrane system, as illustrated in Figure 5.25. The outer membrane contains large pores constructed from three mitochondrial porin molecules. Mitochondrial porin is structurally and functionally analogous to the porin molecules of gram-negative bacteria (e.g., Mannella, 1986). These pores allow molecules of less than 10 kDa to enter the intermembrane space. The outer membrane also contains proteins involved in posttranslational protein import into mitochondria (Section 8.2.4), phospholipid metabolism, and the enzyme monoamine oxidase.

A mitochondrion's inner membrane contains only 20% lipid. PC and PE are imported whereas PG and cardiolipin are synthesized by mitochondria. The inner membrane contains translocators for a variety of compounds such as ATP, ADP,

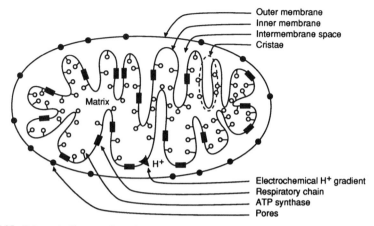

Figure 5.25. Schematic diagram of a mitochondrion. A few of the basic structural elements of mitochondria are indicated. The inner and outer membranes are separated by an intermembrane space. An electrochemical proton gradient is formed across the inner membrane.

glutamate, and fatty acids. The enzyme $NADP^+$–NADPH transhydrogenase is also present in the inner membrane. The inner mitochondrial membrane's most abundant proteins are those participating in oxidative phosphorylation. The respiratory chain of mammals is composed of at least 60 functionally distinct proteins encoded by ~100 genes (Capaldi *et al.*, 1988). These proteins are assembled into four membrane-associated complexes (Table 5.2) that participate in electron transport. The four respiratory complexes are the NADH–ubiquinone reductase (complex I), succinate–ubiquinone reductase (complex II), ubiquinone–cytochrome c reductase (complex III), and cytochrome c oxidase (complex IV). Figure 5.26 shows an SDS-PAGE gel of complexes I through IV. Figure 5.27 shows a schematic view of complexes I, III, IV and the ATP synthase in the mitochondrial inner membrane. The basic features of electron and proton trafficking in this system are also illustrated.

In contrast to bacteriorhodopsin and the photosystems described above, the respiratory chain's enzymes are not generally believed to be spatially organized within inner

Table 5.2.

Components of the Respiratory Chain

Complex	Name	M_r ($\times 10^3$)	Subunits	Prosthetic groups
I	NADH–ubiquinone reductase	850	25 to 28	1 FMN, 5 to 7 Fe–S clusters
II	Succinate–ubiquinone reductase	200	5	1 FAD, 3 Fe–S clusters
III	Ubiquinone–cytochrome c reductase	280	11	1 c and 2 b cytochromes, 2 Fe–S clusters
	Cytochrome c	13	1	1 c cytochrome
IV	Cytochrome c oxidase	208	13	2 a cytochromes

I II III IV

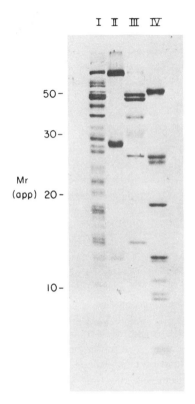

50 –

30 –

Mr
(app) 20 –

10 –

Figure 5.26. SDS-PAGE of respiratory chain proteins. Complexes I through IV of mammalian mitochondria were isolated, solubilized in detergent, and then separated by electrophoresis. The apparent molecular weights are listed on the left. The many different subunits associated with the respiratory chain are apparent. (From Takamiya *et al.*, 1986, *Ann. N.Y. Acad. Sci.* **488**:33.)

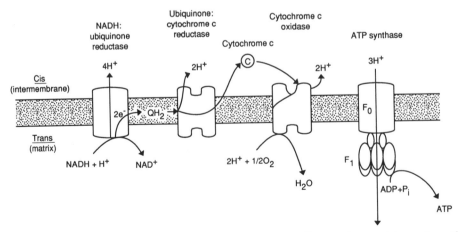

Figure 5.27. Schematic diagram of the mitochondrial inner membrane. Four membrane protein complexes of the mitochondrial inner membrane are shown. An electrochemical transmembrane gradient is generated by NADH dehydrogenase, ubiquinone–cytochrome c reductase, and cytochrome c oxidase. The ATPase utilizes the proton gradient to synthesize ATP.

membranes. Conventional freeze-fracture electron microscopy of mitochondria from higher eukaryotes suggests that these complexes are randomly distributed in membranes. However, there is at least one exception to this rule. Recent quick-freeze deep-etch electron microscopic studies suggest that the NADH–ubiquinone reductase and ATPase within the tubular crista of *Paramecium* are arranged in rows (Allen *et al.*, 1989). It will be important to determine if this supramolecular organization is restricted to *Paramecium*. Furthermore, the lateral organization of complexes II, III, and IV require further study. For example, it is known that complex IV form oligomers; these oligomers could potentially form various sorts of supramolecular organizations. The membrane organization of respiratory chain components remains an active area of investigation.

The protein complexes of the respiratory chain contain tightly bound cofactors, including flavins, hemes, Fe–S clusters, and Cu atoms, that shuttle electrons. The cofactor ubiquinone, which only transiently binds to proteins, carries protons and electrons in mitochondrial membranes. Figure 5.28 shows the temporal sequence of electron transfer events among the respiratory chain's cofactors. Each of the cofactors' redox sites is sequentially reduced as an electron passes along the electron transport chain. Figure 5.29 shows the redox potentials of many of the respiratory chain's cofactors. As seen, NADH is a good electron donor ($E_0' = -320$ mV) and O_2 is a good electron acceptor ($E_0' = +820$ mV). Much of the free energy released at three points along the electron transport chain is conserved as a proton potential. In this section we will examine several features of the respiratory chain.

5.3.2. Complex I: NADH–Ubiquinone Reductase

Several metabolic pathways promote the accumulation of NADH in the mitochondrial matrix. Electrons derived from NADH are imported into mitochondria by inner membrane-bound enzymes. NAD^+ at the inner membrane's matrix surface is reduced to NADH at the expense of cytosolic NADPH by the inner membrane enzyme

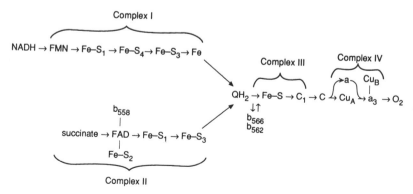

Figure 5.28. Organization of cofactors in the respiratory chain. The likely order of electron flow among the respiratory chain's cofactors is shown.

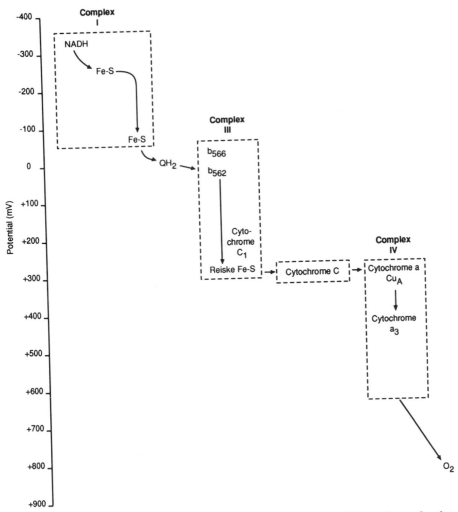

Figure 5.29. Redox potentials associated with the respiratory chain. Redox sites linked to electron flow from NADH to molecular oxygen are shown. The redox potential is listed on the ordinate. The sites associated with respiratory complexes I, III, and IV and cytochrome c are identified by dashed lines.

NAD^+–NADPH transhydrogenase. NADH is generated within the matrix by pyruvate dehydrogenase, the citric acid cycle, and the fatty acid oxidation cycle. NADH provides one important source of electrons for the respiratory chain which begins with complex I, the NADH–ubiquinone reductase.

The NADH–ubiquinone reductase is the respiratory chain's largest and most poorly understood protein complex (M_r = 850,000). It is composed of roughly 25–28 subunits, depending on species and its isolation procedure. One flavin-adenine mononucleotide (FMN) and three iron–sulfur clusters are found in this complex. The NADH–ubiquinone reductase's largest individual component is the NADH dehydrogenase subunit. When these subunits are assembled in the inner membrane, most of the complex's mass is found at the membrane's matrix face.

Recent structural studies have been reported by Brink *et al.* (1987) and Leonard *et al.* (1987). The NADH–ubiquinone reductase extends 9 nm into the matrix whereas it reaches only 1 nm into the intermembrane space. Most of complex I's structure is shown in Figure 5.30. NADH–ubiquinone reductase has at least four major subunits each possessing a diameter of 1.5 nm. Although this model has a square shape, recent studies suggest that another domain lost in earlier isolation procedures gives complex I an "L" shape (Boekema and Leonard, unpublished). A central channel of unknown function is thought to be located in the complex (Boekema *et al.*, 1986), although this could be an artifact of negative staining (Brink *et al.*, 1989). However, a detailed map of its subunits is not available.

Electron transport through the NADH–ubiquinone reductase begins with the NADH dehydrogenase subunit and its associated FMN moiety. The entire pathway of electron transport is NADH$\rightarrow$FMN$\rightarrow$Fe–S$_1$$\rightarrow$Fe–S$_4$$\rightarrow$Fe–S$_3$$\rightarrow$Fe$\rightarrow$S$_2$$\rightarrow$Q. During electron transport the optical absorption of complex I between 350 and 500 nm is greatly diminished. About 50% of this reduction is accounted for by FMN reduction. The

A

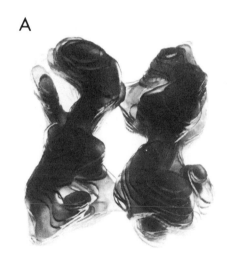

B

Figure 5.30. Two views of a structural model of NADH–ubiquinone reductase. A model of the NADH–ubiquinone reductase was constructed from two-dimensional crystals after negative-stain electron microscopy and image analysis. A view perpendicular to the membrane (panel A) and a lateral perspective (panel B) are shown. (Courtesy of E. Boekema, Groningen University, The Netherlands. Panel A is used with copyright permission of Academic Press. From Boekema *et al.*, *Methods Enzymol.* **126:**344).

remainder is likely due to reduction of nonheme iron atoms. Free radical states associated with the Fe–S clusters have been detected by ESR spectroscopy (Orme-Johnson *et al.*, 1974a,b). Furthermore, ESR has detected bound semiquinone radicals; this indicates that ubiquinone is reduced one electron at a time. Dihydroubiquinone then enters an inner membrane's ubiquinone pool. In addition to reducing ubiquinone, the NADH–ubiquinone reductase also pumps two protons across the inner membrane per electron. Although the mechanism of proton pumping is unknown, it does contribute to a transmembrane proton electrochemical gradient.

5.3.3. Complex II: Succinate–Ubiquinone Reductase

During the citric acid cycle the acetyl group of acetyl CoA is oxidized to CO_2 with the concomitant production of NADH and $FADH_2$. One step of the citric acid cycle is the conversion of succinate to fumarate [Eq. (5.9)]:

$$\text{succinate} + \text{FAD} + 2H^+ \rightarrow \text{fumarate} + FADH_2 \tag{5.9}$$

$$FADH_2 + \text{CoQ} + 2H^+ \rightarrow \text{FAD} + QH_2 \tag{5.10}$$

The mitochondrial inner membrane-associated enzyme succinate–ubiquinone reductase catalyzes the dehydrogenation of succinate followed by the transfer of the two electrons released by succinate to $FADH_2$ and then ubiquinone. The dihydroubiquinone produced by this enzyme enters the inner membrane's ubiquinone pool. Thus, succinate–ubiquinone reductase constitutes a respiratory chain branch that contributes electrons via ubiquinone.

The succinate–ubiquinone reductase (200 kDa) is composed of seven or eight subunits which vary in size from 12 kDa to 70 kDa. Four prosthetic groups associated with the 70- and 27-kDa proteins are required for substrate oxidation. Two iron–sulfur clusters ($Fe–S_1$ and $Fe–S_2$) and a flavin molecule (FAD) are associated with the 70-kDa subunit. One iron–sulfur center ($Fe–S_3$) is associated with the 27-kDa protein. These two proteins of 70 and 27 kDa form the complexes' succinate dehydrogenase activity. A cytochrome b_{558} molecule is also associated with the succinate–ubiquinone reductase. As illustrated in Figure 5.28, electrons flow from succinate→FAD→$Fe–S_1$→$Fe–S_3$→ ubiquinone. The physiological roles of the $Fe–S_2$ and cytochrome b_{558} are unknown. However, the apoprotein component of cytochrome b_{558} is believed to link the succinate dehydrogenase activity to the inner membrane (Girdlestone *et al.*, 1981). The electron transfer reactions through this complex do not result in proton pumping. The product QH_2 does contribute to proton pumping by complexes III and IV of the respiratory chain.

5.3.4. Ubiquinone Molecules Move between Their Redox Sites by Lateral Diffusion

The NADH–ubiquinone reductase and succinate–ubiquinone reductase produce large quantities of reduced ubiquinone (QH_2). Ubiquinone is structurally and functionally homologous to plastoquinone. It links the dehydrogenase activities of the ubiquinone reductases to the respiratory chain's third complex, ubiquinone–cytochrome c reductase. Ubiquinone's covalent structure is shown in Figure 5.14. It is a hydrophobic molecule with a tail composed of ten isoprenoid units. In its reduced dihydroubiquinone form it carries two electrons and two protons.

A variety of biophysical techniques such as ESR, NMR, optical spectroscopy, and calorimetry studies have shown that ubiquinone is buried within the hydrophobic milieu of the mitochondrion's inner membrane and that hydroubiquinone shuttles between its reductases and the ubiquinone–cytochrome c oxidase by lateral diffusion (Schneider *et al.*, 1982). The lateral diffusion of ubiquinone was inferred by fusing phospholipid vesicles with mitochondrial inner membranes. The rate of electron transfer from NADH–ubiquinone reductase to ubiquinone–cytochrome c reductase was measured. As the components of the inner membrane were diluted by the addition of vesicle phospholipids, the rate of electron transfer diminished. This indicates that electron transfer from NADH–ubiquinone reductase to ubiquinone–cytochrome c oxidase is a diffusion-controlled reaction. To understand the mechanism of electron transfer, it is important to determine the identity of the diffusing species. For example, the diffusing species could be free ubiquinone or ubiquinone bound to a protein complex. To distinguish between these possibilities, Schneider *et al.* (1982) also fused ubiquinone-containing lipid vesicles with inner membranes. The variation of electron transport rate with ubiquinone concentration per unit surface area indicates that the unrestrained lateral diffusion of individual ubiquinone molecules is responsible for electron transfer. Therefore, lateral diffusion plays a central role in respiratory electron transfer.

Recently, the lateral diffusion of ubiquinone and an analogue have been directly measured (Gupta *et al.*, 1984; Fato *et al.*, 1985). Gupta *et al.* measured the lateral diffusion coefficient of a fluorescent analogue of ubiquinone in mitochondrial membranes using fluorescence recovery after photobleaching. These investigators found a diffusion coefficient of 3×10^{-9} cm²/sec, rather small for a membrane-bound lipid. Ubiquinone's lateral diffusion in lipid vesicles is much faster. Apparently, additional factors such as low lipid content and ubiquinone–protein interactions decrease its diffusion coefficient in inner membranes. Nonetheless, lateral diffusion is faster than the enzymes' maximal turnover numbers, i.e., lateral diffusion is not the rate-controlling step (Gupta *et al.*, 1984). This contrasts sharply with electron transport in thylakoid membranes where plastoquinone diffusion is believed to be a rate-limiting step. Therefore, lateral diffusion accounts for the rapidity of electron transport observed in mitochondria.

5.3.5. Complex III: Ubiquinone–Cytochrome c Reductase and Its Associated Q Cycle

The reduced ubiquinone molecules generated by complexes I and II are oxidized by complex III, the ubiquinone–cytochrome c reductase. Ubiquinone–cytochrome c reductase contains cytochromes b and c_1; hence, it is also known as the b–c_1 complex. This protein complex oxidizes ubiquinone, translocates protons across the inner membrane, and reduces cytochrome c.

The ubiquinone–cytochrome c reductase is composed of 11 subunits in mammals. However, prokaryotes have only 4 or 5 subunits which presumably play essential roles in electron transfer. The subunits of the mitochondrion's ubiquinone–cytochrome c reductase are encoded by both nuclear and mitochondrial DNA. This protein complex can be divided into three principal regions: a cytoplasmic facing or *cis* domain, a transmembrane region, and a matrix region. The matrix or M domain is composed of two "core" proteins (subunits I and II) and subunit VI. The primary components of the transmembrane region are cytochrome b (i.e., subunit III) and subunit VII. The components of the cytoplasmic domain include cytochrome c_1 (i.e., subunit IV) and the Fe–S protein (i.e., subunit V). The remaining subunits are thought to participate in regulatory interactions. Since the mitochondrial b–c_1 complex is structurally homologous to the b–c_1 complex of photosynthetic bacteria and the b_6–f complex of plants (e.g., Widger *et al.*, 1984), these complexes may exhibit similar electron transport mechanisms.

The subunits of the ubiquinone–cytochrome c reductase dimer are assembled on inner membranes in the shape of an "H" (Leonard *et al.*, 1987). Complex III protrudes 7 nm into the matrix and 3 nm into the intermembrane space, giving it a total height of 15 nm. Its largest dimension parallel to a membrane is 7 nm.

The ubiquinone–cytochrome c reductase possesses three types of prosthetic groups. Two b-type heme groups are associated with cytochrome b. In addition, one c-type heme is found on cytochrome c_1 and an Fe–S center is found on the Fe–S protein. As Figure 5.28 shows, electrons are shuttled along the prosthetic groups of ubiquinone–cytochrome c reductase. We shall return to the mechanism of electron transport later in this section.

The deduced amino acid sequence of ubiquinone–cytochrome c reductase's cytochrome b subunit has been determined (Widger *et al.*, 1984; Saraste, 1984). This protein, encoded by the mitochondrial genome, has nine transmembrane α-helices ($M_r = 42,540$). Cytochrome b possesses binding sites for ubisemiquinone molecules. It has two conserved pairs of His residues: 82–197 and 96–183 (compare with Figure 5.10). These residues are ligands for the two b-type hemes of cytochrome b. The cytochrome b_{562} is associated with the inner membrane's matrix side whereas the b_{566} cytochrome resides at the cytosolic face. Electrons are believed to cross inner membranes via the stacked pair of b-type cytochromes within the protein.

The cytochrome c_1 subunit can be purified from inner membranes in a water-soluble form. It consists of two subunits with masses of 27,924 and 9,175 da. The larger subunit's COOH-terminal region has a short hydrophobic stretch of amino acids that may anchor it to a membrane. The NH$_2$-terminal region is attached to a heme moiety. The smaller subunit has an unusually high content (27%) of glutamic acid. Eight of these Glu residues (positions 5–12) are highly conserved throughout evolution. These negatively charged residues are thought to bind to the positive charges surrounding the exposed edge of cytochrome c's heme (see below).

The iron–sulfur protein of ubiquinone–cytochrome c reductase was first isolated by Rieske; hence, it is sometimes called the Rieske iron–sulfur protein. It is a 24.5-kDa protein containing a single binuclear (2Fe–2S) iron–sulfur cluster. A hydrophobic region presumably anchors the iron–sulfur protein to an inner membrane. The Fe–S cluster exists in a hydrophilic domain near the protein's COOH terminus. This domain contains conserved Cys residues that are ligands for the Fe–S cluster (Beckmann *et al.*, 1987). Its redox state can be detected by ESR spectroscopy. The ubiquinone–cytochrome c reductase may be assembled in inner membranes similar to that previously described for the b–c₁ complex of photosynthetic bacteria (Figure 5.10).

The ubiquinone–cytochrome c reductase has been reconstituted in lipid vesicles (e.g., Beattie and Villalobo, 1982). Under appropriate conditions, the reconstituted enzyme complex participates in electron transfer reactions and proton pumping. Furthermore, the reconstituted complex III displays the expected sensitivities to a panel of inhibitors. When examined by spectroscopic techniques, the reconstituted complex was shown to undergo a conformational change during transport. However, the nature of the conformational change and its relevance to complex III's function are unknown.

The mechanism of electron transport within the ubiquinone–cytochrome c reductase is unknown. Several models of electron transfer have been proposed (Hatefi, 1985; Wikstrom *et al.*, 1981). One of these models is the Q cycle which we have previously mentioned in regard to bacterial and eukaryotic photosynthetic membranes. A schematic diagram illustrating the mitochondrial Q cycle is shown in Figure 5.31, which is very similar to that of Figure 5.24. In this model QH_2 donates one electron to the Fe–S protein and then releases two protons at the intermembrane surface, thus forming a ubisemiquinone anion. The ubisemiquinone anion donates an electron to cytochrome b_{566}. The electron then hops to cytochrome b_{562}. This cytochrome donates an electron to Q, thus re-forming ubisemiquinone at the inner membrane's matrix face. Therefore, cytochrome b's role in the respiratory chain is to facilitate the rapid movement of negatively charged electrons across a membrane's hydrophobic core. Ubisemiquinone takes up two protons and an electron at the matrix face to form QH_2, thus completing the cycle. One proton is thereby transferred across a membrane for each electron shuttled to cytochrome c_1.

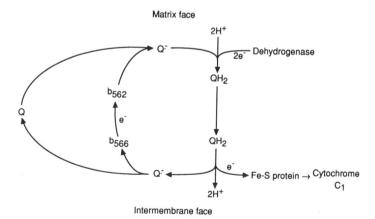

Figure 5.31. The Q cycle. A possible accounting for the flow of electrons and protons during ubiquinone redox reactions is shown.

5.3.6. Cytochrome c

Cytochrome c is a small (13 kDa) protein found in the mitochondrial inter-membrane space. It binds reversibly to the *cis* or cystolic face of the inner membrane during electron transfer from ubiquinone–cytochrome c reductase to cytochrome c oxidase. Mitochondrial cytochrome c is homologous to that of photosynthetic bacteria.

X-ray crystal structures of both oxidized and reduced forms of cytochrome c have been obtained (Mathews, 1984). These two structures are nearly identical. About half of cytochrome c's mass is contained in five short α-helices; no β-sheets are present. Its heme moiety is buried within a hydrophobic pocket. Just an edge of one pyrrole ring is accessible at the protein's surface. This site plays a central role in electron transfer reactions.

Since only a small portion of cytochrome c's heme group is exposed, how can it efficiently transfer electrons from ubiquinone–cytochrome c reductase to cytochrome c oxidase? Specific electrostatic interactions have been found that align the prosthetic groups of cytochrome c with its reductase and oxidase. A series of lysine residues at the protein's surface encircle the heme's exposed edge and play a central role in intermolecular interactions.

The interactions between cytochrome c and its oxidase or reductase can be studied by measuring their binding constants or reaction kinetics (Mathews, 1984). Mutant forms of cytochrome c provided for the first clue that positively charged lysine residues play an important role in cytochrome c's interactions. The loss of lysine residues surrounding the heme's exposed edge decreased the efficiency of cytochrome c's interaction with its oxidase *and* reductase. Lysine's positive charge is also removed by chemical treatments. These experiments showed that the charge of Lys-13 was most important in cytochrome c's interaction with its oxidase and reductase. Its removal alone accounted for a four- to sevenfold reduction in efficiency. The lysine residues participating in cytochrome c's binding to its oxidase are 13, 72, 87, 25, and 8. Similarly, the lysine residues associated with reductase binding are 13, 72, 87, and 27. The inhibition caused by modification of these residues is even more pronounced when the positive charge is chemically altered to a negative charge (Ferguson-Miller *et al.*, 1978). Furthermore, kinetic studies of cytochrome c reactions at differing ionic strengths (i.e., Debye screening lengths) support the electrostatic nature of cytochrome c binding (Smith *et al.*, 1981). The data indicate that the interactions between cytochrome c and its oxidase and reductase are primarily electrostatic in nature.

In addition to ubiquinone–cytochrome c reductase, cytochrome c is also reduced by mitochondrial cytochrome b_5. Mitochondrial cytochrome b_5 is associated with the intermembrane face of a mitochondrion's outer membrane. Although its physiological role is unknown, it is known that it does *not* participate in the respiratory chain. The electrostatic nature of cytochrome c's interactions with its redox partners has been dramatically shown by x-ray crystallographic studies of cytochrome b_5–cytochrome c complexes. The crystal structure of this complex shows a ring of negative charges surrounding the heme of cytochrome b_5 interacting with the positive charges encircling cytochrome c's heme (Mathews, 1984). As mentioned above, the electrostatic binding strategy employed by cytochrome c in the respiratory chain is also used during bacterial photosynthesis.

The cluster of positive charges at cytochrome c's interacting surface creates a large dipole. This dipole helps orient cytochrome c as it enters the electric field of its redox partners. The alignment of positive and negative charges on the surfaces of cytochrome c and its redox partners positions the two heme groups for electron transfer. Although the mechanism of transfer is not certain, quantum mechanical tunneling between the heme groups is believed to be important (Mathews, 1984).

Cytochrome c exists in two forms: bound to membranes via its redox partners and free within the intermembrane space. These two forms are in equilibrium with one another. Membrane-bound fluorescent cytochrome c displays a translational diffusion coefficient of $\geq 2 \times 10^{-9}$ cm²/sec (Gupta et al., 1984). However, these data were necessarily collected at low ionic strengths and low cytochrome c concentrations. Although these conditions are nonphysiological, they support the presence of free and membrane-bound populations of cytochrome c. The diffusion of cytochrome c between its redox sites is faster than the turnover numbers of its associated enzymes.

5.3.7. Complex IV: Cytochrome c Oxidase

The reduced cytochrome c molecules are oxidized by cytochrome oxidase (complex IV), the respiratory chain's terminal enzyme complex. Since cytochrome c oxidase contains the cytochromes a and a_3, it is also known as the aa_3 complex. This protein complex oxidizes cytochrome c, contributes to a transmembrane proton potential, and reduces molecular oxygen to water.

Cytochrome c oxidase is the most thoroughly characterized membrane protein complex of the respiratory chain. In higher animals it is composed of 13 subunits with an aggregate mass of 208 kDa (Table 5.2). These subunits are encoded by both nuclear and mitochondrial genes. Many of the subunits possess hydrophobic sequences of amino acids that span the inner membrane. The cloned gene sequences of subunits I, II, III, IV, and VII predict 10, 2, 6, 1, and 1 transmembrane α-helices, respectively. Subunits I, II, and III are sufficient to catalyze the reduction of molecular oxygen and proton transloca-tion. Functional properties have been assigned to these three subunits. Subunit I is associated with two a-type cytochromes, cytochromes a and a_3. Subunit II possesses two copper atoms, Cu_A and Cu_B. These two subunits collaborate in transporting electrons from cytochrome c to oxygen. Subunit III forms a proton channel across membranes. The remaining subunits are thought to regulate the activity of cytochrome c oxidase.

Two-dimensional crystals of cytochrome c oxidase have been employed to obtain a low-resolution structure by negative-stain electron microscopy (Fuller et al., 1979). As Figure 5.32 shows, cytochrome c oxidase is shaped like a "Y." Its overall height is 12 nm. The two "arms" of the Y extend through the inner membrane. The "stem" extends about 5.5 nm into the intermembrane space. Cytochrome c oxidase functions in dimeric

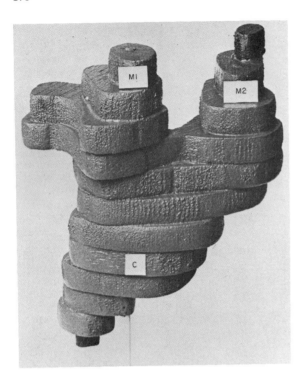

Figure 5.32. A structural model of cytochrome c oxidase. A structural model of cytochrome c oxidase based on negative-stain electron microscopy of two-dimensional crystals is shown. (From Fuller *et al.*, 1979, *J. Mol. Biol.* **134**:305.)

form. When two-dimensional crystals of dimeric complex IV were analyzed, they were found to make contact at their comparatively large intermembrane faces.

The intrinsic absorption properties of complex IV allow transient dichroism experiments to be performed on intact inner membranes (Kawato *et al.*, 1980). These experiments indicate that roughly half of the cytochrome c oxidase complexes are mobile whereas the other half are immobile. The mobile population displays a rotational relaxation time of 300 to 400 μsec. The origin of these differences and their possible physiological relevance are unknown. However, the mobile and immobile populations could correspond to the monomeric and dimeric (or oligomeric) forms.

The first step in cytochrome c oxidase's mechanism of action is its binding to cytochrome c. As described in the preceding section, cytochrome c interacts with its redox sites via electrostatic binding. The role of negative charges at cytochrome c oxidase's surface in promoting cytochrome c binding was established by chemical modification studies (Millett *et al.*, 1982). When exposed carboxylic acid moieties were masked by chemical treatment, the oxidase's function was abolished. Chemical cross-linking studies have shown that cytochrome c binds to negative charges on subunit II of cytochrome c oxidase. When the gene sequences of subunit II were compared among many species, 11 acidic residues were found to be uniformly conserved. Seven or eight of these residues likely interact with positive residues of cytochrome c.

To more precisely define the cytochrome c-to-cytochrome c oxidase interactions, a photoactivatable aryl azide moiety was attached to Lys-13 of cytochrome c and then incubated with cytochrome c oxidase. When photoactivated, the nitrene attached at or near His-161 of subunit II (Bisson *et al.*, 1982). This His residue is likely a ligand for the Cu_A center. A highly conserved negative charge is found at Asp-158. Therefore, subunit II's Asp-158 residue is believed to form an ion pair with Lys-13 of cytochrome c.

After cytochrome c binds to cytochrome c oxidase, it donates an electron to cytochrome a–Cu_A (Hatefi, 1985). ESR studies have indicated that cytochrome a and Cu_A atoms are close to one another but far from cytochrome a_3 and Cu_B (e.g., Scott, 1989). The overall pathway of electron flow between these sites is shown in Figure 5.28. On the basis of kinetic studies, cytochrome a is thought to be the first electron acceptor. However, the proximity of the Cu_A site to cytochrome c's heme suggests that copper might be the first electron acceptor. Electrons are transferred from the cytochrome a–Cu_A site to cytochrome a_3–Cu_B.

Molecular oxygen is reduced at the cytochrome a_3–Cu_B center. Cytochrome a_3 and Cu_B are very close (Scott, 1989). Due to antiferromagnetic coupling, this center generally does not produce an ESR signal in either the oxidized or reduced forms. However, ESR studies have detected the transient appearance of the Cu_B^{2+} state (Wikstrom *et al.*, 1981).

Several models of water reduction consistent with experimental findings have been proposed. The basic features of these models are similar (Wikstrom *et al.*, 1981; Hatefi, 1989). Molecular oxygen is reduced to peroxide when bound to the enzyme's matrix face. One electron is donated by both the nonheme Fe and the Cu atom during this reaction. The peroxide's oxygen–oxygen bond is broken when it acquires another electron. One oxygen atom is simultaneously reduced to water. When the fourth electron arrives, the remaining oxygen atom is reduced to water. The two intermediate electron transfer steps have been linked to proton pumping (Wikstrom, 1989). Electrons then reduce the iron and copper atoms before the next cycle begins.

The enzyme cytochrome c oxidase contributes to the transmembrane proton potential in two fashions. Since this complex catalyzes the reaction:

$$O_2 + 4H^+ + 4e^- \rightarrow 2H_2O \qquad (5.8)$$

at the matrix face, four protons are lost from the matrix side for each oxygen molecule reduced. Furthermore, subunit III participates in the translocation of one proton into the intermembrane space for each electron transported through the chain. Therefore, the reduction of each oxygen molecule is accompanied by a net change of eight protons across the inner membrane. In the next section we will explore how proton potentials are used to make ATP.

Although many of cytochrome c oxidase's subunits are believed to be regulatory in nature, their mechanisms of action are largely unknown. Since the principal product of the respiratory chain is ATP, it likely participates in the regulation of cytochrome c oxidase activity. In an early study, Ferguson-Miller *et al.* (1976) showed that ATP reduced the kinetics of electron transport from cytochrome c to cytochrome c oxidase.

Millimolar ATP concentrations dramatically reduce the affinity of cytochrome c for its oxidase. ATP interacts with subunits IV and VII of cytochrome c oxidase (Bisson *et al.*, 1987). Since these subunits are found at the intermembrane face, they monitor cytosolic ATP levels. ATP triggers a conformational change at the complex's *cis* face. This conformational change reduces the accessibility of the cytochrome c-binding carbonyl groups of subunit II. Consequently, its binding properties are dramatically altered. Therefore, cytosolic ATP levels regulate the efficiency of electron transport to cytochrome c oxidase.

5.4. THE PAYOFF: TRANSDUCING PROTON POTENTIALS INTO ATP

Up to this point we have explored how proton potentials are generated across bioenergetic membranes. As previously mentioned, these proton potentials can be used for tasks such as powering flagellar motion and the uptake of nutrients. However, the primary role of electrochemical proton potentials is the synthesis of ATP, which is required for most energy-dependent biochemical reactions. The energy stored in a membrane's electrochemical proton potential is converted to ATP when protons pass from one membrane face to the other through the ATP synthase complex. ATP synthase is a reversible proton pump. Depending on the prevailing electrochemical proton gradient, this protein complex can hydrolyze ATP (ATPase activity) or synthesize ATP (ATP synthase activity).

As mentioned at the beginning of this chapter, the ATP synthase is a primitive and ubiquitous enzyme complex. It is found in mitochondrial inner membranes, thylakoid membranes of chloroplasts, and bacterial cytoplasmic membranes. All of these ATP synthases are quite similar. They are built from an F_1-ATPase and an F_0 proton channel. In plants these two segments are called CF_1 and CF_0. In this section we will explore how cells use a transmembrane proton potential to make ATP, some of the ATP synthase's properties, and its possible mechanisms of action.

5.4.1. Supramolecular Structure of the ATP Synthase

The ATP synthase has been studied by electron microscopic techniques. The large extra-membrane domain of ATP synthases is associated with the luminal face of mitochondrial inner membranes, the stromal face of thylakoid membranes, and the *cis* face of bacterial cytoplasmic membranes. They appear as membrane-bound "knobs" or "lollipops" about 10 nm in diameter (Figure 5.25). When viewed *en face*, the ATPase possesses pseudo-sixfold symmetry (Akey *et al.*, 1986; Brink *et al.*, 1988). The roughly spherical F_1 subunits are linked to the membrane-bound F_0 subunits by a 4.5-nm stalk composed of the δ and ϵ subunits, which are discussed below. When F_0 is reconstituted into phospholipid or galactolipid vesicles, intramembrane particles 9.5 nm in diameter are observed after freeze-fracture electron microscopy. Since their mass of 90 kDa is insufficient to account for the large size of the particles, it is believed that they are hollow with a channel through the center.

5.4.2. Structural Elements

Plant ATP synthases are composed of eight different proteins. Three integral membrane proteins form a transmembrane channel called the CF_0 domain. These three proteins, I, II, and III, are present at a ratio of 1:2:6 with an aggregate mass of 90 kDa.

The chloroplast's *atpH* gene encodes protein III which binds the inhibitor N,N'-dicyclohexylcarbodiimide (DCCD). This very hydrophobic membrane protein of 8 kDa possesses two transmembrane domains. In membranes it forms a hexameric structure. This structure is believed to form the inner lining of a transmembrane proton channel. Sequencing studies of protein III's bacterial counterpart, protein e, have localized the DCCD binding site to amino acid 65, which is either Asp or Glu depending on species. *E. coli* mutants in which residue 65 is replaced by Gly or Asn are unable to translocate protons, although ATP synthase assembly is normal. Although it is clear that an acidic residue participates in proton migration, the mechanism of proton translocation is unknown. Protein II, a second subunit of 12.5 kDa, locks the hexamer together. Each protein II molecule is thought to possess three transmembrane domains. The two protein II molecules of each CF_0 assembly form a homodimer. In addition to binding protein III hexamers, protein II molecules may also interact with the δ and ϵ subunits of CF_1. The final component of CF_0 is protein I. It is a 15-kDa protein encoded by the chloroplast's *atpF* gene. Protein I possesses one apparent transmembrane domain and a very hydrophilic COOH-terminal domain with over one-third charged amino acids. Since this domain projects into the stroma, it is believed to interact with the CF_1 complex. This idea is supported by the fact that proteolytic cleavage of protein I inhibits CF_0 to CF_1 binding.

The CF_1 complex synthesizes ATP. Since it can be dislodged from membranes by EDTA or high salt, it is a peripheral membrane protein complex. CF_1 is composed of five different proteins with an aggregate mass of 400 kDa. ATP synthases are composed of α, β, γ, δ, and ϵ subunits in the ratio 3:3:1:1:1. In plants they are encoded by the *atpA, B, C, D,* and *E* genes to yield mature proteins of 59, 56, 37, 17.5, and 13 kDa, respectively. The α, β, and γ subunits are homologous among many organisms. Furthermore, the α and β subunits are homologous to one another, suggesting a common ancestral origin. Chemical cross-linking experiments indicate that the α and β subunits are near one another. The three α and β subunits assemble pairwise to form the synthase's "head" region. The δ and ϵ subunits display little homology. Although the $\alpha\beta\gamma$ complex is roughly spherical, it becomes ellipsoidal when the δ subunit is added (Schinkel and Hammes, 1986). These subunits constitute the ATP synthase's head and stalk.

Several sorts of interactions hold the CF_1 complex to the CF_0 baseplate structure. Protein II possesses extended hydrophilic α-helical regions that project into the stroma. These regions may interact with α-helical regions of the δ and ϵ units, thus forming a stalk. The α and β subunits also interact with proteins III and I, respectively.

5.4.3. Functional Interactions

The ATPase's purified subunits can be reconstituted *in vitro*. When assembled in solution, the α, β, and γ subunits constitute a functional ATPase. Reconstitution experiments have shown that the functional interactions among the α, β, and γ subunits are highly conserved. When α, β, and γ subunits from different species are combined *in vitro*, they retain ATPase activity. However, the $\alpha\beta\gamma$ complex alone is unable to bind membranes. The δ and ϵ subunits of the F_1 segment are required for membrane attachment. In particular, membrane binding activity has been localized to the ϵ subunit's NH_2-terminal region (Futai *et al.*, 1989). When reconstituted in liposomes, the F_0 subunit acts as a passive proton conduction pathway. A similar response is observed when the F_1 segment is dislodged from mitochondrial membranes. When δ, ϵ and γ are added to the F_0 segment, the passive proton channel is blocked. This suggests that the α and β subunits do not directly control proton movement. Therefore, the functional attributes of the several components of the ATPase have been confirmed by detailed reconstitution studies.

The proton ATPases of both plants and animals possess six nucleotide binding sites. Three of these sites are catalytic whereas the remainder are noncatalytic in nature. The non-catalytic nucleotide binding sites have been proposed to play structural and/or allosteric roles in ATP formation. These sites are located on the β subunit. They need to be filled to promote the proper assembly of the F_1 complex. The catalytic sites are believed to be located at the interfaces between the α and β subunits. However, the structural details of these sites are unknown.

Although ATP synthase's composition and physiological role are understood, its mechanism of action remains elusive. A variety of models have been proposed to explain ATP synthesis (Mitchell, 1976; Boyer, 1975, 1989; Futai *et al.*, 1989). These models try to explain how an electrochemical proton potential gradient is linked to ATP production. For example, Mitchell has proposed that two protons migrating via the F_0 subunit attack the P_i held in the F_1 segment. This reaction forms water, releases protons, and triggers the addition of P_i to ADP. However, the model least inconsistent with the facts appears to be the conformational coupling model (Boyer, 1989). The F_1 complex undergoes significant conformational changes during its catalytic cycle (e.g., Hartig *et al.*, 1977; Schinkel and Hammes, 1986), although the relevance of these changes to the catalytic mechanism is not certain. Cooperativity between the catalytic sites is an essential feature of ATP synthesis. Positive cooperativity is displayed during ATP synthesis whereas negative cooperativity is found during ADP binding. The catalytic site cycles between at least two distinct conformational states. In one of these conformations ATP is bound tightly to the catalytic site. Protons migrating through the F_0 segment may protonate certain residues of F_1. This triggers a conformational change that releases ATP from the enzyme. ADP and P_i then bind to the catalytic site. The catalytic site then flips back to its original conformation. This conformational change makes ATP formation exergonic. The cycle then repeats itself when ATP is released. It has not yet become possible to develop a quantitative model of ATP generation. Although the ATPase's subunits have been sequenced, its three-dimensional structure remains at a resolution of only 0.9 nm. Until its structure can be interpreted at a molecular level, it will be difficult to synthesize the biochemical, genetic, and biophysical studies of its mechanism.

5.4.4. Free Energy Released by the Proton Potential

Any proposed mechanism of ATP synthesis must be consistent with the free energy available in the proton potential and the free energy required for ATP synthesis. Typically, the proton potential across energized membranes is roughly -200 mV, although occasionally proton potentials of -300 mV have been reported. The free energy required to drive ATP formation in chloroplasts (ΔGp) has been found to be 14 kcal/mole. The free energy released by passing one proton across this potential gradient is quite insufficient to support the synthesis of ATP. However, three protons passing through this potential gradient would provide -600 mV or, equivalently, 16 kcal/mole. Therefore, three protons must be translocated across a membrane to support the synthesis of one ATP molecule. If a 1:1 stoichiometry was employed, the requisite transmembrane potentials would be high enough to disrupt a biological membrane. Therefore, most current models of ATP synthesis utilize a stoichiometry of 3 protons/ATP. However, the study of ATP synthase's mechanism of action has only just begun.

REFERENCES AND FURTHER READING

Proton Pumping

Drachev, L. A., *et al.* 1974. Direct measurement of electric current generation by cytochrome oxidase, H$^+$ ATPase and bacteriorhodopsin. *Nature* **249:** 321–324.

Ferguson, S. J., and Sorgato, M. C. 1982. Proton electrochemical gradients and energy-transduction processes. *Annu. Rev. Biochem.* **51:**185–217.

Harold, F. M. 1986. *The Vital Force: A Study of Bioenergetics.* Freeman, San Francisco.

Hind, G., and Jagendorf, A. T. 1963. Separation of light and dark stages in photophosphorylation. *Proc. Natl. Acad. Sci. USA* **49:**715–722.

Jagendorf, A. T., and Uribe, E. 1966. ATP formation caused by acid–base transition of spinach chloroplasts. *Proc. Natl. Acad. Sci. USA* **55:**170–177.

Mitchell, P. 1961. Coupling of phosphorylation to electron and hydrogen transfer by a chemiosmotic type of mechanism. *Nature* **191:**144–148.

Mitchell, P. 1979. Keilin's respiratory chain concept and its chemiosmotic consequences. *Science* **206:**1148–1159.

Racker, E., and Stoeckenius, W. 1974. Reconstitution of purple membrane vesicles catalyzing light-driven proton uptake and adenosine triphosphate formation. *J. Biol. Chem.* **249:** 662–663.

Raven, J. A., and Smith, F. A. 1976. The evolution of chemiosmotic energy coupling. *J. Theor. Biol.* **57:**301–312.

Wilson, T. H., and Lin, E. C. 1980. Evolution of membrane bioenergetics. *J. Supramol. Struct.* **13:**421–446.

Photosynthetic Membranes

Allen, J. P., *et al.* 1987. Structure of the reaction center from *Rhodobacter sphaeroides* R-26: The protein subunits. *Proc. Natl. Acad. Sci. USA* **84**:6161–6166.

Alt, J., and Herrmann, R. G. 1984. Nucleotide sequence of the gene for pre-apocytochrome f in the spinach plastid chromosome. *Curr. Genet.* **8**:551–557.

Alt, J., *et al.* 1984. Nucleotide sequence of the clustered genes for the 44kd chlorophyll a apoprotein and the "32kd"-like protein of the photosystem II reaction center in the spinach plastid chromosome. *Curr. Genet.* **8**:597–606.

Anderson, J. M. 1986. Photoregulation of the composition, function, and structure of thylakoid membranes. *Annu. Rev. Plant Physiol.* **37**:93–136.

Anderson, J. M., and Andersson, B. 1988. The dynamic photosynthetic membrane and regulation of solar energy conversion. *Trends Biochem. Sci.* **13**:351–355.

Andersson, B., and Anderson, J. M. 1980. Lateral heterogeneity in the distribution of chlorophyll–protein complexes in the thylakoid membranes of spinach chloroplasts. *Biochim. Biophys. Acta* **593**:427–440.

Ashby, M. K., *et al.* 1987. Cloning, nucleotide sequence and transfer of genes for the B800–850 light harvesting complex of *Rhodobacter sphaeroides*. *FEBS Lett.* **213**:245–248.

Barber, J. 1987. Photosynthetic reaction centres: A common link. *Trends Biochem. Sci.* **12**: 321–326.

Barber, J., *et al.* 1987. Characterisation of a PSII reaction center isolated from the chloroplasts of *Pisum sativum*. *FEBS Lett.* **220**:67–73.

Braiman, M. S. 1986. Resonance Raman methods for proton translocation in bacteriorhodopsin. *Methods Enzymol.* **127**:587–597.

Brenton, J., *et al.* 1986. Femtosecond spectroscopy of excitation energy transfer and initial charge separation in the reaction center of the photosynthetic bacterium *Rhodopseudomonas viridis*. *Proc. Natl. Acad. Sci. USA* **83**:5121–5125.

Brudvig, G. W., *et al.* 1989. Mechanism of photosynthetic water oxidation. *Annu. Rev. Biophys. Biophys. Chem.* **18**:25–46.

Brunisholz, R. A, *et al.* 1985. The light harvesting polypeptides of *Rhodopseudomonas viridis*: The complete amino acid sequences of B1015-α, B1015-β and B1015-γ. *Biol. Chem. Hoppe-Seylers* **366**:87–98.

Butt, H. J., *et al.* 1989. Aspartic acids 96 and 85 play a central role in the function of bacteriorhodopsin as a proton pump. *EMBO J.* **8**:1657–1663.

Chitnis, P. R., *et al.* 1986. Assembly of the precursor and processed light-harvesting chlorophyll a/b protein of *Lemma* into the light harvesting complex II of barley etiochloroplasts. *J. Cell Biol.* **102**:982–988.

Darr, S. C., *et al.* 1986. Monoclonal antibodies to the light-harvesting chlorophyll a/b protein complex of photosystem II. *J. Cell Biol.* **103**:733–740.

Deisenhofer, J., and Michel, H. 1989. The photosynthetic reaction center from the purple bacterium *Rhodopseudomonas viridis*. *Science* **245**:1463–1473.

Deisenhofer, J., *et al.* 1985. Structure of the protein subunits in the photosynthetic reaction centre of *Rhodopseudomonas viridis* at 3Å resolution. *Nature* **318**:618–624.

Dollinger, G., *et al.* 1986. Fourier transform infrared difference spectroscopy of bacteriorhodopsin and its photoproducts regenerated with deuterated tyrosine. *Biochemistry* **25**:6524–6533.

Dracheva, S. M., *et al.* 1988. Electrogenic steps in the redox reactions catalyzed by photosynthetic reaction centre complex from *Rhodopseudomonas viridis*. *Eur. J. Biochem.* **171**:253–264.

Dunahay, T. G., and Staehelin, L. A. 1985. Isolation of photosystem I complexes from octyl glucoside/sodium dodecyl sulfate solubilized spinach thylakoids. Characterization and reconstitution into liposomes. *Plant Physiol.* **78**:606–613.

Engelhardt, H., *et al.* 1983. Electron microscopy of photosynthetic membranes containing bacteriochlorophyll b. *Arch. Microbiol.* **135**:169–175.

Feher, G., *et al.* 1975. Endor experiments on chlorophyll and bacteriochlorophyll in vitro and in the photosynthetic unit. *Ann. N.Y. Acad. Sci.* **244**:239–259.

Fish, L. E., *et al.* 1985. Two partially homologous light-inducible maize chloroplast genes encoding polypeptides of the P700 chlorophyll a-protein complex of photosystem I. *J. Biol. Chem.* **260**:1413–1421.

Fowler, G. J. S., *et al.* 1992. Genetically modified photosynthetic antenna complexes with blueshifted absorbance bands. *Nature* **355**:848–850.

Fragata, M., *et al.* 1984. Lateral diffusion of plastocyanin in multilamellar mixed bilayers studied by fluorescence recovery after photobleaching. *Biochemistry* **23**:4044–4051.

Gabellini, N. 1988. Organization and structure of the genes for the cytochrome b/c_1 complex in purple photosynthetic bacteria. A phylogenetic study describing the homology of the b/c_1 subunits between prokaryotes, mitochondria, and chloroplasts. *J. Bioenerg. Biomembr.* **10**:59–83.

Gabellini, N., and Sebald, W. 1986. Nucleotide sequence and transcription of the *fbc* operon from *Rhodopseudomonas sphaeroides* (note: subsequently identified as *Rhodobacter capsulatus*). *Eur. J. Biochem.* **154**:569–579.

Gabellini, N., *et al.* 1989. Reconstitution of cyclic electron transport and photophosphorylation by incorporation of the reaction center, cytochrome bc_1 complex and ATP synthase from *Rhodobacter capsulatus* into ubiquinone-10/phospholipid vesicles. *Biochim. Biophys. Acta* **974**:202–210.

Glazer, A. N., and Melis, A. 1987. Photochemical reaction centers: Structure, organization, and function. *Annu. Rev. Plant. Physiol.* **38**:11–45.

Gounaris, K., *et al.* 1986. The thylakoid membranes of higher plant chloroplasts. *Biochem. J.* **237**:313–326.

Hackett, N. R., *et al.* 1987. Structure–function studies on bacteriorhodopsin. V. Effects of amino acid substitution in the putative helix F. *J. Biol. Chem.* **262**:9277–9284.

Hasselbacher, C. A., and Dewey, T. G. 1986. Changes in retinal position during the bacteriorhodopsin photocycle: A resonance energy transfer study. *Biochemistry* **25**:6236–6243.

Hauska, G., *et al.* 1983. Comparative aspects of quinol-cytochrome c/plastocyanin oxidoreductases. *Biochim. Biophys. Acta* **726**:97–133.

Heinemeyer, W., *et al.* 1984. Nucleotide sequence of the clustered genes for apocytochrome b_6 and subunit 4 of the cytochrome b/f complex in the spinach plastid chromosome. *Curr. Genet.* **8**:543–549.

Henderson, R., *et al.* 1990. Model for the structure of bacteriorhodopsin based on high-resolution electron cryo-microscopy. *J. Mol. Biol.* **213**:899–929.

Hinshaw, J. E., and Miller, K. R. 1989. Localization of light-harvesting complex II to the occluded surfaces of photosynthetic membranes. *J. Cell Biol.* **109**:1725–1731.

Izawa, S., and Good, N. E. 1966a. Effect of salts and electron transport on the conformation of isolated chloroplasts. I. Light scattering and volume changes. *Plant Physiol.* **41**:533–543.

Izawa, S., and Good, N. E. 1966b. Effect of salts and electron transport on the conformation of isolated chloroplasts. II. Electron microscopy. *Plant Physiol.* **41**:544–552.

Jay, F., and Wildhaber, I. 1988. Structure of the photosynthetic unit of *Rhodopseudomonas viridis.* *Eur. J. Cell Biol.* **46**:227–232.

Jay, F., *et al.* 1984. The preparation and characterization of native photoreceptor units from the thylakoids of *Rhodopseudomonas viridis.* *EMBO J.* **3**:773–776.

Kirmaier, D., and Holten, D. 1987. Primary photochemistry of reaction centers from the photosynthetic purple bacteria. *Photosynth. Res.* **13**:225–260.

Kirmaier, C., *et al.* 1991. Charge separation in a reaction center incorporating bacteriochlorophyll for photoactive bacteriopheophytin. *Science* **251**:922–927.

Knapp, E. W., *et al.* 1985. Analysis of optical spectra from single crystals of *Rhodopseudomonas viridis* reaction centers. *Proc. Natl. Acad. Sci. USA* **82**:8463–8467.

Kohorn, B. P., *et al.* 1986. Functional and mutational analysis of the light-harvesting complex chlorophyll a/b protein of thylakoid membranes. *J. Cell Biol.* **102**:972–981.

Kok, B., *et al.* 1970. Cooperation of charges in photosynthetic O_2 evolution. I. A linear four step mechanism. *Photochem. Photobiol.* **11**:457–475.

Kuhlbrandt, W., and Wang, D. N. 1991. Three-dimensional structure of plant light-harvesting complex by electron crystallography. *Nature* **350**:130–134.

Kyle, D. J., *et al.* 1983. Lateral mobility of the light harvesting complex in chloroplast membranes controls excitation energy distribution in plants. *Arch. Biochem. Biophys.* **222**:527–541.

Li, J. 1985. Light harvesting chlorophyll a/b-protein: Three dimensional structure of a reconstituted membrane lattice in negative stain. *Proc. Natl. Acad. Sci. USA* **82**:386–390.

Lyon, M. K., and Miller, K. R. 1985. Crystallization of the light-harvesting chlorophyll a/b complex within thylakoid membranes. *J. Cell Biol.* **100**:1139–1147.

McDonnel, A., and Staehelin, L. A. 1980. Adhesion between liposomes mediated by the chlorophyll a/b light harvesting complex isolated from chloroplast membranes. *J. Cell Biol.* **84**:40–56.

McElroy, J. D., *et al.* 1969. On the nature of the free radical formed during the primary process of bacterial photosynthesis. *Biochim. Biophys. Acta* **172**:180–183.

Marinetti, T., *et al.* 1989. Replacement of aspartic residues 85, 96, 115, or 212 affects the quantum yield and kinetics of proton release and uptake by bacteriorhodopsin. *Proc. Natl. Acad. Sci. USA* **86**:529–533.

Michel, H., *et al.* 1985. The "heavy" subunit of the photosynthetic reaction centre from *Rhodopseudomonas viridis*: Isolation of the gene, nucleotide and amino acid sequence. *EMBO J.* **4**:1667–1672.

Michel, H., *et al.* 1986. The "light" and "medium" subunits of the photosynthetic reaction centre from *Rhodopseudomonas viridis*: Isolation of the genes, nucleotide and amino acid sequence. *EMBO J.* **5**:1149–1158.

Mullet, J. E. 1983. The amino acid sequence of the polypeptide segment which regulated membrane adhesion (grana stacking) in chloroplasts. *J. Biol. Chem.* **258**:9941–9948.

Mullet, J. E., and Arntzen, C. J. 1980. Stimulation of grana stacking in a model membrane system. Mediation by a purified light-harvesting pigment-protein complex from chloroplasts. *Biochim. Biophys. Acta* **589**:100–117.

Murphy, D. J. 1986. The molecular organization of the photosynthetic membranes of higher plants. *Biochim. Biophys. Acta* **864**:33–94.

Nanba, O., and Satoh, K. 1987. Isolation of a photosystem II reaction center consisting of D-1 and D-2 polypeptides and cytochrome b-559. *Proc. Natl. Acad. Sci. USA* **84**:109–112.

Norris, J. R., *et al.* 1971. Electron spin resonance of chlorophyll and the origin of signal I in photosynthesis. *Proc. Natl. Acad. Sci. USA* **68**:625–628.

Oesterhelt, D., and Stoeckenius, W. 1973. Functions of a new photoreceptor membrane. *Proc. Natl. Acad. Sci. USA* **70**:2853–2857.

Oh-oka, H., *et al.* 1986. Complete amino acid sequence of 33 kDa protein isolated from spinach photosystem II particles. *FEBS Lett.* **197**:63–66.

Rees, D. C., *et al.* 1989. The bacterial photosynthetic reaction center as a model for membrane proteins. *Annu. Rev. Biochem.* **58**:607–633.

Reilly, P., and Nelson, N. 1988. Photosystem I complex. *Photosynth. Res.* **19**:73–84.

Rogner, M., *et al.* 1987. Size, shape and mass of the oxygen-evolving photosystem II complex from the thermophilic cyanobacterium *Synechococcus* sp. *FEBS Lett.* **219**:207–211.

Rothschild, K. J. 1986. Fourier transform infrared studies of an active proton transport pump. *Methods Enzymol.* **127**:343–353.

Rutherford, A. W., and Heathcote, P. 1985. Primary photochemistry in photosystem I. *Photosynth. Res.* **6**:295–316.

Stark, W., *et al.* 1984. The structure of the photoreceptor unit of *Rhodopseudomonas viridis*. *EMBO J.* **3**:777–783.

Stoeckenius, W., and Bogomolni, R. A. 1982. Bacteriorhodopsin and related pigments of halobacteria. *Annu. Rev. Biochem.* **52**:587–616.

Tadros, M. H., *et al.* 1984. Isolation and complete amino acid sequence of the small polypeptide from light-harvesting pigment-protein complex I (B870) of *Rhodopseudomonas capsulata*. *Eur. J. Biochem.* **138**:209–212.

Wechsler, T., *et al.* 1985. The complete amino acid sequence of a bacteriochlorophyll a binding polypeptide isolated from the cytoplasmic membrane of the green photosynthetic bacterium *Chloroflexus aurantiacus*. *FEBS Lett.* **191**:34–38.

Weyer, K. A., *et al.* 1987. Amino acid sequence of the cytochrome subunit of the photosynthetic reaction centre from the purple bacterium *Rhodopseudomonas viridis*. *EMBO J.* **6**:2197–2202.

Williams, J. C., *et al.* 1986. Primary structure of the reaction center from *Rhodopseudomonas sphaeroides*. *Proteins* **1**:312–325.

Youvan, D. C., and Marrs, B. 1984. Molecular genetics and the light reactions of photosynthesis. *Cell* **39**:1–3.

Zuber, H. 1986. Structure of light-harvesting antenna complexes of photosynthetic bacteria, cyanobacteria and red algae. *Trends Biochem. Sci.* **11**:414–419.

Mitochondrial Membranes: The Respiratory Chain

Allen, R. D., *et al.* 1989. An investigation of mitochondrial inner membranes by rapid-freeze deep-etch techniques. *J. Cell Biol.* **108**:2233–2240.

Beattie, D. S., and Villalobo, A. 1982. Energy transduction by the reconstituted b–c_1 complex from yeast mitochondria. *J. Biol. Chem.* **257**:14745–14752.

Beckmann, J. D., *et al.* 1987. Isolation and characterization of the nuclear gene encoding the Rieske iron–sulfur protein (RIP1) from *Saccharomyces cerevisiae*. *J. Biol. Chem.* **262**:8901–8909.

Beinert, H., and Albracht, S. P. J. 1982. New insights, ideas and unanswered questions concerning iron–sulfur clusters in mitochondria. *Biochim. Biophys. Acta* **683**:245–277.

Bisson, R., *et al.* 1982. Mapping of the cytochrome c binding site on cytochrome c oxidase. *FEBS Lett.* **144**:359–363.

Bisson, R., *et al.* 1987. ATP induces conformational changes in mitochondrial cytochrome c oxidase. Effect on the cytochrome c binding site. *J. Biol. Chem.* **262**:5992–5998.

Boekema, E. J., *et al.* 1986. Preparation of two-dimensional crystals of complex I and image analysis. *Methods Enzymol.* **126**:344–353.

Brink, B., *et al.* 1987. The structure of NADH: ubiquinone oxidoreductase from beef-heart mitochondria: Crystals containing an octameric arrangement of iron–sulfur protein fragments. *Eur. J. Biochem.* **166**:287–294.

Brink, J., *et al.* 1989. Computer image analysis of two-dimensional crystals of beef heart NADH: ubiquinone oxidoreductase fragments. II. Comparison of frozen hydrated and negatively stained specimens. *Ultramicroscopy* **27**:91–100.

Capaldi, R. A. 1982. Arrangement of proteins in the mitochondrial inner membrane. *Biochim. Biophys. Acta* **694**:291–306.

Capaldi, R. A., *et al.* 1988. Complexity and tissue specificity of the mitochondrial respiratory chain. *J. Bioenerg. Biomembr.* **20**:291–311.

Deatherage, J. F., *et al.* 1982. Three-dimensional structure of cytochrome c oxidase vesicles in negative stain. *J. Mol. Biol.* **158**:487–499.

De Haan, M., *et al.* 1984. The biosynthesis of the ubiquinol–cytochrome c reductase complex in yeast: DNA sequence analysis of the nuclear gene coding for the 14-kDa subunit. *Eur. J. Biochem.* **138**:169–177.

Fato, R., *et al.* 1985. Measurement of the lateral diffusion coefficients of ubiquinones in lipid vesicles by fluorescence quenching of 12-(9-anthroyl)stearate. *FEBS Lett.* **179**:238–242.

Ferguson-Miller, S., *et al.* 1976. Correlation of the kinetics of electron transfer activity of various eukaryotic cytochromes c with binding to mitochondrial cytochrome c oxidase. *J. Biol. Chem.* **251**:1104–1115.

Ferguson-Miller, S., *et al.* 1978. Definition of cytochrome c binding domains by chemical modification. III. Kinetics of reaction of carboxydinitrophenyl cytochromes c with cytochrome c oxidase. *J. Biol. Chem.* **253**:149–159.

Fuller, S. P., *et al.* 1979. Structure of cytochrome c oxidase in deoxycholate-derived two-dimensional crystals. *J. Mol. Biol.* **134**:305–327.

Girdlestone, J., *et al.* 1981. Interaction of succinate–ubiquinone reductase (complex II) with (arylazido)phospholipids. *Biochemistry* **20**:152–156.

Gupta, S., *et al.* 1984. Relationship between lateral diffusion, collision frequency, and electron transfer of mitochondrial inner membrane oxidation–reduction components. *Proc. Natl. Acad. Sci. USA* **81**:2606–2610.

Hamamoto, T, *et al.* 1985. Direct measurement of the electrogenic activity of o-type cytochrome oxidase from *Escherichia coli* reconstituted into planar lipid bilayers. *Proc. Natl. Acad. Sci. USA* **82**:2570–2573.

Hatefi, Y. 1985. The mitochondrial electron transport and oxidative phosphorylation system. *Annu. Rev. Biochem.* **54**:1015–1069.

Henderson, R., *et al.* 1977. Arrangement of cytochrome oxidase molecules in two-dimensional vesicle crystals. *J. Mol. Biol.* **112**:631–648.

Kawato, S., *et al.* 1980. Cytochrome oxidase rotates in the inner membrane of intact mitochondria and submitochondrial particles. *J. Biol. Chem.* **255**:5508–5510.

Leonard, K., *et al.* 1987. Three-dimensional structure of NADH: ubiquinone reductase (complex I) from *Neurospora* mitochondria determined by electron microscopy of membrane crystals. *J. Mol. Biol.* **194**:277–286.

Lorusso, M., *et al.* 1989. Effect of papain digestion on polypeptide subunits and electron-transfer pathways in mitochondrial $b-c_1$ complex. *Eur. J. Biochem.* **179**:535–540.

Mannella, C. A. 1986. Mitochondrial outer membrane channel (VDAC, porin) two-dimensional crystals from *Neurospora*. *Meth. Enzymol.* **125**:595–611.

Mathews, F. S. 1984. The structure, function and evolution of cytochromes. *Prog. Biophys. Mol. Biol.* **45**:1–56.

Matsushita, K., *et al.* 1983. Reconstitution of active transport in proteoliposomes containing cytochrome o oxidase and lac carrier protein purified from *Escherichia coli. Proc. Natl. Acad. Sci. USA* **80**:4889–4893.

Millett, F., *et al.* 1982. Cytochrome c is cross-linked to subunit II of cytochrome c oxidase by a water-soluble carbodiimide. *Biochemistry* **21**:3857–3862.

Muller, M., *et al.* 1982. Selective labeling and rotational diffusion of the ADP/ATP translocator in the inner mitochondrial membrane. *J. Biol. Chem.* **257**:1117–1120.

Orme-Johnson, N. R., *et al.* 1974a. Electron paramagnetic resonance detectable electron acceptors in beef heart mitochondria: Reduced diphosphopyridine nucleotide ubiquinone reductase segment of the electron transfer system. *J. Biol. Chem.* **249**:1922–1927.

Orme-Johnson, N. R., *et al.* 1974b. Electron paramagnetic resonance detectable electron acceptors in beef heart mitochondria: Ubihydroquinone–cytochrome c reductase segment of the electron transfer system and complex mitochondrial fragments. *J. Biol. Chem.* **249**:1928–1939.

Racker, E., and Kandrach, A. 1973. Partial resolution of the enzymes catalyzing oxidative phosphorylation. Reconstitution of the third segment of oxidative phosphorylation. *J. Biol. Chem.* **248**:5841–5847.

Salemme, F. P. 1977. Structure and function of cytochromes c. *Annu. Rev. Biochem.* **46**:299–329.

Saraste, M. 1984. Location of haem-binding sites in the mitochondrial cytochrome b. *FEBS Lett.* **166**:367–372.

Schneider, H., *et al.* 1982. Lateral diffusion of ubiquinone during electron transfer in phospholipid- and ubiquinone-enriched mitochondrial membranes. *J. Biol. Chem.* **257**:10789–10793.

Scott, R. A. 1989. X-ray absorption spectroscopic investigations of cytochrome c oxidase structure and function. *Annu. Rev. Biophys. Biophys. Chem.* **18**:137–158.

Senior, A. E. 1983. Secondary and tertiary structure of membrane proteins involved in proton translocation. *Biochim. Biophys. Acta* **726**:81–95.

Smith, H. T., *et al.* 1981. Electrostatic interaction of cytochrome c with cytochrome c_1 and cytochrome oxidase. *J. Biol. Chem.* **256**:4984–4990.

Takamiya, S., *et al.* 1986. Mitochondrial myopathies involving the respiratory chain: A biochemical analysis. *Ann. N.Y. Acad. Sci.* **488**:33–43.

Wendoloski, J. J., *et al.* 1988. Molecular dynamics of a cytochrome c–cytochrome b_5 electron transfer complex. *Science* **238**:794–797.

Widger, W. R., *et al.* 1984. Sequence homology and structural similarity between cytochrome b of mitochondrial complex III and the chloroplast b_6–f complex: Position of the cytochrome b hemes in the membrane. *Proc. Natl. Acad. Sci. USA* **81**:674–678.

Wikstrom, M. 1989. Identification of the electron transfers in cytochrome oxidase that are coupled to proton-pumping. *Nature* **338**:776–778.

Wikstrom, M., *et al.* 1981. Proton-translocating cytochrome complexes. *Annu. Rev. Biochem.* **50**:623–655.

Wingfield, P., *et al.* 1979. Membrane crystals of ubiquinone: cytochrome c reductase from Neurospora mitochondria. *Nature* **280**:696–697.

Wuttke, D. S., *et al.* 1992. Electron-tunneling pathways in cytochrome c. *Science* **256**:1007–1009.

Young, I. C., *et al.* 1981. Nucleotide sequence coding for the respiratory NADH dehydrogenase of *Escherichia coli. Eur. J. Biochem.* **116**:165–170.

Zhang, Y.-Z., *et al.* 1988. Orientation of the cytoplasmically made subunits of beef heart cytochrome c oxidase determined by protease digestion and antibody binding experiments. *Biochemistry* **27**:1389–1394.

ATP Synthesis

Akey, C. W., *et al*. 1986. Electron microscopy of single molecules and crystals of F_1-ATPases. *Methods Enzymol*. **126**:434–446.

Amzel, L. M., and Pedersen, P. L. 1983. Proton ATPases: Structure and mechanism. *Annu. Rev. Biochem*. **52**:801–824.

Amzel, L. M., *et al*. 1982. Structure of the mitochondrial F_1 ATPase at 9 Å resolution. *Proc. Natl. Acad. Sci. USA* **79**:5852–5856.

Boyer, P. D. 1975. A model for conformational coupling of membrane potential and proton translocation to ATP synthesis and to active transport. *FEBS Lett*. **58**:1–6.

Boyer, P. D. 1989. A perspective of the binding change mechanism for ATP synthesis. *FASEB J.* **3:** 2164–2178.

Brink, J., *et al*. 1988. Electron microscopy and image analysis of the complexes I and V of the mitochondrial respiratory chain. *Electron Microsc. Rev*. **1**:175–199.

Futai, M., *et al*. 1989. ATP synthase (H^+ ATPase): Results by combined biochemical and molecular biological approaches. *Annu. Rev. Biochem*. **58**:111–136.

Hartig, P. R., *et al*. 1977. 5-Iodoacetamidofluorescein-labeled chloroplast coupling factor 1: Conformational dynamics and labeling-site characterization. *Biochemistry* **16**:4275–4282.

Hinkle, P. C., *et al*. 1991. Mechanistic stoichiometry of mitochondrial oxidative phosphorylation. *Biochemistry* **30**:3576–3582.

McCarty, R. E., and Hammes, G. G. 1987. Molecular architecture of chloroplast coupling factor 1. *Trends Biochem. Sci*. **12**:234–237.

Mitchell, P. 1976. Possible molecular mechanisms of the protonmotive function of cytochrome systems. *J. Theor. Biol*. **62**:327–367.

Morschel, E., and Staehelin, L. A. 1983. Reconstitution of cytochrome b_6–f and CF_0–CF_1 ATP synthetase complexes into phospholipid and galactolipid liposomes. *J. Cell Biol*. **97**:301–310.

Nelson, N. H., *et al*. 1980. Biosynthesis and assembly of the proton-translocating adenosine triphosphatase complex from chloroplasts. *Proc. Natl. Acad. Sci. USA* **77**:1361–1364.

Pedersen, P. L., and Carafoli, E. 1987. Ion motive ATPases. II. Energy coupling and work output. *Trends Biochem. Sci*. **12**:186–189.

Schinkel, J. E., and Hammes, G. G. 1986. Chloroplast coupling factor 1: Dependence of rotational correlation time on polypeptide composition. *Biochemistry* **25**:4066–4071.

Wakabayaski, T., *et al*. 1975. Structure of ATPase (coupling factor TF) from thermophilic bacterium. *J. Mol. Biol*. **117**:515–519.

Chapter 6

Transport across Membranes

Membrane transport processes are utilized by all living cells. Transport is required for the accumulation of nutrients and elimination of wastes. Membrane transport steps have been identified in most biological events. For example, it occurs during the formation of proton electrochemical potentials, the uptake of saccharides and amino acids, the endocytotic internalization of macromolecules and particles, and during oxygen transport in respiration. The widespread nature of membrane transport phenomena is illustrated in Table 6.1, which lists transport processes presented in other chapters. We will classify transport phenomena into two broad categories based on the size of the translocated species. The migration of individual ions and molecules is termed molecular transport whereas the movement of large molecules or particles is called macromolecular and bulk transport. We shall begin this chapter's discussion of transport by considering the movements of ions and molecules.

6.1. MOLECULAR TRANSPORT

Figure 6.1 and Table 6.2 summarize the variety of molecular transport systems found in biological membranes. These systems are categorized into two broad classes: those that require energy and those that do not. When an energy source is not required, molecules enter or leave cells according to their electrochemical potential gradient. For example, when a neutral substance is present at a higher concentration outside a cell, the substance will tend to move down its chemical gradient into a cell. This process is strictly diffusional in character. Hydrophobic compounds simply diffuse across a bilayer membrane. However, hydrophilic molecules encounter great resistance in passing across a membrane's hydrophobic core. Membrane-bound channels or carriers dramatically enhance a hydrophilic compound's rate of transport by a process called facilitated diffusion. During facilitated diffusion, an ion or molecule moves from one face of a membrane to the other. All of the processes just described, with or without the intervention of a carrier or channel, do not require an energy supply and cannot move compounds against their electrochemical potential gradients.

The right side of Figure 6.1 shows three fundamental types of molecular transport

Table 6.1.

Additional Examples of Membrane Transport Processes

Chapter 4	Movement through gap junctions
Chapter 5	Electron and proton transport in bioenergetic membranes
Chapter 7	Ion transport through ACh receptors
Chapter 8	Co- and post-translational movement of proteins
	Phospholipid translocators (flippases)
	Secretory transport
Chapter 9	P-glycoproteins translocate chemotherapeutic drugs out of tumor cells

that require energy sources. The first type of energy-requiring transport shown is group translocation. During group translocation a substrate is always covalently modified during transit. An energy source, such as phosphoenolpyruvate, is used to transport and covalently modify a substrate. Primary active transport directly utilizes an energy supply without modifying its substrate. Finally, energy in the indirect form of an electrochemical gradient can also be used to drive active transport; this is known as secondary active transport. During secondary active transport and, occasionally, facilitated diffusion, two molecules are simultaneously transported. For example, the far right-hand side of Figure 6.1 illustrates two molecules (A and B) passing across a membrane. Energy stored in the electrochemical gradient of molecule B is used to transport molecule A. When two molecules are simultaneously transported across a membrane in the same direction, it is called symport. Alternatively, if two molecules are

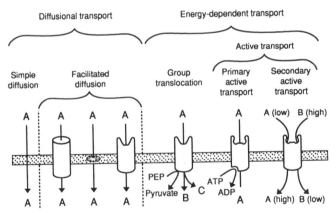

Figure 6.1. Mechanisms of molecular biological membrane transport. Several energy-independent and energy-dependent mechanisms of membrane transport are presented. Diffusion of molecules down their electrochemical potential gradients is shown on the left. Simple diffusion and several forms of facilitated diffusion are illustrated. The right side shows three broad classes of energy-dependent transport that includes group translocation, primary and secondary active transport. A, B, and C represent transported ions or molecules.

Table 6.2.
Selected Solute Transport Mechanisms

Solute	Polarity	Cell type[a]	
		Mammalian	*E. coli*
Na^+	Out	Primary AT	Secondary AT
Ca^{2+}	Out	Primary AT	Secondary AT
K^+	In	Primary AT	Secondary AT
Glucose	In	Secondary AT	Group translocation
Disaccharide	In	—	Secondary AT
Amino acid	In	Secondary AT	Secondary AT
cAMP	Out	Primary AT	Secondary AT

[a]AT, active transport. Secondary active transport in mammalian cells is linked to sodium cotransport whereas proton cotransport occurs during secondary active transport in *E. coli*.

simultaneously translocated in opposite directions, it is referred to as antiport. We will now explore mechanisms of molecular membrane transport.

6.1.1. Simple Diffusion

Simple diffusion is just as its name describes. A molecule on one side of a membrane enters or "dissolves" in a membrane and then exits a membrane. Once a molecule has entered a bilayer, it could leave via either membrane face. For example, if a transmembrane concentration gradient of a neutral molecule is present, there will be an overall net molecular flux across a bilayer until the same concentration is reached on both sides. When the concentrations at each face are equal, molecules are still entering the membrane, but the number of molecules entering and leaving each face are equivalent. We will refer to this process as "simple diffusion" instead of the frequently encountered "free diffusion." Free diffusion across membranes is misleading since it really is not free at all because membranes retard a solute's rate of diffusion thousands of fold. "Simple" diffusion is also preferred since it draws a sharp contrast with the more complex mechanism of facilitated diffusion.

The ability of a molecule to traverse a membrane by simple diffusion is determined by hydrophobicity, size, and charge. Bangham *et al.* (1965) defined some of the elementary physical features of molecules transported by simple diffusion across well-defined membranes. These investigators measured the efflux of radioactive ions or water from PC liposomes. Encapsulated potassium cations were retained with liposomes for a period of days. On the other hand, labeled water crossed the membranes so fast that its rate could not be measured with standard biochemical methods. These liposome membranes were not complicated by the presence of carbohydrates, oligopeptides, proteins, or multiple lipid phases, which are encountered in biological membranes.

Bangham's results provide us with two key clues toward understanding cell membranes: (1) charged molecules cross membranes *very* slowly and (2) small uncharged molecules, such as water, can quickly pass through bilayers.

The influence of molecular size on cell permeability is schematized in Figure 6.2, panel a. All of the compounds shown are capable of forming hydrogen bonds. The ether–water partition coefficients, which are a rough measure of their hydrophobicity, are given in parentheses. As seen, cell membrane permeability rapidly decreases as molecular size increases.

The ability of hydrophobic compounds to cross bilayers has been known for many years. Figure 6.2, panel b shows the membrane permeability of a variety of compounds plotted against their lipid solubility or partition coefficient. A solute's permeability is the ratio of its transmembrane flux to its concentration gradient. As this graph reveals, permeability is roughly proportional to lipid solubility. This relationship between simple diffusion and lipid solubility is known as Overton's rule.

In contrast to the transport mechanisms described below, simple diffusion lacks stereochemical requirements. Although the chemical properties of a solute are important, their spatial positions are not. This is reasonable since specific chemical interactions, such as binding to an active site, do not participate in simple diffusion. Furthermore, simple diffusion is not saturable. Since specific translocators are not involved in this type of transport, they cannot be saturated. This is illustrated in Figure 6.3. The rate of transport is only dependent on the solute's concentration.

A given solute's rate of simple diffusion is affected by a membrane's phospholipid composition. Early studies showed that fluid membranes are more rapidly crossed by simple diffusion than are solid membranes. This has been confirmed by measuring water transport as a function of temperature. An abrupt change in transport occurs when the phase transition temperature is exceeded. Apparently, solutes can more readily permeate the disorganized environment of a fluid membrane.

Although simple diffusion may sound inelegant, it is the most fundamental transport process. In addition to water, oxygen and carbon dioxide are transported across membranes by simple diffusion. Most drugs and some low-molecular-weight toxins are transported into cells via simple diffusion. The endoplasmic reticulum makes harmful compounds more hydrophobic, thus allowing them to escape from cells by simple diffusion.

6.1.2. Water Movement across Membranes Regulates Cell Volume

In the preceding section we learned that water can cross lipid bilayers by simple diffusion. As previously mentioned, the tendency of molecules to diffuse across membranes can be quantitated by the permeability coefficient. The diffusion permeability coefficient (P_d) of lipid bilayers is about 10 μm/sec. The permeability coefficient of a membrane can also be determined by establishing an osmotic pressure gradient across the membrane using an impermeant solute and then measuring the water flux (P_f). For model lipid bilayers, $P_d = P_f$, indicating that water crosses membranes by simple diffusion alone. However, for many cells such as erythrocytes, $P_d < P_f$ (Macey, 1984).

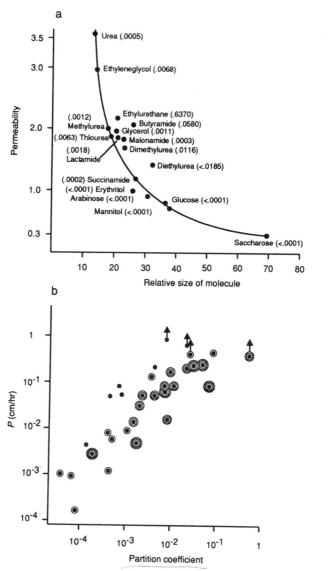

Figure 6.2. Simple diffusion. Panel a illustrates the effect of molecular size on simple diffusion across *Beggiatoa* membranes. As the relative molecular size increases, permeability rapidly decreases. A third variable, the ether–water partition coefficient, is listed by each compound in parentheses to account for hydrophobicity. Panel b shows the relationship between partition coefficient of a molecule (abscissa) and its penetration (P; ordinate) into *Chara*. The sizes of the data points, drawn as circles, are roughly proportional to molecular weight. As the partition coefficient, or solubility in a hydrocarbon environment, increases, the penetration into cells increases. [This material was originally published in *Planta* **1:**1 (1925) and *Trans. Faraday Soc.* **33:**987 (1937) and subsequently modified by A. Giese, *Cell Physiology*, Saunders (1973) and G. Karp, *Cell Biology*, McGraw–Hill (1984). Used with permission of the Royal Society of Chemistry, Springer-Verlag, A. Giese, and McGraw–Hill.]

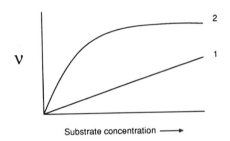

Substrate concentration ⟶

Figure 6.3. Concentration dependence of diffusional transport across membranes. Trace 1 shows the rate or velocity (v) of a hypothetical case of simple diffusion as a function of solute concentration. Trace 2 shows the transport behavior of a solute undergoing facilitated diffusion. As the solute concentration rises, the transport sites become saturated; the rate can increase no further.

For example, P_d = 52 μm/sec and P_f = μm/sec for erythrocytes. This indicates that water crosses erythrocyte membranes by simple diffusion and by flowing through a channel or opening.

Physiologists debated the presence of water channels for some time because of potential artifacts of unstirred layers near membranes. Such unstirred layers would decrease the apparent value of P_d. Therefore, it was necessary to confirm the presence of water "channels" by two independent routes. First, the activation energy of water transport was measured. Cell membranes had a lower activation energy than did model membranes. Second, the addition of covalent channel inhibitors increased the activation energy for water transport and decreased P_f without affecting P_d. This indicates that water is able to cross membranes by a pathway other than simple diffusion.

The membrane molecules apparently responsible for heightened water transport in erythrocytes are band 3 and band 4.5 (Benga *et al.*, 1986). This has been determined using radiolabeled mercurials that bind to these proteins in a 1:1 stoichiometry. This makes sense because both band 3 and band 4.5 (the glucose transporter) transport hydrophilic substances during their normal course of events. Therefore, it should not be surprising that small molecules (water) leak across these hydrophilic membrane channels when an osmotic pressure gradient is imposed.

Net water movement across membranes is regulated by osmotic pressure. In general, animal cells are in osmotic equilibrium with their surroundings. This is not generally true for plant cells since they have cell walls that can withstand large hydrostatic pressures. The more vulnerable animal cells must therefore carefully regulate their internal ionic concentration and water content to ensure a constant cell volume.

Cells contain many large negatively charged polymers such as proteins and nucleic acids. These charged polymers are trapped within a cell's plasma membrane. These trapped or nondiffusible anions cause a redistribution of permeant ions across biological membranes (the Donnan effect). At *equilibrium*, anions (A) and cations (C) redistribute across a membrane according to:

$$[A_o][C_o] = [A_i][C_i] \tag{6.1}$$

where the subscripts refer to the outer (o) and inner (i) compartments. Although the Donnan effect potentiates the redistribution of ions, it generally does not reach equilibrium in cells. Because of the redistribution of ions, the *sums* of the external and internal ionic concentrations are *not* equal ($A_o + C_o \neq A_i + C_i$). Since osmotic pressure is dependent on the total number of solute molecules in each compartment, a net osmotic pressure gradient is caused by nondiffusible anions. Animal cells must diminish this osmotic pressure gradient to survive.

One might imagine that membrane-impermeable extracellular ions could counterbalance the effects of nondiffusible intracellular anions. However, both *in vitro* and *in vivo* experiments have shown that extracellular anions are unimportant in regulating cell volume (Macknight and Leaf, 1986). A cell maintains a steady state away from Donnan equilibrium by pumping Na^+ and K^+ ions across its membrane. During each cycle the Na^+/K^+-ATPase pumps three Na^+ out of a cell followed by the uptake of two K^+. By moving Na^+ out of a cell, a net reduction in ionic strength is obtained (thereby reducing the transmembrane osmotic pressure gradient). In a very real sense, the Na^+/K^+-ATPase makes Na^+ the *inwardly* impermeant ion that counterbalances the bound (outwardly impermeant) ions within a cell. Although the regulation of cell volume is not yet understood completely, it seems clear that membrane pumps and leaks play a central role in determining animal cell volume.

A cell's steady-state distribution of ions can be perturbed by metabolic inhibitors or toxins. These changes often allow cells to move toward Donnan equilibrium. If the Na^+/K^+-ATPase is inhibited, membrane channels will allow a redistribution of ions. This in turn will drive water into a cell, causing its volume to increase. Ultimately, the equilibration will cause cells to swell and then burst as Donnan equilibrium is approached. Erythrocytes can resist this so-called colloid-osmotic swelling by a cell volume-gated potassium efflux. When an erythrocyte's volume slowly or transiently expands, these volume-sensitive pores allow potassium to escape down its concentration gradient, thus decreasing the ionic strength and osmotic pressure inside a cell. A cell's ability to resist swelling can be overwhelmed by the introduction of membrane pores. Membrane pores allow ions to rapidly redistribute across a membrane. The low tensile strength of an erythrocyte's membrane is quickly exceeded, causing the membrane to rupture and its cytoplasm to spill out.

6.1.3. Facilitated Diffusion

We have learned that small and/or hydrophobic compounds can cross a bilayer by simple diffusion. But most nutrients are neither small nor hydrophobic. Some nutrients enter cells by the process of facilitated diffusion. During facilitated diffusion, molecules are transported in an energy-independent fashion "down" their electrochemical potential gradient. Facilitated diffusion requires a membrane-bound oligopeptide or protein to shield a hydrophilic substrate from a membrane's hydrophobic core. A translocator participating in facilitated diffusion catalyzes the dissipation of a neutral substrate's concentration gradient. In the absence of a concentration gradient, no net flux of substrate is observed. However, living cells generally maintain concentration gradients by metabolizing molecules upon entry, thus accounting for the biological utility of facilitated diffusion.

6.1.3.1. Properties and Kinetics of Facilitated Diffusion

In contrast to simple diffusion, facilitated diffusion is saturable because membranes contain a finite number of translocators. Figure 6.3 compares simple and facilitated diffusion across membranes. The rate or velocity of transport (V) is plotted against substrate concentration (S). Simple diffusion is linearly related to substrate concentration whereas facilitated diffusion is a hyperbolic function. When these translocators are tested with various substrates, they are found to display chemical and stereochemical specificity.

The dose–response behavior of facilitated diffusion, as exemplified by Figure 6.3, trace 2, is described by a Michaelis–Menten treatment. The flux, J, is related to substrate concentration according to the relationship:

$$J = \frac{SV_{max}}{K_s + S} \tag{6.2}$$

where V_{max} is the maximum rate of transport and K_s is the substrate concentration when $J = V_{max}/2$. The kinetic behavior of translocators parallels that of enzymes. For example, irreversible inhibitors decrease V_{max} by destroying some translocation sites. Competitive inhibitors decrease K_s by blocking the active site and thereby decreasing the apparent affinity. In contrast to the high affinities generally expressed by membrane receptors, membrane translocators have relatively low affinities ($K_d \approx 10$ mM).

We have already described in detail several examples of membrane molecules catalyzing facilitated diffusion. This type of translocator is represented by a class of oligopeptides known as ionophores (Section 2.2). Ionophores facilitate the diffusion of hydrophilic solutes across membranes by directly carrying molecules across a bilayer or by forming a pore. We will now focus on facilitated translocation mediated by membrane proteins.

6.1.3.2. Erythrocyte Band 3: The Ping-Pong Model of Ion Exchange

Throughout this book we have used erythrocyte band 3 as a "model" integral membrane protein; lessons learned from the study of this protein could be applied when investigating other membrane proteins. For example, in Section 2.3.5.2 we discussed the structure and physiological setting of band 3. In Sections 3.1.3 and 3.2.1 we described band 3's carbohydrate attachments and transmembrane links to the membrane skeleton. In this section we will discuss how band 3 facilitates the diffusion of anions across membranes.

Band 3 is encoded by the *AE1* (anion exchange-1) gene. It is just one member of a family of homologous proteins that participate in anion movements; most cells possess band 3-like molecules. As mentioned in Chapter 2, it exchanges Cl^- and HCO_3^- anions across erythrocyte membranes, although other anions can also pass through band 3 to various degrees. In addition to productive exchanges of Cl^- and HCO_3^-, self-exchange of $Cl^- \leftrightarrow Cl^-$ and $HCO_3^- \leftrightarrow HCO_3^-$ also occur.

Although band 3 has been crystallized, its structure is not yet available. By analogy with other proteins such as bacteriorhodopsin, its membrane-bound domain is believed to assume a cylindrical shape. Positive charges at its aqueous surface attract anions whereas charged groups inside the membrane catalyze transmembrane anion movement. But how does it happen?

Although band 3's mechanism of anion translocation has not been established definitively, the best current model consistent with most experimental observations is the Ping-Pong model (Gunn and Frohlich, 1979; Macara and Cantley, 1981; Jay and Cantley, 1986; Jennings, 1989). The central feature of the Ping-Pong model is that anions are transported one at a time back and forth across a membrane, like a Ping-Pong ball. Figure 6.4 shows a fanciful illustration of what the Ping-Pong model might be like on a molecular level (just a single ping is shown). In this model an ion binds to a transport site at one face of a membrane (it could be either the *cis* or *trans*). The anion moves across the membrane by a series of small conformational changes in which salt bridges are exchanged among the diffusing anion and bound ionic residues lining the protein's transport pathway. The anion is then released at the opposite membrane face (Figure 6.4). This is one-half of a cycle. Another anion *must* now move in the opposite direction to complete the cycle. The series of salt bridge exchanges is reversed to move another anion in the opposite direction, thus returning band 3 to its original state. This model requires the presence of two stable conformations that do not interconvert in the absence of an anion.

The charged intramembrane residues participating in salt bond exchange during transport have not been completely worked out, in part because there are so many to choose from. Amino acid residues known to participate in transport include Lys, Arg, Glu, and His. Site-directed mutagenesis of murine Lys-558 followed by expression in *Xenopus* oocytes has shown that this site is responsible for binding the covalent transport inhibitor dicyclohexylcarbodiimide (DCCD) but does not participate in transport (Bartel *et al.*, 1989). An analysis of band 3's deduced amino acid sequence reveals many candidate residues (Kopito and Lodish, 1985), which might be profitably explored using site-directed mutagenesis.

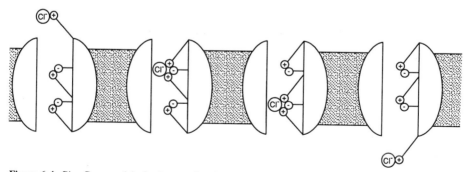

Figure 6.4. Ping-Pong model of anion translocation across erythrocyte membranes. The transport of a Cl⁻ anion across a membrane is illustrated. The salt bridges exchange partners as an anion moves across the membrane. Only one-half of a cycle is shown.

6.1.3.3. Erythrocyte Glucose Transporter: Molecular Characterization and
        Mechanism

Almost all mammalian cells take up glucose by facilitated diffusion. In addition, the apical surfaces of certain kidney and intestinal cells use sodium-linked secondary active transport to internalize glucose. The erythrocyte glucose transporter catalyzes facilitated diffusion of glucose. Since erythrocytes do not have mitochondria, they must generate ATP from glycolysis. Because of their simplicity, erythrocytes also serve as a prime model of facilitated glucose transport.

The erythrocyte's glucose transporter has been identified as a diffuse 55-kDa band after SDS-PAGE. This integral membrane protein migrates as at least part of band 4.5 on these gels. After glycosidase treatment, it migrates as a sharper 46-kDa band. Cytochalasin B (Figure 3.18), which inhibits F-actin formation, inhibits the erythrocyte glucose transporter. The inhibitory activity of cytochalasin B is due to direct binding to the transporter since: (1) cytochalasin B labels the 55-kDa band in a 1:1 stoichiometry and (2) it inhibits the function of reconstituted glucose transporters. The glucose transporter is purified by solubilization of erythrocyte ghosts with octylglucoside followed by column chromatography. This transporter preparation was judged to be pure since it migrated as a single band during SDS-PAGE and bound cytochalasin B in a 1:1 molar fashion. The purified protein retained its ability to facilitate the diffusion of glucose through membranes when reconstituted in lipid vesicles (e.g., Chen *et al.*, 1986; Wheeler and Hinkle, 1985).

Mueckler *et al.* (1985) have cloned a gene for the human erythrocyte glucose transporter (M_r = 54,117), which is one member of a family of homologous genes. Hydropathy analysis of the deduced amino acid sequence suggests the presence of 12 transmembrane domains. Seven of these domains are composed of only hydrophobic residues. The five remaining helices contain both nonpolar and polar residues such as serine, threonine, glutamine, and asparagine. These polar residues are positioned such that they could form amphipathic α-helices. This suggests that the polar residues might form a cavity that could hydrogen bond with glucose. Circular dichroism and Fourier transform infrared spectroscopy (Chin *et al.*, 1987) indicate that: (1) most of the glucose transporter's mass is α-helical (~80%) and (2) these helices are oriented perpendicular to a bilayer's plane. Therefore, it is reasonable to propose the presence of numerous transmembrane α-helices.

The glucose transporter's NH_2- and COOH-termini are located at the *cis* face of an erythrocyte's membrane. This receives direct support from proteolysis and labeling studies. It finds indirect support from its deduced amino acid sequence since Asn-45 is apparently the only site available for *N*-linked carbohydrate attachment and must therefore be at the *trans* face. Other integral membrane proteins that participate in molecular membrane transport such as band 3 and bacteriorhodopsin have charged amino acids buried within a bilayer. However, these membrane proteins participate in ionic exchanges or net ionic fluxes and therefore may require ionic residues within their structures to mediate ionic movements. In contrast, the facilitated diffusion of neutral glucose molecules apparently does not require intramembrane charges.

A simple model of glucose transport has been proposed based on kinetic studies (Wheeler and Hinkle, 1985). Direct experimental studies (Chin *et al.*, 1987; Gorga and

Lienhard, 1982) have shown that the glucose transporter alters its conformation when glucose is bound. Also, it is quite unlikely that the structure described in the preceding paragraphs rotates *through* membranes (or "flip-flops") during transport. Therefore, an alternating conformation model of transport is preferred. Two conformations corresponding to an outward or *trans*-facing binding site (C_t) and an inner or *cis*-facing binding site (C_c) have been proposed (Wheeler and Hinkle, 1985). The glucose (G) transport scheme can be written as:

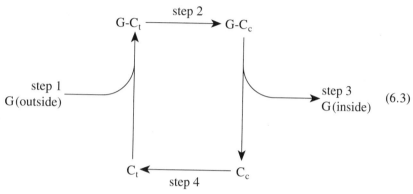

This equation illustrates the net inward movement of glucose. The steps participating in the outward movement of glucose, such as that found in liver cells during glycogenolysis, are identical except that the arrowheads are pointed in the opposite directions. Depending on its concentration gradient, glucose can be translocated from the exterior to the interior or from the interior to the exterior of a cell. In addition to net flux, the glucose transporter can exchange intracellular glucose for extracellular glucose or vice versa. Radiotracer experiments have shown that exchange reactions are faster than net transfer reactions. This observation suggests that step 4 [Eq. (6.3)] is the slow step in the reaction pathway. This $C_t \rightarrow C_c$ or $C_c \rightarrow C_t$ conformational change limits the rate of net flux but is bypassed during exchange (step $1 \rightarrow 3$ and $3 \rightarrow 1$).

Barnett *et al.* (1973a,b) have used glucose analogues to define several structure–activity relationships of the glucose transporter's active site. When deoxy and fluoride-substituted analogues were tested, they indicated that hydroxyls at C-1, C-3, and C-6 positions of glucose interact with its transporter. A hydrophobic cleft is postulated to be near positions C-4 and C-6 based on the binding of glucose analogues possessing bulky substituents. These observations led Barnett *et al.* (1973b) to suggest that the transporter opens near the vicinity of C-1 and closes behind C-6. This conformational change could be mediated by a swing in the position of helix M9 (Walmsley, 1988).

The molecular pathway of glucose binding and translocation remains open to speculation. Further detailed structural information of the native and active transporter is needed. Furthermore, site-directed mutagenesis studies focusing on the hydrophilic residues in conjunction with reconstitution should aid in dissecting this transport pathway. An analysis of deoxyglucose analogues binding to site-specific translocator mutants may allow specific substrate–transporter interactions to be identified.

The glucose transporter is regulated by several mechanisms. One way to regulate the amount of glucose uptake is to alter the number of glucose transporters present in a membrane. For example, insulin ligand–receptor complexes stimulate the delivery of preformed glucose translocators from cytoplasmic vesicles to plasma membranes (Section 7.4). The glucose transporter may also be regulated by its phospholipid environment (Tefft *et al.*, 1986). These investigators found that phospholipid head group and acyl chain composition were primary factors in determining the K_s and turnover number of reconstituted translocators. However, when lipids and translocators are co-reconstituted, one cannot separate the effects of lipids on the reconstitution *per se* from their effect on the protein's membrane-bound functional activity. In other words, the lipids could have destroyed transport function before the membranes were formed. It seems likely that lipid composition does affect K_s and turnover number since these are complex functions of temperature (Tefft *et al.*, 1986). It may be useful to reconstitute the transporter in a well-characterized membrane. With a known baseline, one could then alter the composition of this system by using phospholipid exchange proteins or by allowing cholesterol to partition into the membranes.

6.1.4. Group Translocation: The Prokaryotic Phosphoenolpyruvate-Dependent Sugar Translocator

The hallmark of group translocation is the covalent modification of a substrate during passage across a membrane. Kundig *et al.* (1964) first discovered the phosphoenolpyruvate-dependent sugar transporter (phosphotransferase system; PTS) in *E. coli*. The PTS has since become the best understood group translocation mechanism.

Saccharide group translocation systems are widely distributed among prokaryotes. Both gram-positive and -negative microbes use PTS transport systems. Facultative anaerobes such as *Escherichia*, *Salmonella*, and *Staphylococcus* and obligate anaerobes (e.g., *Clostridium*) utilize PTSs to internalize many sugars. There are claims that non-PTS group translocation mechanisms may participate in disaccharide uptake in higher eukaryotes, although these data are not compelling. However, it is certain that the PTS plays a central role in the metabolic activity of prokayotes (Higgins, 1989).

The PTS is comprised of four distinct proteins or protein complexes (Table 6.3). In the presence of Mg^{2+}, these proteins catalyze the reaction:

$$PEP + sugar_{(out)} \rightarrow sugar\text{-}phosphate_{(in)} + pyruvate \qquad (6.4)$$

where PEP is phosphoenolpyruvate. PEP (**1**) is a high-energy intermediate generated during glycolysis:

$$CH_2{=}C{-}\overset{\overset{\displaystyle O}{\|}}{C}{-}O^-$$

$$\underset{\underset{\displaystyle O}{\|}}{\overset{\overset{\displaystyle O}{|}}{{}^-O{-}P{-}O^-}}$$

(**1**)

Table 6.3.
Components of the Glucose Phosphotransferase
System of *E. coli*

Component	Gene	Sugar specificity	M_r	Position
EI	*ptsI*	No	70,000	Soluble
HPr	*ptsH*	No	9,500	Soluble
EIII	*crrA*	Yes	20,000	Peripheral
EII		Yes	60,000	Integral

PEP donates its phosphoryl group to a His residue of PTS's enzyme I (EI), yielding pyruvate and phosphoenzyme. Several lines of evidence indicate that EI functions in a dimeric or oligomeric state. One phosphate is bound per monomer of EI. Since EI has a higher K_m for PEP (200 μM) than HPr (5 μM), PEP concentration may regulate this transport system's activity. Phosphorylated EI transfers the phosphoryl moiety to HPr (Figure 6.5). HPr (heat-resistant protein; $M_r = 9500$) is phosphorylated on the N-1 atom of a histidine's imidazole ring. These first two enzymes are general cytoplasmic phosphate carriers that are common to all of a cell's PTSs. The HPr phosphoenzyme can donate its phosphate to any of the several sugar-specific enzyme III (EIII) molecules.

EIII is a peripheral membrane protein that associates with enzyme II (EII), an integral membrane protein. In conjunction with EII, EIII transfers the phosphoryl moiety of PEP to a sugar. There is no evidence to support the formation of an EII–phosphate intermediate. A glucose molecule is transformed to glucose-6-phosphate upon entering the cytoplasm. This is particularly convenient since: (1) phosphorylated glucose is unable to leave a cell by simple diffusion and (2) glucose-6-phosphate is the first intermediate formed during glycolysis. This mechanism is efficient since both transport and glucose phosphorylation are catalyzed by each PEP molecule.

The deduced amino acid sequence of the EII molecule that participates in mannitol transport has been determined (Lee and Saier, 1983). The NH_2-terminal domain contains seven transmembrane α-helices, as suggested by hydropathy analysis. In analogy to the glucose transporter, there are several hydrophilic residues that may participate in forming a channel. In addition, there are three charged residues associated with these

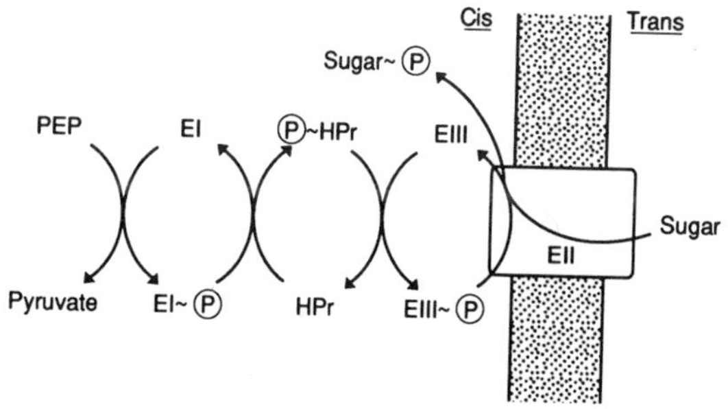

Figure 6.5. Transfer of the phosphoryl moiety during group translocation. The phosphoryl moiety of phosphoenolpyruvate is sequentially transferred to EI, HPr, EIII, and then to a sugar via EII.

hydrophobic helices. These charged groups may be required to trigger phosphate transfer to a sugar. Its large intracellular COOH-terminal domain may interact with EIII.

In the preceding discussions, we have learned that eukaryotes generally transport glucose by facilitated diffusion via a single protein type whereas *E. coli* uses several proteins and an energy supply to obtain the same result. It may seem odd that higher eukaryotes, which are much more evolved than prokaryotes, use a *simpler* means of glucose transport. This apparent paradox is resolved by considering the cells' environments. Metazoans regulate glucose availability by monitoring blood glucose levels whereas prokaryotes have absolutely no control over their nutrient supply. Therefore, a complex mechanism of nutrient uptake became an unnecessary complication in animals.

6.1.5. Primary Active Transport: The Sodium/Potassium Pump

During active transport an energy source such as ATP, electron transport, or electrochemical gradients is used to move ions or molecules against their electrochemical potential gradients. As mentioned previously, all cells must control their internal ionic milieu. The principal means of regulating Na^+, K^+, Ca^{2+}, Mg^{2+}, and sometimes H^+ is by direct ATP-dependent pumping mechanisms. The Na^+/K^+-ATPase is one member of a family of E_1E_2 membrane pumps. The resting pumps are phosphorylated by ATP. The phosphorylated pump E_1–P alters its conformation to E_2–P to allow vectorial ion translocation. In this section we will explore the Na^+/K^+-ATPase, a well-studied but incompletely understood ionic pump.

6.1.5.1. The Pump's Physiological Setting

The Na^+/K^+-ATPase is present in almost all animal cell membranes. Its primary function is to maintain Na^+ and K^+ gradients across plasma membranes. This is achieved by pumping Na^+ out of a cell while K^+ is moved into a cell. The importance of sodium/potassium pumping extends to many areas of membrane and cell biology. The Na^+/K^+-ATPase plays an important role in generating a transmembrane potential. Voltage swings, brought about by intercellular signaling events such as nerve impulses, rely on fluctuations in these ion gradients. The Na^+ gradient is employed by other membrane-bound translocators to regulate cytosolic pH and transport nutrients into cells. The heightened intracellular K^+ level is required for many enzyme activities. The importance of Na^+/K^+ pumping is underscored by the fact that a typical cell uses about one-third of its ATP production just to maintain this gradient.

6.1.5.2. Structural Properties of the Na^+/K^+-ATPase

The Na^+/K^+-ATPase is randomly distributed about the periphery of most cells. However, in some polarized neuronal, exocrine, and epithelial cells, the Na^+/K^+ pump is asymmetrically distributed at a cell's surface. Nelson and Veshnock (1987) have shown that the Na^+/K^+-ATPase binds to the cytoskeletal protein ankyrin. Therefore, the cytoskeleton may participate in restricting the pump to specific membrane domains in some cell types.

The Na^+/K^+-ATPase is composed of at least two integral membrane protein subunits: the α and β subunits. The presence of a third γ subunit remains controversial. The α and β subunits have masses of 100 and 40 kDa, respectively. Figure 6.6 shows a highly schematic model of the α and β chains which are associated in a 1:1 stoichiometry. Reconstitution experiments in lipid vesicles and cDNA transfection experiments indicate that both the α and β chains are required for transport.

The α subunit is expressed in at least two isoforms, α_1 and α_2, which can be distinguished by SDS-PAGE. A third gene, α_3, has been discovered, although its functional expression is uncertain. Human genomic cDNA libraries have revealed the presence of two additional α subunit genes or pseudogenes. The physiological role(s) of these isoforms is not known, although different tissues can vary in their Na^+/K^+ pumping stoichiometries. It is possible that the α subunit's physical isoforms account for its physiological variability.

The α subunit binds ATP and ouabain, an inhibitor of the Na^+/K^+-ATPase. The α subunit presumably undergoes conformational changes that conduct sodium and potas-

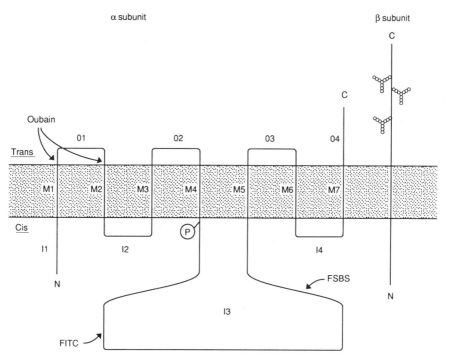

Figure 6.6. Schematic model of the Na^+/K^+-ATPase. A stick figure model of the Na^+/K^+-ATPase based on its deduced amino acid sequence is given. The α and β subunits are shown. The positions of the ouabain, FITC (fluorescein isothiocyanate), and FSBS [5'-(p-fluorosulfonyl)benzoyl adenosine] binding sites are indicated. The transmembrane domains are labeled M1 through M7. The outer and inner domains are labeled O1 to O4 and I1 to I4, respectively. The carbohydrate chains of the β subunit are represented by a series of circles.

sium ions across a plasma membrane. Although the β subunit is required for transport, its function is uncertain. It is apparently required for the proper insertion and structural stability of the α subunit in a membrane. The β subunit also interacts with the α subunit during cation translocation. This is suggested by studies showing that anti-β-subunit antibodies inhibit functional activity.

The deduced amino acid sequences of several α subunits from different species have been obtained (e.g., Kawakami *et al.*, 1985). As expected for a transport protein, its sequence suggests the presence of seven transmembrane domains (M1 to M7 in Figure 6.6). The M4 domain contains the negatively charged residue Glu-334. This anionic group and possibly other groups near M5 and M6 may participate in transporting cations. The NH_2-terminus is located at a membrane's *cis* face. The third intracellular domain (I3) is known to be linked to transport activity. This domain possesses a consensus ATP-binding region and the aspartic acid residue (Asp-376) that becomes phosphorylated during active transport. Since this residue is relatively near M4, it may be able to affect the transmembrane domains. The I3 loop also contains binding sites for some inhibitors. These include Lys-507 and Lys-725 which bind fluorescein isothiocyanate and 5'-(*p*-fluorosulfonyl)benzoyl adenosine, respectively (Ohta *et al.*, 1986). As mentioned above, the reagent ouabain inhibits the pump's functional activity. Early studies indicated that ouabain's binding site is located at the pump's *trans* face. By comparing the α subunit's sequence from species sensitive and resistant to ouabain, candidate sites for ouabain binding were identified. Site-directed mutagenesis studies have shown that Gln-111 and Asn-122 are important participants in ouabain binding (Price and Lingrel, 1988). These residues are located on the O1 outer loop between the M1 and M2 transmembrane domains. However, it is not known how these sites affect transport.

6.1.5.3. Mechanism of Action

Several lines of evidence now indicate that the Na^+/K^+-ATPase pumps three sodium ions out of a cell followed by the uptake of two potassium ions (Forbush, 1984; Sachs, 1986). In an important study, Forbush followed ionic movements during a single cycle of the pump. A simple mechanistic scheme of the pump's action based on the Albers–Post model is shown in Figure 6.7 (Blostein, 1989). Three sodium ions bind to the translocator's E_1 conformation at its *cis* face. The translocator becomes phosphorylated at Asp-376, thus forming the E_1–P intermediate and moving the sodium ions to an occluded (or hidden) location (perhaps in the middle of the pump) (Forbush, 1987). This potentiates the formation of the E_2–P conformation which releases the sodium ions at a membrane's *trans* face. After the sodium ions are released, the translocator can become dephosphorylated to yield the E_2 and then E_1 conformers or it can bind potassium. After two potassium ions bind at the *trans* face, they become occluded, ATP binds, and then the potassium ions are released from the *cis* face. The enzyme, now in the E_1·ATP conformation, is ready to repeat the cycle.

At present the active site of the Na^+/K^+-ATPase can only be imagined. Experiments have shown that the translocator is highly specific for sodium. However, Rb^+,

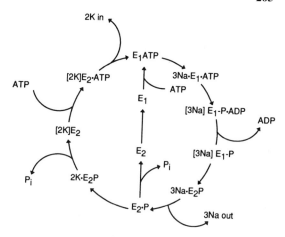

Figure 6.7. Albers–Post model of ion translocation. A possible model of ion translocation by the Na^+/K^+ pump is shown. The pump's two conformations, E_1 and E_2 are indicated. When the ion is listed in brackets, it is occluded. (Adapted from Blostein, 1989, *Curr. Opin. Cell Biol.* **1:**746.)

Li^+, and other cations can substitute for potassium. This suggests that the active conformations of the enzyme impose quite different physical constraints on the bound ions.

6.1.6. Secondary Active Transport: The Lac Permease of *E. coli*

As briefly mentioned above, secondary active transport requires an indirect source of energy, such as proton gradient. In 1963, Mitchell proposed that a proton gradient drives the transport of β-galactosides across cell membranes. This proposal was one corollary of his chemiosmotic hypothesis. Although it was ignored for several years, this mechanism of transport is now rigorously established. We will discuss the lac permease, an excellent model of proton symport.

6.1.6.1. Lac Permease

The lac permease is an integral membrane protein of the *E. coli* inner membrane that translocates lactose from the periplasmic space into the cytoplasm. It is one of three structural genes encoded by the *lac* operon. These three structural genes, genes *Z*, *Y*, and *A*, encode β-galactosidase, lac permease, and thiogalactoside transacetylase, respectively. The β-galactosidase cleaves lactose upon entry whereas the function of the transacetylase is uncertain. In the presence of lactose, these genes are expressed to stimulate its uptake and utilization.

In Section 5.2 we learned how electron transport creates an electrochemical proton gradient across membranes. In addition to ATP production, this gradient is used by bacteria, mitochondria, and chloroplasts to drive the uptake of solutes. Lac permease catalyzes the symport of a proton and a lactose molecule in a 1:1 stoichiometry.

6.1.6.2. Identification, Characterization, Cloning, and Structural Biology of Lac Permease

The modern quantitative analysis of lac permease began with the cloning of the *lacY* gene in 1980. The *lacY* gene of the lactose operon was cloned in recombinant plasmids (Buchel *et al.*, 1980). This development was important for two reasons: (1) it revealed the permease's deduced amino acid sequence and (2) it could be overexpressed in *E. coli*, thus providing a large, convenient, and physiologically relevant supply of permease.

Lac permease is comprised of 417 amino acids (M_r = 46,504). Most of these residues (71%) are nonpolar, making it one of the most hydrophobic proteins yet observed. Almost 19% of these hydrophobic residues are phenylalanine; a striking observation of unknown significance. At neutral pH the permease has a net positive charge. The clustering of hydrophobic amino acids suggests 12 transmembrane domains (Figure 6.8). Antibody binding experiments have shown that both the NH_2- and COOH-termini are located at the cytoplasmic membrane's *cis* face. Hence, there is an even number of transmembrane domains. Circular dichroism and Raman spectroscopy indicate that lac permease is rich (~80%) in α-helical structure (Foster *et al.*, 1983; Vogel *et al.*, 1985). The high α-helical content is consistent with a large fraction of this hydrophobic protein's mass being embedded within a bilayer as transmembrane α-helices. These transmembrane helices are arranged as a cylinder in membranes.

The functional dissection of the lac permease began with the study of *E. coli* mutants (Mieschendahl *et al.*, 1981). These early experiments showed that the NH_2-terminal region of the permease was unimportant in its physiological activity. These sequences could be replaced by those coding for β-galactosidase with no effect on

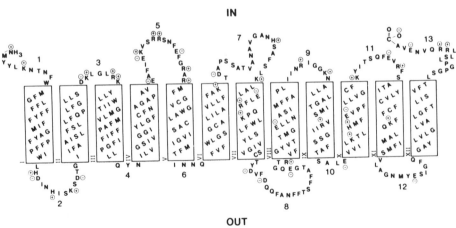

Figure 6.8. Structural model of lactose permease. A model of lactose permease based on its deduced amino acid sequence is shown. The identities of the amino acids are given in the single-letter code. The charged residues are so indicated. (Reprinted with permission from *Biochemistry* **26**:2071, copyright © 1987 by the American Chemical Society.)

transport. Mutations between residues 191 and 360 were found to influence the affinity of the protein for protons and/or lactose. We shall return to the functional analysis of lac permease in the following section.

When overexpressed in *E. coli*, lac permease accumulates to about 3% of the total membrane protein. This greatly simplified its purification and characterization. Solubilization in the detergent octylglucoside does not damage the translocator; it retains functional activity when reconstituted in lipid vesicles. Lac permease is photoaffinity labeled by 4-nitrophenyl-α-D-galactopyranoside (NPG; **2**):

(2)

In the absence of light, NPG acts as a competitive inhibitor of lactose transport. When illuminated, it becomes covalently bound to the permease and irreversibly inhibits transport. Prior to, but not after, illumination it can be displaced by other saccharides that bind to this translocator. This effect is specific for lac permease since the membrane's electrochemical gradient and proline transport are unaffected. When analyzed by SDS-PAGE, NPG was found to label a 33-kDa protein. The difference between this apparent molecular mass and the deduced molecular mass cannot be accounted for by posttranslational modifications of the permease (Ehring *et al.*, 1980). Due to the permease's unusually hydrophobic character, it likely binds more SDS than do typical proteins, thus increasing its charge density and leading to an anomalously high migration rate and low apparent molecular mass.

The lac permease is purified from overexpressing bacteria by several negative selection steps yielding a homogeneous band after SDS-PAGE. Lipid vesicles containing the permease are prepared by detergent dialysis. The permease is uniformly distributed in the vesicles' membranes as shown by freeze-fracture microscopy.

Several very important observations have been made with these reconstituted membranes. These reconstitution experiments: (1) unambiguously confirmed the identity of this structural protein (or structural gene product) as *the* functional lac permease and (2) showed that this single protein expresses all of the attributes necessary to catalyze lactose symport. These two observations may sound like trivial footnotes to the uninitiated. But, as we will see in the next chapter, it can be extraordinarily difficult to reach these elementary conclusions in some membrane systems. These reconstituted membranes will catalyze the accumulation of lactose in the presence of an electrochemical proton gradient (Kaback, 1986; Matsushita *et al.*, 1983). In fact, the turnover number and K_m (~16 mM) of these reconstituted membranes are indistinguishable from those found for cell membrane vesicles. Therefore, these synthetic membranes provide a vital link between membrane structures and their physiological function.

Lac permease's saccharide binding site has been characterized by fluorescence spectroscopy (Mitaku *et al.*, 1984). A fluorescent dansyl galactoside molecule was employed. The fluorescence contributions of free, nonspecifically bound, and specifically bound dansyl galactoside can be distinguished using appropriate controls. When bound to the permease's active site, changes in dansyl galactoside's emission spectrum and fluorescence lifetime indicate that this site is hydrophobic. Although the active site is hydrophobic, it remains accessible to the solvent since aqueous quenchers diminish the fluorescence of dansyl galactoside. Quenchers bound to fatty acyl groups do not affect specifically bound dansyl galactoside. Therefore, its hydrophobic binding site must be sequestered away from the lipid environment. In contrast, fluorophores covalently bound to the permease were accessible to lipid-associated quenchers. Therefore, the permease's active site is thought to be a hydrophobic pocket near its center.

Structural studies are beginning to outline lac permease's shape. When prepared for freeze-fracture electron microscopy, the liposome-reconstituted permease appears as 5.0-nm intramembrane particles. No surface structures are revealed by etching these samples; this suggests that most of its mass is located inside a bilayer. During favorable shadowing conditions, a cleft across most intramembrane particles is seen. These results have been extended by negative-stain electron microscopy of one- and two-dimensional crystalline arrays of lac permease (Li and Tooth, 1987). Monomers within the arrays measure 5.0×3.0 nm within the plane of the membrane. Furthermore, image analysis consistently revealed the presence of a cleft along the center of the protein. This cleft likely corresponds to the permease's saccharide binding site. A similar cleft has been identified as the saccharide binding site in a periplasmic L-arabinose binding/translocation protein in x-ray diffraction studies (Gilliland and Quiocho, 1981).

Structural flexibility is necessary for lac permease's transport function (Dornmair and Jahnig, 1989). When reconstituted in phospholipid vesicles, lac permease's ability to translocate lactose is very sensitive to a bilayer's phase. Although fluid membranes support transport, lactose translocation is dramatically decreased in solid-phase membranes. NMR spectroscopy has shown a membrane phase-dependent reduction in the permease's flexibility which in turn affects its ability to undergo the conformational changes required for transport.

6.1.6.3. Molecular Mechanism of Transport

Our understanding of lac permease's mechanism of action has improved dramatically in recent years (Kaback, 1987). Most of our recent insights into the permease's transport mechanism come from site-directed mutagenesis studies. This sort of experiment definitely reveals what is not important. However, one can never be absolutely certain about what is important. For example, changing an amino acid could affect protein function by removing or just moving a critical functional group required for activity. Alternatively, an amino acid alteration could change the secondary or tertiary structure or some other parameter which diminishes transport apart from any direct modification of a catalytic site. This potential problem can be minimized by testing a hypothesis from several points of view. However, this approach is generally a big improvement over earlier chemical modification experiments.

Table 6.4.

Site-Directed Mutagenesis Studies of *Lac* Permease[a]

Amino acid position	Wild-type residue	Mutant residue	% of wild-type's transport rate
154	Cys	Gly	0
		Ser	10
		Val	30
117	Cys	Ser	70
148	Cys	Ser	100
		Gly	25
176	Cys	Ser	80
333	Cys	Ser	100
35	His	Arg	100
39	His	Arg	100
205	His	Arg	0
		Asn	100
		Gln	100
322	His	Arg	0[b]
		Asn	0
		Gln	0
325	Glu	Ala	0

[a]Modified from and reprinted with permission from *Biochemistry* **26:**2071, copyright ©1987 by the American Chemical Society.
[b]Retains facilitated diffusion activity.

Table 6.4 shows the results of several representative site-directed mutagenesis studies of lac permease. Early covalent modification studies suggested that a Cys residue was required for permease function. To test the role of Cys in transport, each of the permease's eight Cys residues were separately modified by site-directed mutagenesis followed by construction of a plasmid vector and transformation of a *lac Y⁻* strain of *E. coli*. When these transformants were tested for transport activity, only Cys-154 was found to be required for transport. Its role is restricted to transport since saccharide binding was unaffected by this mutation. Although Cys-154 is critical for transport, it probably does not participate in proton binding or movement since a Val-154 mutant retains 30% transport activity.

Since chemical modification experiments suggested that His residues participate in symport, each of the permease's four His residues was altered by site-directed mutagenesis (Kaback, 1987). These experiments showed that His-35 and His-39 are not important in transport (Table 6.4). However, His-205 and His-322 both participate in transport, but in different ways. The mutation His-205 to Arg-205 abolishes transport activity although the positive charge at this position is conserved. However, this mutation does not conserve the hydrogen bonding properties of histidine's imidazole ring. The amide nitrogens of Asn and Glu can adopt spatial positions that match the N-1 and N-3 imidazole nitrogens of His. The permease retains normal activity when His-205 is replaced by Asn or Glu (Puttner *et al.*, 1986). This suggests that His-205 participates in forming a hydrogen bond required for transport.

Proton:saccharide symport is abolished by the His-322-to-Arg-322 mutation (Figure 6.9). Although this mutant transporter no longer catalyzes secondary active trans-

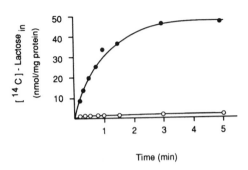

Figure 6.9. Kinetics of lactose transport in an *E. coli* test system. The kinetics of ^{14}C-labeled lactose uptake were measured. An *E. coli* strain deficient in lactose transport was transformed with plasmids containing wild-type and mutant genes encoding the lactose permease. The wild-type data are given as solid circles. Mutants containing Ala-325, Asn-322, Arg-322, or Glu-322 are indicated as open circles. (Redrawn with permission from *Biochemistry* **26**:2071, copyright © 1987 by the American Chemical Society.)

port, it does act as a facilitated translocator! This suggests that the imidazole ring is required for active transport driven by a proton gradient. Furthermore, symport is not retained by Asn-322 and Glu-322, indicating that hydrogen bonding is likely not important. Therefore, protonation of His-322 is required for active translocation of lactose across membranes. Protonation of His-322 is believed to be the first mechanistic step in lac permease-catalyzed symport.

When the importance of His-322 was recognized, it immediately raised the question: how does this residue move a proton across a membrane? An inspection of the permease's sequence reveals a negatively charged glutamic acid moiety at position 325. When part of an α-helix, this amino acid is in an optimal position to form an ion pair with His-322 (Figure 2.27). This suggests that Glu-325 may participate in a charge-relay-type system with His-322 (Kaback, 1987). To test this idea, Glu-325 was mutated to Ala-325. When tested for transport activity, this mutant permease displayed no secondary transport activity but it did participate in facilitated diffusion. A preliminary model based on Kaback's results suggests that His-322 binds protons at a membrane's *cis* face followed by their release from Glu-325 at the *trans* face. This proton pathway provides the energy required for secondary active transport.

6.2. MACROMOLECULAR AND BULK TRANSPORT

In the preceding section we examined how ions and small molecules move across membranes. Macromolecules and inert particles are simply too big to diffuse across a bilayer. In general, preformed proteins cross bilayers during membrane fusion (e.g., secretion) or fission (e.g., pinocytosis) events (Section 8.1). [However, under special conditions, immature proteins pass across membranes during their synthesis or translocation into organelles (Section 8.2).] For example, during pinocytosis macromolecules in the extracellular fluid phase are transferred into the cytoplasm via pinocytotic vesicles which bud from a plasma membrane. On the other hand, during exocytosis molecules or macromolecules within storage vesicles fuse with a plasma membrane and thereby release their contents into the extracellular environment. The remainder of this chapter will consider the mechanisms of macromolecular or bulk transport.

6.2.1. Endocytosis

During endocytosis soluble and particulate substances in the extracellular environment are brought into a cell's cytoplasm. This primitive function is widespread among eukaryotes—from amoebae to people. Soluble extracellular materials are internalized by pinocytosis whereas particulate objects are engulfed during phagocytosis. The internalized materials are surrounded by a bilayer membrane derived from the plasma membrane. These endocytotic vesicles and vacuoles then interact with lysosomes that digest most of the engulfed materials. In a few cases, endocytosis is used to transport materials from one location to another within an organ. For example, proteins are transported across epithelial cell layers by endocytosis at one side of the cell layer and exocytosis from the opposite side.

Endocytosis plays many important roles in cell physiology. The most primitive function performed by endocytosis is cell nutrition. This is particularly evident in microorganisms such as amoebae. Phagocytic cells of metazoans endocytose and then destroy infectious agents, thus providing host defense. This includes, for example, the wandering phagocytes of invertebrates and a human's specialized phagocytic cells. The transfer of protective antibodies from mother to offspring relies on endocytosis and transport across epithelial cell barriers. Endocytosis also participates in the control of metabolism and the expression of cell surface receptors. We shall now consider a few endocytotic processes in greater detail.

6.2.1.1. Pinocytosis

Classically, pinocytosis is defined as the uptake of fluid by a cell (cell drinking). Of course, cells do not really *need* to drink. As we learned in Section 6.1, water equilibrates across membranes by simple diffusion. When living cells are observed by phase-contrast microscopy, it just looks like water interiorization. Molecules and macromolecules contained within this fluid or adsorbed to a membrane's *trans* face are brought into a cell. These forms of pinocytosis participate both in nutrient uptake and in cell regulatory pathways.

As mentioned above, both fluid-phase and adsorptive pinocytosis take place. During fluid-phase pinocytosis solutes dissolved in a cell's environment are internalized. In contrast, during adsorptive pinocytosis macromolecules first bind to a plasma membrane followed by their pinocytotic uptake. These mechanistic differences in uptake result in quite distinct biological properties. The rate of adsorptive endocytosis saturates at high substrate levels. Since there are a finite number of receptors for adsorptive endocytosis, the endocytotic rate cannot be increased after all of the receptors are bound to ligands. In contrast, when the solute concentration is raised during fluid-phase pinocytosis, a cell pinocytoses the same *volume* but more solute molecules. Consequently, the rate of fluid-phase endocytosis is not saturable. Another ramification of these mechanistic differences is that fluid-phase pinocytosis is inefficient. Adsorptive pinocytosis concentrates solute particles at a membrane invagination whereas no such concentration occurs, by definition, during fluid-phase uptake.

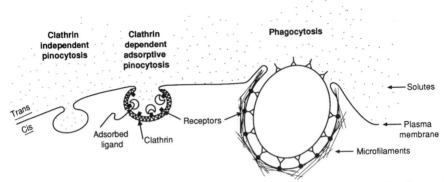

Figure 6.10. Illustrations of three endocytotic mechanisms. Clathrin-independent and -dependent pinocytosis are illustrated on the left. The adsorptive endocytosis of a ligand is shown during the clathrin-dependent pinocytosis. Receptor-dependent phagocytosis is shown on the right.

Figure 6.10 shows a schematic illustration of fluid-phase pinocytosis, adsorptive pinocytosis, and phagocytosis. Clathrin and actin are always associated with adsorptive pinocytosis and phagocytosis, respectively. As Figure 6.10 suggests, some fluid-phase molecules are internalized with the adsorbed molecules during adsorptive pinocytosis. Many scientists believe that fluid-phase pinocytosis occurs during both clathrin-dependent and -independent pinocytosis. However, the existence of clathrin-independent pinocytosis has not been rigorously proven (Hubbard, 1989).

Fluid-phase endocytosis is quantitated by adding soluble tracers to the incubation media. The tracers are often labeled with a radioactive atom or fluorescent group. For example, radioactive sucrose and inulin and fluorescent lucifer yellow are frequently employed pinocytotic tracers. Another frequently employed pinocytotic tracer is the glycoprotein enzyme horseradish peroxidase. Although horseradish peroxidase offers the advantage of parallel cytochemical studies and biochemical measurements, it is complicated by the fact that its carbohydrate chain is specifically bound by some cell membranes. To ensure that only fluid-phase endocytosis is measured, the tracers must not bind to membranes. The apparent volume of solution internalized by cells should be independent of the tracer's concentration. If the apparent level of fluid-phase endocytosis is dependent on the tracer's concentration, then the tracer may be binding to the cell's membrane or stimulating pinocytotic activity.

During pinocytosis the plasma membrane invaginates and then buds off to form an intracellular vesicle. These vesicles are generally about 150 to 300 nm in diameter. Consequently, pinocytotic vesicles have a very high surface-to-volume ratio. Moreover, pinocytosis results in the internalization of large quantities of surface membrane. Macrophages can turn over an area equivalent to their cell surface area in 30 min; this is accompanied by the uptake of a volume equivalent to 25% of the cell's volume every hour (Steinman *et al.*, 1978). In comparison, fibroblasts pinocytose a volume equivalent to 3% of their volume each hour whereas other cells do not pinocytose at all.

Plasma membrane internalized by pinocytosis is not destroyed; most of its constituents are recycled back to the cell surface. The high pinocytotic rates of macrophages and certain amoebae could not be supported by membrane biosynthesis alone. Most pinocytosed membrane components flow back to the plasma membrane, sometimes via the Golgi or *trans* Golgi network.

Several properties of fluid-phase endocytosis have been established. Fluid-phase endocytosis is completely inhibited by incubation at 4°C. Experiments with drugs such as colchicine have indicated that microtubules play no apparent role in fluid-phase pinocytosis. However, fluid-phase pinocytosis is perturbed or abolished by various metabolic inhibitors and reagents that affect cytoplasmic pH and potassium levels. However, it has been difficult to sort out the contributions of the clathrin-dependent and - independent modes of fluid-phase pinocytosis. The following section will describe our current understanding of clathrin-dependent pinocytosis.

6.2.1.2. The LDL Receptor: A Transport Receptor for Adsorptive Pinocytosis

LDL binds to its receptor in a saturable, high-affinity fashion. In Section 2.3.6 we discussed the LDL receptor's deduced amino acid sequence. In comparing the deduced amino acid sequences of the receptor with a ligand, apolipoprotein B, a molecular mechanism of ligand-to-receptor binding has been suggested. As schematized in Figure 2.25, the LDL receptor possesses a cysteine-rich region near its NH_2-terminus. This region of the receptor contains eight homologous negatively charged repeats. A clustering of *negative* charges in each repeat is found in the sequence Asp-Cys-X-Asp-Gly-Ser-Asp-Glu. Apoprotein B-100, the protein component of LDL, possesses two regions containing several closely spaced *positive* residues with the sequence Arg-X-X-Arg-Lys-Arg-X-X-Arg/Lys (Yang *et al.*, 1986; Knott *et al.*, 1986). At least one of these regions from about amino acid 3350 to 3380 is known to bind to LDL receptors. Stable receptor binding likely results from the complementary pairing of the ligand's and receptor's charge clusters.

About 70% of the cell surface LDL receptors are localized at clathrin coated pits. As shown in Figure 4.22, LDL binds to the outer surface of coated pits. A coated pit then pinches off from a plasma membrane to form a coated vesicle (Figure 6.11). As the coated vesicle becomes acidified, LDL dissociates from its receptor. Many of the free LDL receptors contained within a coated vesicle are then recycled to the plasma membrane via a vesicular compartment. An LDL receptor may be recycled over 100 times during its 20-hr lifetime (Pathak *et al.*, 1988). After fusion with lysosomes, an endosome's contents are digested by the action of lysosomal enzymes. LDL's constituent protein and cholesteryl ester components are hydrolyzed. Cholesteryl esters yield free fatty acids and cholesterol. Cholesterol partitions out of secondary lysosomes into the cytosol and other organelles.

6.2.1.3. Molecular Cues for Adsorptive Pinocytosis

We are just beginning to unravel the complex molecular reactions that lead to adsorptive pinocytosis. In the preceding section, we described the recognition events

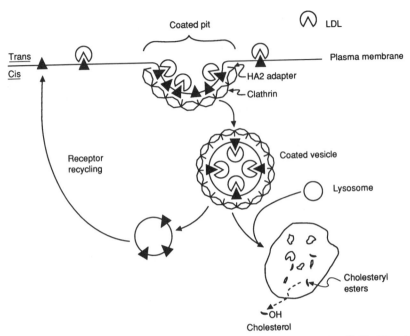

Figure 6.11. Adsorptive pinocytosis and processing of LDL. The physiological roles of LDL (⋀) and LDL receptors (▲) are illustrated. Cholesterol-carrying LDL is internalized via adsorptive pinocytosis. The LDL and its constituent cholesteryl esters are delivered to the endolysosomal compartment. The LDL receptors are recycled to the plasma membrane.

at the *trans* face and their physiological consequences. This section will describe some of the events at a membrane's *cis* face that contribute to adsorptive pinocytosis.

Studies of LDL endocytosis provided the first clue in identifying the molecular events responsible for pinocytotic function. A hypercholesterolemia patient with the J.D. mutation displayed dramatically reduced adsorptive pinocytosis of LDL. When this patient's LDL receptor cDNA was examined, a tyrosine was missing from its cytoplasmic COOH-terminal tail (Davis *et al.*, 1986). This indicates that a tyrosine residue at a membrane's *cis* face plays a central role in triggering endocytosis.

Lazaraovits and Roth (1988) constructed three variants of the influenza viruses' COOH-terminal tail by site-directed mutagenesis. Tyr residues were inserted separately at three locations along the short ten-residue stretch of amino acids at a membrane's *cis* face. All of these mutant membrane proteins underwent normal synthesis and processing. However, one of these mutant proteins was rapidly endocytosed and then recycled to the plasma membrane. Therefore, the presence and location of a tyrosine residue is a critical determinant of the endocytotic trafficking of membrane proteins.

Although an aromatic residue, especially Tyr, near a protein's cytoplasmic juxta-membrane surface is necessary for rapid internalization (Chen *et al.*, 1990), a great deal of sequence heterogeneity is observed among endocytotic receptors (e.g., LDL, transferrin, and mannose-6-phosphate receptors). Furthermore, studies with chimeric receptors indicate that these cytoplasmic regions can be exchanged between receptors without affecting endocytosis (Collawn *et al.*, 1991). Although the primary structure is not well conserved, the secondary structure appears to be a key to triggering endocytosis (Collawn *et al.*, 1990; Bansal and Gierasch, 1991). Peptide conformation studies have shown the internalization signal to be an aromatic residue within a β-turn in the juxtamembrane region. This is supported by the observation that adding a Tyr residue within a turnlike sequence at glycophorin's cytoplasmic juxtamembrane region triggers its uptake via coated pits (Ktistakis *et al.*, 1990).

The cytoplasmic COOH-terminal tyrosine and β-turn region of transport receptors interact with cytoskeletal proteins. Pearse (1988) has shown that HA-2 adaptors interact with the cytoplasmic tails of several endocytotic receptors. HA-2 is comprised of three peripheral membrane proteins, adaptins (100 kDa), AP50, and AP17 (Section 3.2.2.9). The HA-2 protein complex is sufficient to assemble coated vesicles *in vitro*. HA-2 adaptors also bind to clathrin triskelions, thus forming links between plasma membranes and clathrin assemblies.

6.2.1.4. Phagocytosis

Phagocytosis is a complex process which leads to the engulfment of large particles such as intact microorganisms. The first step during phagocytosis is the recognition of a target particle. The surfaces of a phagocyte and a target adhere tightly to one another during recognition. Specific receptors at a phagocyte's surface often mediate particle recognition/adhesion. For example, cell surface receptors for the Fc domain of antibody molecules and the RGD tripeptide (Section 7.5) mediate antibody- and complement-dependent recognition and adhesion of mammalian phagocytes. In some instances phagocytosis is receptor-independent. For example, when smoke particles enter the lung they are phagocytosed by alveolar macrophages. The physical properties of the target and macrophage surfaces likely contribute to adherence. Although nonspecific binding sites for these particles may exist, the presence of purported "nonspecific receptors" (Silverstein *et al.*, 1977) has never been borne out by their biochemical isolation and seems to be a contradiction in terms (Section 7.1.1).

The first step in phagocytosis is the receptor-dependent or -independent adherence of a target particle to a phagocytic cell's surface (e.g., Section 7.5). Phagocytic cells generally have many surface folds or ruffles (so-called redundant membrane; e.g., Figure 4.23). Surface folds and an intracellular membrane pool provide the surface area required to surround target particles during their internalization. *De novo* membrane synthesis appears to be unimportant during phagocytosis. Actin polymerizes beneath the phagocyte's plasma membrane, thereby forming pseudopods that surround a target. The importance of microfilaments in phagocytosis is supported by the fact that cytochalasin B abolishes phagocytosis whereas colchicine has no effect. The polymerization of actin requires ATP; this may account for the ability of metabolic inhibitors to block phagocytosis. When advancing pseudopods cover a target they fuse to form an intact plasma membrane and a phagosome.

One of the most fully studied examples of phagocytosis is macrophage-mediated antibody-dependent phagocytosis. Antibody molecules bind to antigens expressed at a target's surface. The Fc stem of an antibody then binds to Fc receptors at a phagocyte's surface. This perturbation of the phagocyte's surface must be communicated to the cytosol to initiate a cascade of events eventually leading to phagocytosis and target destruction. Unfortunately, just how this transmembrane signaling takes place is not known.

The first event detected during phagocytic recognition is membrane depolarization. Although membrane depolarization has been detected in both receptor-dependent and -independent phagocytosis, its physiological significance is uncertain. Fc receptors accumulate at sites of antibody-coated targets (Petty *et al.*, 1989). Since the Fc receptor (type II) has only one transmembrane domain, its ion channel activity might be due to its ability to cluster at these sites. The binding of phagocytes to antibody-coated surfaces triggers actin recruitment near a separate plasma membrane protein of the integrin supergene family (Zhou *et al.*, 1992), thus suggesting that Fc receptors communicate their ligand binding status to an integrin. This may contribute to the assembly of microfilaments about a target. Fc receptors have been shown to trigger phospholipase activity, changes in cytosolic Ca^{2+} levels, and protein kinase activity. However, no consensus viewpoint for phagocytic membrane signaling and internalization has emerged.

6.2.2. Exocytosis

In addition to the internalization of extracellular substances, cells often secrete molecules and macromolecules into their extracellular environment. This is accomplished in two ways. First, the secretory material can be biosynthesized near a plasma membrane and cotranslationally exported. Prokaryotes, for example, use this sort of macromolecular transport to secrete substances that inhibit the growth of competing microorganisms. In Section 8.2 we will discuss how polypeptides are individually translocated across membranes. Eukaryotes, especially metazoans, require the *rapid* release of many compounds. These rapid secretory events are accomplished by the bulk transport (and therefore simultaneous release) of many thousands of molecules via exocytosis. The secreted compounds are initially stored in secretory granules. When secretory granules fuse with a plasma membrane, their contents are released into the extracellular environment.

Exocytotic transport phenomena are classified as either regulated or constitutive. During regulated exocytosis substances are biosynthesized and then stored within intracellular granules. When an appropriate stimulus is received by a cell, secretory granules fuse with a plasma membrane. For example, when the appropriate surface receptors are stimulated, mast cells release histamine, neurons release acetylcholine, chromaffin cells release adrenergic ligands, and pancreatic cells release insulin. Some cellular material is continuously secreted; this is constitutive exocytosis. For example, hepatocytes and fibroblasts continuously release materials into the extracellular environment, albeit at a lower rate than regulated exocytosis. Regulated exocytosis is better understood because experimentalists can turn it on or off to study how it works. Exocytosis has three distinct steps: approach, membrane–membrane adherence, and fusion. In this section we will explore the exocytotic events as they relate to membrane transport; the role of constitutive exocytosis in membrane biosynthesis is reserved for Section 8.3.

6.2.2.1. Approach

After an appropriate exocytotic stimulus has reached the cytoplasm, the secretory granules must be moved into position near a plasma membrane. Cytoskeletal components are generally believed to be responsible for granule translocation (Vale, 1987). Microtubule-depolymerizing drugs such as colchicine inhibit most types of exocytotic phenomena. However, some exocytotic events are not inhibited, suggesting that both microtubule-dependent and -independent granule translocation systems are present.

The events participating in microtubule-dependent regulated exocytosis can be lightly sketched. Often, an appropriate signal is generated by a plasma membrane ligand–receptor interaction. In the following chapter we will discuss several classes of ligand–receptor interactions in detail. This signal is generally linked to PI turnover and calcium fluxes. The transmembrane signaling activates microtubule "motors." Individual secretory granules likely possess receptors for microtubule motors. Microtubule motors, such as kinesin, use ATP to move unidirectionally along a microtubule. This role for ATP accounts for its absolute requirement during exocytosis. Therefore, transmembrane signaling leads to motor activation, perhaps via phosphorylation reactions, which causes granules to approach plasma membranes.

6.2.2.2. Adhesion and Fusion

The adhesion and fusion of secretory vesicles with a plasma membrane are morphologically distinct processes. During adhesion the granule and plasma membranes come into intimate contact. A series of intermediate structures (Section 8.1.2.4) are formed which culminate in the two membranes joining to form one.

How do secretory granules adhere to a plasma membrane's *cis* face? There is no unambiguous answer to this question since several factors may be involved. The site of granule–membrane fusion may have specific arrangements of intramembrane particles (e.g., Figure 4.20, panel A). These intramembrane particles may represent docking proteins that potentiate adhesion. The production of DAG and release of Ca^{2+} will stimulate physical mechanisms of membrane adherence since: (1) Ca^{2+} decreases electrostatic repulsion and increases ion-correlation effects (a type of van der Waals force), (2) sites of PI turnover, which are rich in DAG, will have little hydration and thus allow the membranes to come into intimate molecular contact, and (3) the *cis* faces of the granular and plasma membranes are rich in PE and PS, two lipids with a tendency to form nonbilayer phases that lead to fusion (Section 8.1.2.4).

After adhesion and the formation of intermediate lipid structures, the system is thought to spontaneously form one continuous bilayer membrane. The first structures detected after fusion begins are small pores. These pores have been identified by patch clamp techniques (Breckenridge and Almers, 1987). Initially, an unstable pore of roughly 3 nm is formed between the granule and plasma membrane. Within a few milliseconds the small pore dilates, thus triggering the release of granule contents. As the pore dilates it becomes visible by electron microscopy as a narrow channel (50 to 100 nm in diameter) between a granule and the extracellular environment (Chandler and Heuser, 1980). Figure 6.12 shows the formation of a pore during mast cell secretion.

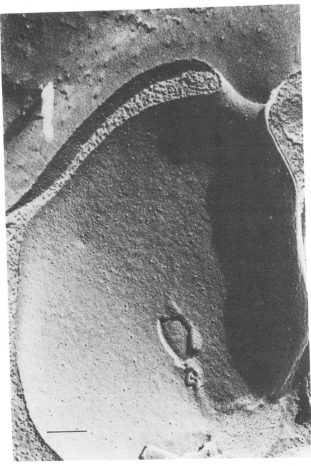

Figure 6.12. Membrane fusion during mast cell exocytosis. Mast cells were stimulated to release histamine and then quick-frozen (Section 4.2.7). An etchable pore is seen at the top. Intramembrane particles can be seen in the region of fusion. Bar = 0.2 μm. (Micrograph produced by Dr. John E. Heuser of Washington University School of Medicine, St. Louis.)

These structures can be etched thereby showing that they are filled with water. Therefore, during secretion small electrically conductive pores are formed which rapidly enlarge to release granule contents.

REFERENCES AND FURTHER READING

Molecular Transport

Bangham, A. D., *et al.* 1965. Diffusion of univalent ions across the lamellae of swollen phospholipids. *J. Mol. Biol.* **13:**238–252.

Barnett, J. E. G., *et al.* 1973a. Structural requirements for binding to a sugar-transport system of the human erythrocyte. *Biochem. J.* **131:**211–221.

Barnett, J. E. G., *et al.* 1973b. An explanation of the asymmetric binding of sugars to the human erythrocyte sugar-transport system. *Biochem. J.* **135:**539–541.

Bartel, D., *et al.* 1989. Identification by site-directed mutagenesis of Lys-558 as the covalent attachment site of dihydroDIDS in the mouse erythroid band 3 protein. *Biochim. Biophys. Acta* **985:**355–364.

Benga, G., *et al.* 1986. p-(Chloromercuri)benzenesulfonate binding by membrane proteins and the inhibition of water transport in human erythrocytes. *Biochemistry* **25:**1535–1538.

Bibi, E., and Kaback, H. R. 1992. Functional complementation of internal deletion mutants in the lactose permease of *Escherichia coli*. *Proc. Natl. Acad. Sci. USA* **89:**1524–1528.

Bieseler, B., *et al.* 1985. Topological studies of lactose permease of *Escherichia coli* by protein sequence analysis. *Ann. N.Y. Acad. Sci.* **456:**309–325.

Birnbaum, M. J., *et al.* 1986. Cloning and characterization of a cDNA encoding the rat brain glucose-transporter protein. *Proc. Natl. Acad. Sci. USA* **83:**5784–5788.

Blostein, R. 1989. Ion pumps. *Curr. Opin. Cell Biol.* **1:**746–752.

Buchel, D. E., *et al.* 1980. Sequence of the lactose permease gene. *Nature* **283:**541–545.

Carrasco, N., *et al.* 1986. Lac permease of *Escherichia coli*: Histidine-322 and glutamic acid-325 may be components of a charge-relay system. *Biochemistry* **25:**4486–4488.

Chen, C. C., *et al.* 1986. Human erythrocyte glucose transporter: Normal asymmetric orientation and function in liposomes. *Proc. Natl. Acad. Sci. USA* **83:**2652–2656.

Chin, J. J., *et al.* 1987. Structural basis of human erythrocyte glucose transporter function in proteoliposome vesicles. *Proc. Natl. Acad. Sci. USA* **84:**4113–4116.

Deuticke, B., and Haest, C. W. M. 1987. Lipid modulation of transport proteins in vertebrate cell membranes. *Annu. Rev. Physiol.* **49:**221–235.

Dornmair, K., and Jahnig, F. 1989. Internal dynamics of lactose permease. *Proc. Natl. Acad. Sci. USA* **86:**9827–9831.

Ebner, R., and Lengeler, J. W. 1988. DNA sequence of the gene *scrA* encoding the sucrose transport protein EnzymeIIScr of the phosphotransferase system from enteric bacteria: Homology of the EnzymeIIScr and EnzymeIIBgl proteins. *Mol. Microbiol.* **2:**9–17.

Ehring, R., *et al.* 1980. *In vitro* and *in vivo* products of *E. coli* lactose permease gene are identical. *Nature* **283:**537–540.

Forbush, B. 1984. Na$^+$ movement in a single turnover of the Na pump. *Proc. Natl. Acad. Sci. USA* **81:**5310–5314.

Forbush, B. 1987. Rapid release of ^{42}K or ^{86}Rb from two distinct transport sites on the Na$^+$,K$^+$-pump in the presence of P$_i$ or vanadate. *J. Biol. Chem.* **262:**11116–11127.

Foster, D. L., *et al.* 1983. Structure of the *lac* carrier protein of *Escherichia coli. J. Biol. Chem.* **258:**31–34.

Gilliland, G. L., and Quiocho, F. A. 1981. Structure of the L-arabinose-binding protein from *Escherichia coli* at 2.4 Å resolution. *J. Mol. Biol.* **146:**341–362.

Gorga, F. R., and Lienhard, G. E. 1982. Changes in the intrinsic fluorescence of the human erythrocyte monosaccharide transporter upon ligand binding. *Biochemistry* **12:**1905–1908.

Gunn, R. B., and Frohlich, O. 1979. Asymmetry in the mechanism for anion exchange in human red blood cell membranes. Evidence for reciprocating sites that react with one transported anion at a time. *J. Gen. Physiol.* **74:**351–374.

Higgins, C. F. 1989. Bacterial membranes. *Curr. Opin. Cell Biol.* **1:**701–705.

Jay, D., and Cantley, L. 1986. Structural aspects of the red cell anion exchange protein. *Annu. Rev. Biochem.* **55:**511–538.

Jennings, M. L. 1989. Structure and function of the red cell anion transport protein. *Annu. Rev. Biophys. Biophys. Chem.* **18:**397–430.

Kaback, H. R. 1986. Active transport in *Escherichia coli*: Passage to permease. *Annu. Rev. Biophys. Biophys. Chem.* **15:**279–319.

Kaback, H. R. 1987. Use of site-directed mutagenesis to study the mechanism of a membrane transport protein. *Biochemistry* **26:**2071–2076.

Kawakami, K., *et al.* 1985. Primary structure of the α-subunit of *Torpedo californica* ($Na^+ + K^+$) ATPase deduced from cDNA sequence. *Nature* **316:**733–736.

Kopito, R. R., and Lodish, H. F. 1985. Structure of the murine anion exchange protein. *J. Cell Biochem.* **29:**1–17.

Kundig, W., *et al.* 1964. Phosphate bound to histidine in a protein as an intermediate in a novel phosphotransferase system. *Proc. Natl. Acad. Sci. USA* **52:**1067–1073.

Lee, C. A., and Saier, M. H. 1983. Mannitol-specific enzyme II of the bacterial phosphotransferase system. III. The nucleotide sequence of the permease gene. *J. Biol. Chem.* **258:** 10761–10767.

Leonard, J. E., and Saier, M. H. 1983. Mannitol-specific enzyme II of the bacterial phosphotransferase system. *J. Biol. Chem.* **258:**10757–10760.

Li, J., and Tooth, P. 1987. Size and shape of the *Escherichia coli* lactose permease measured in filamentous arrays. *Biochemistry* **26:**4816–4823.

Macara, I. G., and Cantley, L. C. 1981. Interactions between transport inhibitors at the anion binding sites of the band 3 dimer. *Biochemistry* **20:**5095–5105.

Macey, R. I. 1984. Transport of water and urea in red blood cells. *Am. J. Physiol.* **246:**C195–C203.

Macknight, A. D. C., and Leaf, A. 1986. Regulation of cell volume. In: *Physiology of Membrane Disorders* (T. E. Andreoli *et al.*, eds.), Plenum Press, New York, pp. 311–328.

Matsushita, K., *et al.* 1983. Reconstitution of active transport in proteoliposomes containing cytochrome o oxidase and *lac* carrier protein purified from *Escherichia coli. Proc. Natl. Acad. Sci. USA* **80:**4889–4893.

Mieschendahl, M., *et al.* 1981. Mutations in the *lacY* gene of *Escherichia coli* define functional organization of lactose permease. *Proc. Natl. Acad. Sci. USA* **78:**7652–7656.

Mitaku, S., *et al.* 1984. Localization of the galactoside binding site in the lactose carrier of *Escherichia coli. Biochim. Biophys. Acta* **776:**247–258.

Mueckler, M., *et al.* 1985. Sequence and structure of a human glucose transporter. *Science* **229:** 941–945.

Nelson, W. J., and Veshnock, P. J. 1987. Ankyrin binding to ($Na^+ + K^+$) ATPase and implications for the organization of membrane domains in polarized cells. *Nature* **328:**533–536.

Nicholson, B., *et al.* 1987. Two homologous protein components of hepatic gap junctions. *Nature* **329:**732–734.

Ohta, T., *et al.* 1986. The active site structure of the Na$^+$/K$^+$ transporting ATPase: Location of the 5′(p-fluorosulfonyl) benzoyladenosine binding site and soluble peptides released by trypsin. *Proc. Natl. Acad. Sci. USA* **83**:2071–2075.

Price, E. M., and Lingrel, J. B. 1988. Structure–function relationships in the Na$^+$,K$^+$-ATPase α-subunit: Site-directed mutagenesis of glutamine-11 to arginine and asparagine-122 to aspartic acid generates a ouabain-resistant enzyme. *Biochemistry* **27**:8400–8407.

Puttner, I. B., *et al.* 1986. *Lac* permease of *Escherichia coli*: Histidine-205 and histidine-322 play different roles in lactose/H$^+$ symport. *Biochemistry* **25**:4483–4485.

Sachs, J. R. 1986. The order of addition of sodium and release of potassium at the inside of the sodium pump of the human red cell. *J. Physiol. (London)* **381**:149–168.

Tefft, R. E., *et al.* 1986. Reconstituted human erythrocyte sugar transporter activity is determined by bilayer lipid head groups. *Biochemistry* **25**:3709–3718.

Tsuchiya, T., *et al.* 1985. Melibiose-cation cotransport system of *Escherichia coli*. *Ann. N.Y. Acad. Sci.* **456**:326–341.

Villegas, R., and Villegas, G. M. 1981. Nerve sodium channel incorporation in sodium vesicles. *Annu. Rev. Biophys. Bioeng.* **10**:387–419.

Vogel, H., *et al.* 1985. The structure of the lactose permease derived from Raman spectroscopy and prediction methods. *EMBO J.* **4**:3625–3631.

Walmsley, A. R. 1988. The dynamics of the glucose transporter. *Trends Biochem. Sci.* **13**:226–231.

Weigel, N., *et al.* 1982. Sugar transport by the bacterial phosphotransferase system. Isolation and characterization of enzyme I from *Salmonella typhimurium*. *J. Biol. Chem.* **257**:14461–14469.

Wheeler, T. J., and Hinkle, P. C. 1985. The glucose transporter of mammalian cells. *Annu. Rev. Physiol.* **47**:503–517.

Wright, J. K., *et al.* 1975. Lactose: H$^+$ carrier of *Escherichia coli*: Kinetic mechanism, purification, and structure. *Ann. N.Y. Acad. Sci.* **456**:326–341.

Wright, J. K., *et al.* 1986. Molecular aspects of sugar: ion cotransport. *Annu. Rev. Biochem.* **55**:225–248.

Macromolecular and Bulk Transport

Bansal, A, and Gierasch, L. M. 1991. The NPXY internalization signal of the LDL receptor adopts a reverse-turn conformation. *Cell* **67**:1195–1201.

Breckenridge, L. J., and Almers, W. 1987. Currents through the fusion pore that forms during exocytosis of a secretory vesicle. *Nature* **328**:814–827.

Chandler, D. E., and Heuser, J. E. 1980. Arrest of membrane fusion events in mast cells by quick-freezing. *J. Cell Biol.* **86**:666–674.

Chen, W.-J., *et al.* 1990. NPXY, a sequence often found in cytoplasmic tails, is required for coated pit-mediated internalization of the low density lipoprotein receptor. *J. Biol. Chem.* **265**:3116–3123.

Collawn, J. F., *et al.* 1990. Transferrin receptor internalization sequence YXRF implicates a tight turn as the structural recognition motif for endocytosis. *Cell* **63**:1061–1072.

Collawn, J. F., *et al.* 1991. Transplanted LDL and mannose-6-phosphate receptor internalization signals promote high-efficiency endocytosis of the transferrin receptor. *EMBO J.* **10**:3247–3253.

Davis, C. G., *et al.* 1986. The J. D. mutation in familial hypercholesterolemia: Amino acid substitution in cytoplasmic domain impedes internalization of LDL receptors. *Cell* **45**:15–24.

Goldstein, J. L., *et al*. 1979. Coated pits, coated vesicles, and receptor-mediated endocytosis. *Nature* **279**:679–685.

Griffin, F. M., *et al*. 1976. Studies on the mechanism of phagocytosis II. The interaction of macrophages with anti-immunoglobulin IgG-coated bone marrow-derived lymphocytes. *J. Exp. Med*. **144**:788–809.

Holtzman, E. 1989. *Lysosomes*. Plenum Press, New York.

Hubbard, A. L. 1989. Endocytosis. *Curr. Opin. Cell Biol*. **1**:675–683.

Knott, T. J., *et al*. 1986. Complete protein sequence and identification of structural domains of human apolipoprotein B. *Nature* **323**:734–738.

Ktistakis, N. T., *et al*. 1990. Characteristics of the tyrosine recognition signal for internalization of transmembrane surface glycoproteins. *J. Cell Biol*. **111**:1393–1407.

Lazarovits, J. and Roth, M. 1988. A single amino acid change in the cytoplasmic domain allows the influenza virus hemagglutinin to be endocytosed through coated pits. *Cell* **53**:743–752.

Pathak, R. K., *et al*. 1988. Immunocytochemical localization of mutant low density lipoprotein receptors that fail to reach the Golgi complex. *J. Cell Biol*. **106**:1831–1841.

Pearse, B. M. F. 1988. Receptors compete for adaptors found in plasma membrane coated pits. *EMBO J*. **7**:3331–3336.

Petty, H. R., *et al*. 1989. Cell surface distribution of Fc receptors II and III on living human neutrophils before and during antibody-dependent cellular cytotoxicity. *J. Cell Physiol*. **141**:598–605.

Rodman, J. S., *et al*. 1986. The membrane composition of coated pits, microvilli, endosomes, and lysosomes is distinct in the rat kidney proximal tubule cell. *J. Cell Biol*. **102**:77–87.

Silverstein, S. C., *et al*. 1977. Endocytosis. *Annu. Rev. Biochem*. **46**:669–722.

Vale, R. D. 1987. Intracellular transport using microtubule-based motors. *Annu. Rev. Cell Biol*. **3**:347–378.

Vega, M. A., and Strominger, J. L. 1989. Constitutive endocytosis of HLA class I antigens requires a specific portion of the intracytoplasmic tail that shares structural features with other endocytosed molecules. *Proc. Natl. Acad. Sci. USA* **86**:2688–2692.

Wilson, C., *et al*. 1991. Three-dimensional structure of the LDL receptor-binding domain of human apolipoprotein E. *Science* **252**:1817–1822.

Yang, C.-Y., *et al*. 1986. Sequence, structure, receptor-binding domains and integral repeats of human apolipoprotein B-100. *Nature* **323**:738–742.

Zhou, M.-J., *et al*. 1992. Surface-bound immune complexes trigger transmembrane proximity between complement receptor type 3 and the neutrophil's cortical microfilaments. *J. Immunol*. **148**:3550–3553.

Chapter 7

Receptors and Responses

Cells communicate with other cells and with their environment. In many tissues neighboring cells communicate with each other via electrical connections through gap junctions. Cells communicate over relatively long distances using diffusible substances or ligands as messengers. When ligands bind to receptors at a cell's surface, the receptors trigger the release of second messengers that transmit a ligand's signal to intracellular biochemical pathways. In this chapter we will learn how receptors are studied, the major classes of membrane receptors, and how a ligand's signal is transduced into the cytoplasm to affect a physiological response.

7.1. RECEPTORS: AN INTRODUCTION

Receptors are specific macromolecules or oligomers that recognize and bind specific ligands. Both soluble and membrane-bound receptors are found within cells. For example, the insulin receptor is a membrane glycoprotein whereas the estrogen receptor is a soluble protein complex. This chapter will only consider membrane receptors.

7.1.1. When Is a Receptor Really a Receptor?

Several criteria must be met to identify a substance as a receptor. Most importantly, the putative receptor must generate a physiological response. A membrane site that simply sticks to a substance is called a binding site. In addition, a ligand–receptor interaction should display the following features: appropriate affinity, saturable number of sites, reversibility, stereospecificity, and temporal specificity. A ligand must have an affinity for its receptor which will allow it to bind and stimulate a response at a physiologically relevant concentration. Furthermore, membranes must possess a finite number of receptors. For small ligands such as acetylcholine and epinephrine, their receptors should display specificity for a ligand's stereochemical configuration. Although it may seem obvious, kinetic studies should show that a ligand binds prior to a receptor-mediated response.

7.1.2. Quantitating Ligand–Receptor Interactions

Although Scatchard analysis was originally developed for the study of colloids, it has provided an essential tool in the analysis of ligand–receptor interactions. In the simplest case, ligands (L) and receptors (R) interact at equilibrium according to:

$$L + R \overset{K_d}{\leftrightarrows} LR \tag{7.1}$$

where

$$K_d = \frac{[L][R]}{[LR]} \tag{7.2}$$

Substituting with $R_{tot} = [R] + [LR]$ and rearranging yields

$$\frac{[LR]}{[L]} = -\frac{1}{K_d}[LR] + \frac{R_{tot}}{K_d} \tag{7.3}$$

which is plotted in the form $y = mx + b$ (Figure 7.1). The values of K_d and R_{tot} are determined by inspection of the Scatchard plot. Specific membrane receptors are usually high affinity ($K_d < 10^{-9}$ M) and low capacity ($R_{tot} \approx 1000$ to 10,000). Nonspecific

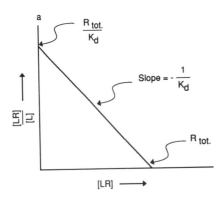

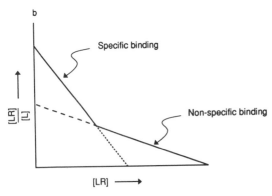

Figure 7.1. Scatchard plots. Hypothetical Scatchard plots illustrating data interpretation are shown. In both panels the ordinate shows [LR]/[L] (bound over free) and the abscissa gives [LR] (bound). Panel a is a graphical representation of Eq. (7.3); a single class of noninteracting receptors is present. Panel b shows a biphasic Scatchard plot. This sort of plot might be encountered in the presence of specific and nonspecific binding. The nonspecific binding is exaggerated for clarity.

binding is generally of low affinity and high capacity. When nonspecific binding takes place, a Scatchard plot becomes biphasic with a line of low slope extending to high LR values (Figure 7.1, panel b).

Scatchard analyses are usually performed using radiolabeled ligands. The amount of membrane-bound ligand is measured in the absence and presence of a 1000-fold excess of unlabeled (or "cold") ligand. This cold ligand should saturate the high-affinity binding sites; any residual membrane-associated radioactivity is probably due to nonspecific binding sites. When the amount of radioactivity in the presence of cold ligand is subtracted from that in its absence, only the specifically bound activity should remain. This type of analysis provides excellent results for a single population of noninteracting receptors.

In some cases Scatchard plots are nonlinear. One simple explanation for this type of behavior is the presence of multiple types of specific receptors. In addition, one receptor type could exhibit cooperative binding leading to a variable value of K_d. One method to analyze cooperative interactions of ligands with receptors is provided by Hill's equation (Hill, 1913). Consider an equilibrium where n ligands interact with a receptor:

$$n\text{L} + \text{R} \overset{K_d}{\rightleftharpoons} \text{RL}_n \tag{7.4}$$

The implicit assumption in this equation is that all n ligands bind simultaneously. This represents a highly cooperative binding reaction. In this case

$$K_d = \frac{[\text{R}][\text{L}]^n}{[\text{RL}_n]} \tag{7.5}$$

Since the fractional occupancy (Y) is

$$Y = \frac{[\text{R}]}{[\text{R}_{\text{tot}}]} \tag{7.6}$$

and $[\text{R}_{\text{tot}}] = [\text{R}] + [\text{RL}_n]$, then

$$1 - Y = \frac{[\text{RL}_n]}{\text{R}_{\text{tot}}} \tag{7.7}$$

To derive the Hill equation, insert Eqs. (7.6) and (7.7) into (7.5) and then take logs to remove the exponent. This gives:

$$\log \frac{Y}{1 - Y} = \log \frac{1}{K_d} + n \log [\text{L}] \tag{7.8}$$

When data are plotted in this form, the Hill coefficient n is obtained as the slope. Since a receptor could transiently bind 1 to $n - 1$ ligands, the Hill plot is often slightly nonlinear. The minimum number of interacting sites is determined from the maximum slope of the plot (Levitzki, 1984).

Figure 7.2 illustrates the differences between noncooperative and cooperative ligand–receptor interactions. Panel a shows the hyperbolic and sigmoidal curves obtained for noncooperative binding and positive cooperativity, respectively. Although the abscissa shows the amount of ligand bound, it could also represent a physiological

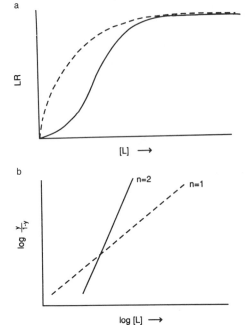

Figure 7.2. Plots of cooperative binding. Hypothetical plots of noncooperative (– – –) and positive cooperative (———) ligand binding are shown. Panel a shows the dose-dependent binding of ligands to receptors. A noncooperative response is hyperbolic whereas the cooperative response is sigmoidal. Hill plots showing hypothetical values of $\log Y/(1 - Y)$ (ordinate) and $\log L$ (abscissa) are given for noncooperative and positive cooperative binding (panel b). The increased slope of the Hill plot should be noted.

response to ligand binding such as ion conductance or muscle contraction (Section 7.2.2 discusses these two approaches using a specific example). Panel b of Figure 7.2 shows Hill plots of these two hypothetical examples. The noncooperative and cooperative cases yield Hill coefficients of 1 and 2, respectively. Most ligand–receptor interactions can be appropriately analyzed using Scatchard and Hill plots.

7.1.3. Principles of Receptor Binding

Ligands and receptors bind to one another due to the formation of intermolecular interactions including ionic bonds, hydrogen bonds, van der Waals forces, and hydrophobic forces. Steric factors and hydration pressure also contribute to ligand–receptor binding and its control. Later in this chapter we will specify some likely interactions that lead to ligand-to-receptor binding.

In a classic paper, Berg and Purcell (1977) defined some of the physical parameters affecting a ligand's interaction with its receptor. The microscopic world is very different from the macroscopic environment that we directly sense. Since molecules have almost no mass, their motion is impeded by a solvent's viscosity, not their inertia. Brownian motion, which is a manifestation of thermal fluctuations in a solvent, is significant for small particles (<0.5 μm) and a dominant factor in molecular motion. In the context of membrane receptors, our concern is: what are the elementary physical limitations concerning ligand–receptor interactions?

A few generally applicable results of Berg and Purcell (1977) will now be summarized. First, a ligand that comes within a few microns of a cell surface will almost always eventually bind to the cell. This is true because cells act as nearly perfect ligand absorbers although only a fraction of a percent of their surfaces is covered by a particular receptor. The reason cells absorb ligands so well is that a ligand in the vicinity of a cell surface will collide with the surface hundreds of times before it leaves for good. Consequently, there is an excellent chance that a ligand will eventually find its receptor. Another conclusion related to this point is that the optimum number of receptors at a cell's surface is roughly the ratio of a cell's diameter to the effective size of the receptor. For example, a 10-μm cell with a 1-nm receptor would require only 10,000 receptor copies to display optimal sensitivity to ligand. Transport receptors are most efficient when they are spread uniformly about a cell's perimeter. If they were all localized in a large cluster, the local concentration of ligand to be transported would be substantially reduced in the area of the cluster, thus decreasing its transmembrane flux into a cell. These principles are generally true. However, in certain specialized physiological niches, they may not be usefully applied. For example, at a postsynaptic membrane, speed is far more important than efficiency. To minimize response time, many receptors are clustered together and pointed toward the ligand source.

Some receptors require clustering to transmit a biological signal. For example, receptor clustering may be required to form an ion channel or to propagate a conformational change. The signal amplification provided by clustering can be as great as $\sim 10^{20}$ (DeLisi, 1981). As outlined in the previous paragraph, the measured ligand affinities of receptor clusters increase while a cell's sensitivity to ligand decreases. Therefore, the spatial distribution of receptors on a cell is expected to have important consequences on receptor behavior based on purely physical grounds.

7.1.4. Summary of Receptor Classes

The great variety of receptors found in living organisms can be classified into six major groups (Table 7.1). This receptor nomenclature system, originally introduced by Racker (1985), is useful because it organizes receptors according to their structural, mechanistic, and/or functional attributes. This chapter will discuss channel, RGC, polypeptide hormone, and adherence receptors. Each receptor's transmembrane signaling mechanism is integrated into the discussion of the receptor proper since signaling is only physiologically important in the context of a receptor. Transport receptors, such as lac permease and the LDL receptor, were discussed in the preceding chapter.

Since there are so many kinds of receptors, we have chosen only a few types of receptors to discuss in detail. These include the nicotinic acetylcholine receptor, the β-adrenergic receptor, and the insulin receptor. Three broad categories of adherence receptors are discussed: RGD, neural cell adhesion molecule-like, and carbohydrate

Table 7.1.
Racker's Receptor Nomenclature System[a]

Type	Section	Receptor examples	Ligand	Properties
Channel receptors	7.2	Nicotinic acetylcholine	Same	Multiple transmembrane domains that form channel
RGC receptors	7.3	Adrenergic	Epinephrine	Three subunits: R = recognition, G = G binding, C = catalytic
Polypeptide hormone receptors	7.4	Insulin	Same	Often one transmembrane domain; ligand binds *trans* face, enzyme or effector site *cis*
Transport receptors	6.1	Lactose	Same	Moves ligand into cell
	6.2	LDL	Same	
Adherence receptors	7.6	RGD	Fibronectin	Recognizes ligand to change cell structure, metabolism, development, etc.
Drug and toxin receptors		GM$_1$	Cholera toxin	Promotes change in cell physiology, such as disease

[a]This scheme of receptor classification was first proposed by Racker (1985).

receptors. These families of adherence receptors were chosen because they illustrate a broad variety of recognition mechanisms.

Our emphasis on a narrow selection of receptors covers all of the fundamental types of receptors. Since mechanisms of action are often shared among members of the same receptor family, breadth of training is not sacrificed for brevity.

7.2. CHANNEL RECEPTORS: THE NICOTINIC ACETYLCHOLINE RECEPTOR

Acetylcholine (ACh) is widely distributed throughout nature including both pro-karyotes and eukaryotes. Although ACh has no known function in prokaryotes, in higher eukaryotes it binds to membrane-bound cholinergic receptors. The presence of ACh excites or inhibits the activity of cells possessing cholinergic receptors. Cholinergic receptors are classified as muscarinic or nicotinic according to their pharmacologic and kinetic properties. Muscarinic receptors require several seconds to respond; their activity is evoked by muscarine, a drug found in certain poisonous mushrooms. In contrast, nicotinic cholinergic receptors respond in only a few milliseconds and are triggered by nicotine, a drug found in tobacco. In this section we will focus on the nicotinic cholinergic receptor, a model channel receptor that displays features common to many other channel receptors including the glycine receptor and γ-aminobutyric acid (GABA) receptor.

7.2.1. Structure–Function Correlations of ACh and Its Nicotinic Receptor

Due to their inherent complexity, the analysis of membrane receptors was limited until the late 1960s to pharmacological studies of their agonists, which potentiate receptor function, and antagonists, which block receptor function. The properties of "receptive sites" were inferred from the structural features of those compounds that bound to the receptor. These correlations between a ligand's physical attributes and its ability to evoke a physiological response are called structure–activity relationships.

Structure–activity relationships for ACh (**1**)

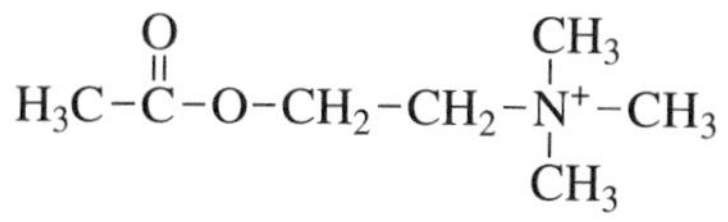

$$H_3C-\overset{\overset{\displaystyle O}{\|}}{C}-O-CH_2-CH_2-\overset{\overset{\displaystyle CH_3}{|}}{\underset{\underset{\displaystyle CH_3}{|}}{N^+}}-CH_3$$

(**1**)

were developed in the 1950s. A series of ACh analogues possessing small structural changes were synthesized. These studies found that a positive charge is required for ligand-to-receptor binding. ACh fits into a cavity at the receptor's surface. When one methyl group is removed from the tertiary amino, no change in receptor triggering is observed. However, if two methyl groups are removed or if a methyl is replaced by an ethyl, the ACh analogue does not function properly. Apparently, ACh fits into a negatively charged receptor cavity that has just enough room to accommodate two methyl groups. At the molecule's other end, the carbonyl group hydrogen bonds with the receptor while the neighboring methyl group helps hold it in place.

The ACh structure–activity relationships have been confirmed and extended by studying the three-dimensional structures of its agonists and antagonists. Chothia and Pauling (1970) determined the crystal structures of several cholinergic agonists. Due to its small size, ACh can adopt only a few conformations. However, it is difficult to predict which of these conformations participates in receptor binding. This problem can be addressed by examining the crystal structures of specific agonists such as muscarine (**2**) and nicotine (**3**).

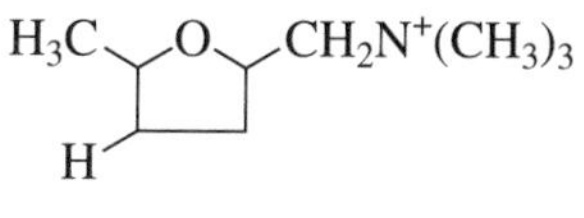

$$H_3C \diagdown \diagup O \diagdown \diagup CH_2N^+(CH_3)_3$$
$$H \diagup$$

(**2**)

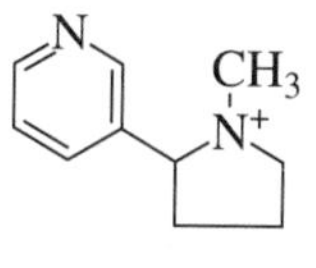

(**3**)

These molecules have ring structures that severely limit the ligands' possible conformations. The range of possible agonist conformations is further restricted by the known ionic and hydrogen bonding interactions of the receptor with ACh. By comparing the agonists' conformations with the covalent structure of ACh, the conformation of ACh on muscarinic and nicotinic receptors can be deduced. Figure 7.3 shows the crystal structures of muscarine and nicotine. The distance between the cationic group and the hydrogen bond acceptor of muscarine and nicotine are 0.44 and 0.59 nm, respectively. When the relevant groups of ACh are arranged in the same spatial pattern, their probable receptor-bound conformations become apparent (Figure 7.3).

The nicotinic ACh receptor is the most thoroughly characterized receptor. This success is due to the availability of great quantities of receptors from electric eels' electric organs and specific high-affinity toxins. A typical neuromuscular synapse has about 4×10^7 ACh receptors. In contrast, a single synapse from an electric eel has 2×10^{11} receptors. The eels most frequently studied are in the genus *Torpedo*. The purification and electrophysiological characterization of ACh receptors have been aided by using toxins isolated from venoms. The venom of the banded krait contains the protein α-bungarotoxin. This toxin, which can paralyze the snake's prey, binds tightly to nicotinic ACh receptors and blocks their activation. When linked to immobilized supports, α-bungarotoxin is used to rapidly purify nicotinic receptors.

The nicotinic ACh receptor is a pentameric transmembrane glycoprotein of 250 kDa. It is composed of two α subunits (40 kDa), and one each of the β (50 kDa), γ (60 kDa), and δ (65 kDa) chains. Cross-linking studies indicate that the two α subunits are at each side of the β subunit. All of these subunits are glycosylated with high-mannose-type carbohydrate chains. The α-subunit has an *N*-linked high-mannose chain at Asn-143. The α subunit's NH_2-terminal sequence of amino acids (6 to 85) acts as an autoantigenic domain which triggers the disease myasthenia gravis. In addition to their high-mannose chains, the γ and δ subunits also possess complex carbohydrates (Poulter *et al.*, 1989). Each α chain possesses one ACh receptor site and one α-bungarotoxin binding site. In its native membrane environment the ACh receptor exits as a dimer. The dimers in turn are clustered at postsynaptic membranes.

The ACh receptor is shaped like a cylindrical funnel (Figures 4.10 and 7.4). The subunits' long NH_2-terminal domains form the top of a funnel that extends 7 nm outward from a membrane's *trans* face. Each of the receptor's subunits is believed to be arranged

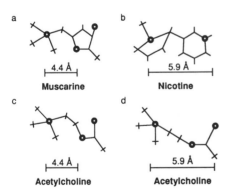

a   Muscarine   4.4 Å   b   Nicotine   5.9 Å

c   Acetylcholine   4.4 Å   d   Acetylcholine   5.9 Å

Figure 7.3. Structures of receptor agonists. The crystal structures of the cholinergic agonists muscarine and nicotine are shown. The distances between the cationic moiety and H-bond acceptor are shown. The deduced binding site conformations of acetylcholine bound to muscarinic and nicotinic receptors are given. [Redrawn from Okamato, 1980, in: *Advanced Cell Biology* (L. Schwartz and M. Azar, eds.), copyright © 1980 by Van Nostrand–Reinhold.]

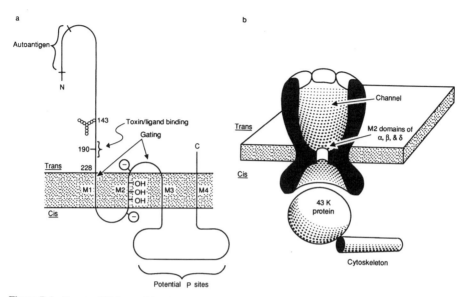

Figure 7.4. Structural biology of the nicotinic acetylcholine receptor. The macromolecular and supramolecular structures of the ACh receptor are presented. Panel a is a schema of the ACh receptor's α subunit. The transmembrane domains M1, M2, M3, and M4 are indicated. In addition to sites responsible for glycosylation (small circles), autoantigenic activity, and ligand/toxin binding, potential sites for phosphorylation and gating are also shown. Negative charges and hydroxyls that may participate in transport function are identified. Panel b shows a cross section of the pentameric receptor. The M2 domains of α, β, and γ subunits line the transmembrane region of the channel. Peripheral proteins associated with a membrane's *cis* face are illustrated.

about its central channel like staves of a barrel (Figure 7.4). The deduced amino acid sequences of the subunits have revealed several important facets of the receptor's structure. The subunits' primary structures are homologous to one another. There is now general agreement that each subunit traverses a membrane four times, as Figure 7.4 schematically shows. These transmembrane domains are called M1 through M4 in the NH_2- to COOH-terminal direction. The primary structures of the M2 domains suggest that they form amphipathic α-helices. Since this domain's sequence is highly conserved among all four subunits, the five M2 domains likely provide a moderately hydrophilic channel across a membrane's hydrophobic core. This role for the M2 domain is suggested by several quite independent experimental results. First, noncompetitive inhibitors that block the channel bind to Ser-262 of the M2 domain. Each M2 domain possesses three hydrophilic serine residues. These residues line the channel and hydrogen bond with water molecules. If these serine residues are replaced by alanine or phenylalanine using site-directed mutagenesis, the channel's ability to conduct a current is selectively decreased. Moreover, receptor binding by QX-222, a local anesthetic, is diminished when the serines are replaced (Leonard *et al.*, 1988). The M2 domains' serine residues play an important role in the receptor's channel function.

The deduced amino acid sequences of the M2 domains revealed clusters of negative charges at each end of this transmembrane domain. If these charge clusters are reduced or eliminated by site-directed mutagenesis, the modified ACh receptors display dramatically reduced conductance (Imoto *et al.*, 1988). The decrease in conduction is roughly proportional to the number of charges removed. The structural and electrophysiological data suggest that the negative charges of each M2 domain form anionic rings at both ends of the receptor's transmembrane channel. These anionic rings enhance the rate of ion movement through the channel by attracting cations into the channel's vicinity.

7.2.2. Ligand–Receptor Binding

There are two fundamentally different approaches used to quantitate ligand–receptor interactions. As we emphasized above, the amount of radiolabeled ligand bound to a membrane can be measured as a function of ligand concentration. This dose-binding curve is an *equilibrium* property of a receptor and it reflects how well a ligand *binds*. This experiment can also be conducted by measuring the response of a membrane (i.e., current flow) as a function of ligand concentration. Since an ACh receptor opens very briefly, this dose–response curve may not reflect the receptor's equilibrium properties. Dose–response curves are directly linked to *conductance* but are not necessarily closely tied to equilibrium binding properties. The K_d values determined by binding and response measurements are indistinguishable for many ligand–receptor combinations; however, this is not the case for ACh receptors.

Figure 7.5, panel a shows the dose-dependent binding of [^{3}H]-ACh to *Torpedo* membranes (Fels *et al.*, 1982). As the concentration of ACh is raised, the amount of bound ACh increases in a hyperbolic fashion. When these data are plotted according to Eq. (7.3), a nonlinear Scatchard plot (panel b) is obtained. The curve's convex shape suggests that the receptor's two ACh binding sites interact in a weak, positive cooperative fashion. The interaction between each ligand and the receptor was analyzed according to:

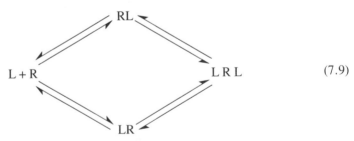

$$\text{(7.9)}$$

where RL, LR, and LRL are the two monoliganded and diliganded forms of the ACh receptor. This showed that the binding sites had distinct dissociation constants of $K_d = 25$ and 8 nM (Fels *et al.*, 1982). These studies indicate that the binding sites of the receptor's two α subunits possess distinguishable equilibrium binding properties and interact cooperatively.

The binding properties of ACh to cell membranes have also been studied using

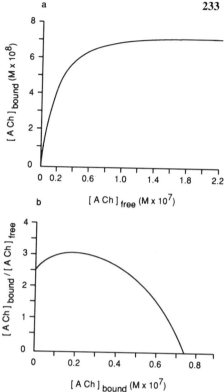

Figure 7.5. ACh binding to *Torpedo* membranes. Radiolabeled ACh binding to isolated *Torpedo* membranes was measured. Panel a shows the dose-dependent binding of ACh to membranes. The amount of bound ligand (abscissa) is shown versus ligand added (ordinate). Ligand cooperativity is not obvious in this plot. Panel b shows a Scatchard plot of the data given in panel a. The Scatchard plot is nonlinear, suggesting cooperativity among subunits. (Redrawn from Fels *et al.*, 1982, *Eur. J. Biochem.* **127**:31–38.)

functional assays. These experiments measure a cell's response, such as a change in transmembrane potential, as a function of agonist concentration. These ligand binding curves are sigmoidal in shape (Changeux and Podleski, 1968), suggesting strong cooperative interactions between subunits. When these data are analyzed, they yield K_d values roughly 1000-fold larger ($K_d = 3 \times 10^{-6}$ M) than those found in the dose-binding studies. The detailed reasons for the apparent differences in K_d and cooperativity are not certain. They are likely due to the distinction between receptor properties required for binding and triggering. For example, the equilibrium binding experiment represents a *composite* behavior of ACh receptors: its measured K_d includes contributions from at the least the resting, open, and inactivated receptor states. In contrast, the K_d measured in *response* studies represents only the affinity associated with channel opening.

Another factor contributing to these differences in apparent receptor affinity is revealed by their cooperativity. Physiological measurements have yielded a Hill coefficient of nearly two, suggesting that two ACh molecules must bind to trigger channel opening (Changeux *et al.*, 1981). The biochemical binding experiment measures how one ACh molecule binds to one of its binding sites. Since two ACh molecules must be bound to trigger opening, the channel's behavior reflects the activities of both sites. Similarly, the dissociation of either ligand is believed to cause channel closing. Therefore, the apparent dissociation constant in dose–response measurements must be greater than in binding studies.

7.2.3. The Receptor's Ligand Binding Site

Our molecular understanding of how ACh binds to its receptor has dramatically improved in the past decade. Early affinity labeling studies showed that the ACh and α-bungarotoxin binding sites were extremely close or identical to one another and that they are near a sulfhydryl group. When the sequence of the α subunit became available (Noda *et al.*, 1982), one could reasonably guess that the ACh and α-bungarotoxin binding sites were near one of the Cys residues at positions 128, 142, 192, and 193. Several lines of evidence now indicate that the binding site(s) includes Cys residues 192 and 193. The amino acid sequence from residue 187 to 196 is Trp-Val-Tyr-Tyr-Thr-Cys-Cys-Pro-Asp-Thr. When a fusion protein containing α-subunit residues 166 to 200 is expressed by *E. coli* transformants, these bacteria are able to specifically bind toxin (Gershoni, 1987). A synthetic oligopeptide including α-subunit residues 185 to 196 was synthesized by Neumann *et al.* (1986). This oligopeptide specifically bound α-bungarotoxin with high affinity. Furthermore, binding was specifically inhibited by various cholinergic ligands including ACh, consistent with the proximity of these two binding sites. The rank order potency of the cholinergic ligands to block α-bungarotoxin binding to synthetic oligopeptides paralleled their ability to block binding to intact receptors. The disulfide between residues 192 and 193 was found to be essential for binding activity. Further oligopeptide and recombinant DNA work in this area should be directed toward integrating the molecular structure–activity relationships with the newly acquired structural information. The negatively charged residue Asp-195 may form an ionic bond with the tertiary amino group of ACh. Similarly, the hydroxyl group of Thr-195 or Thr-191 may provide a hydrogen bond donor for the carbonyl group of ACh. The rather odd geometry brought about by the -Cys-Cys-Pro- sequence may contribute to a ligand binding cavity. Further studies should clarify these possibilities.

7.2.4. The Receptor's Channel Activity

ACh receptors initiate an action potential at cell membranes. However, we do not know how ACh binding causes the channel's gate to open. A Pro residue in M1 and the stretch of amino acids between M2 and M3 have been suggested to participate in channel gating. The ACh binding sites and the gate may communicate via conformational changes. The cooperative interactions among subunits indicate that at least some conformational changes are occurring. Conformational changes upon ligand binding have been observed by physical and chemical techniques. This evidence includes both changes in the receptor's intrinsic fluorescence and its reactivity with certain covalent probes (Cohen and Changeux, 1975). These conformational changes presumably transmit ligand binding information to the channel's gate.

When the ACh receptor's gate opens, ions pass through its hydrophilic pore. The most permeant species are Na^+ and K^+, although Ca^{2+} and Mg^{2+} can also cross, but to a lesser extent. This depolarizes a membrane leading to impulse propagation along an axon via the voltage-gated Na^+ and K^+ channels. Although the channel is fairly nonspecific, divalent cations larger than about 0.8 nm can get stuck near the negatively charged "rings" at the transmembrane pore's opening, thus interfering with conduction. Cationic anesthetics are also thought to interact in this area of the receptor. Patch-clamp studies have shown that under normal conditions the conductivity of the channel is about 25 pS. When reconstituted into lipid bilayers, each receptor displays all of the pharmacologic, kinetic, and conductance properties found in living cell membranes. This indicates that these components are sufficient to form functionally active receptors. Although ACh binds to ACh receptors reconstituted into any lipid environment, channel activity is only retained in the presence of cholesterol and anionic lipids (Fong and McNamee, 1986). This suggests that an appropriate fluidity and head group composition are required to maintain function. The large internal membrane potentials generated by anionic lipids may influence the orientation and action of charged gates and rings within the receptor.

7.2.5. A 43-kDa Protein Mediates ACh Receptor Clustering

ACh receptors aggregate into clusters at synapses. Although several proteins co-cluster with ACh receptors, a 43-kDa myristoylated protein appears to be a primary determinant of receptor clustering. Early experiments showed that this 43-kDa peripheral membrane protein co-purified with receptors. Chemical cross-linking studies demonstrated that the 43-kDa protein was closely associated with the ACh receptor's β subunit. When synaptic membranes were treated with a pH 11 buffer, which dissociates the 43-kDa protein, a dramatic increase in receptor mobility was observed. The apparent role of the 43-kDa protein in controlling receptor distribution was directly tested using transfection experiments. The ACh receptor's α, β, γ, and δ subunits were transfected into fibroblasts where they demonstrated ligand-gated channel activity (Phillips *et al.*, 1991a,b). These ACh receptors were uniformly distributed at fibroblast surfaces. However, when the 43-kDa protein was expressed by this transfected cell line, ACh receptors formed clusters. These results show that the 43-kDa protein is a key participant in receptor clustering. Interestingly, when the 43-kDa protein is transfected into cells in the *absence* of ACh receptors, it retains its ability to form clusters at the *cis* face of plasma membranes. An analysis of several mutant 43-kDa proteins showed that the NH_2- and COOH-termini were important in membrane attachment. The central region contains leucine zipper consensus sequences, which coil two α-helices together, that may form links between two copies of the 43-kDa protein or between the 43-kDa and other proteins. When the central region of the 43-kDa protein was deleted, it associated with membranes but was unable to interact with receptors, thus suggesting that this region is important in receptor binding. In view of their pH-dependent association, it seems likely that basic amino acids participate in the 43-kDa protein-to-β-subunit interaction.

7.2.6. Regulatory Phosphorylation of Nicotinic Receptors

As previously mentioned (Section 2.3.1), membrane proteins can be covalently modified in several ways including glycosylation, methylation, acetylation, fatty acylation, and phosphorylation. The nicotinic ACh receptor is phosphorylated on seryl and tyrosyl residues *in vivo*. Phosphorylation sites are located on a large intracellular domain between the M3 and M4 helices (Figure 7.4, panel a). The receptor's functional activity is affected by phosphorylation of these sites.

At least three different protein kinases participate in ACh receptor phosphorylation. A cAMP-dependent protein kinase (PKA) phosphorylates seryl residues in the γ and δ subunits. A calcium- and phospholipid-dependent protein kinase (PKC) phosphorylates seryl residues of the α and δ chains. A third kinase, an endogenous protein tyrosine kinase, phosphorylates tyrosyl residues of the β, γ, and δ chains. Phosphorylation of either seryl or tyrosyl residues enhances the rate of receptor desensitization or inactivation in the presence of ligand (Huganir *et al.*, 1986; Hopfield *et al.*, 1988; Hemmings *et al.*, 1989). This suggests that at least three protein kinase signaling pathways regulate a cell's sensitivity to ACh. Recently, Miles *et al.* (1989) have identified an extracellular peptide that increases nicotinic receptor phosphorylation via PKA. Although the role(s) of receptor phosphorylation in cell function is unknown, regulatory desensitization may contribute to short- and/or long-term cell memory.

7.3. RGC RECEPTORS: THE β-ADRENERGIC RECEPTORS

RGC membrane receptors are a broad class of structurally and mechanistically similar proteins. They all contain three common elements: a regulatory component (R), a guanine nucleotide binding component (G), and a catalytic component (C). Included among these receptors are the β-adrenergic receptors, the 5-hydroxytryptamine receptor, the muscarinic ACh receptor, and rhodopsin. In this section we will discuss β-adrenergic receptors, a prototypic RGC receptor.

7.3.1. Physiological Setting of Adrenergic Receptors

Adrenergic receptors are a family of membrane proteins that recognize and trigger cellular responses to epinephrine and norepinephrine (Figure 7.6). Depending on their anatomical location, epinephrine and norepinephrine act as neurotransmitters like ACh or as hormones like insulin. Epinephrine and norepinephrine are also known as adrenaline and noradrenaline, respectively. They are two members of a class of compounds called catecholamines. Catecholamines contain catechol (**4**) and a positively charged amine.

(**4**)

During stressful conditions epinephrine and norepinephrine, which are stored in the adrenal medula's chromaffin cells, are released into the bloodstream. These hor-

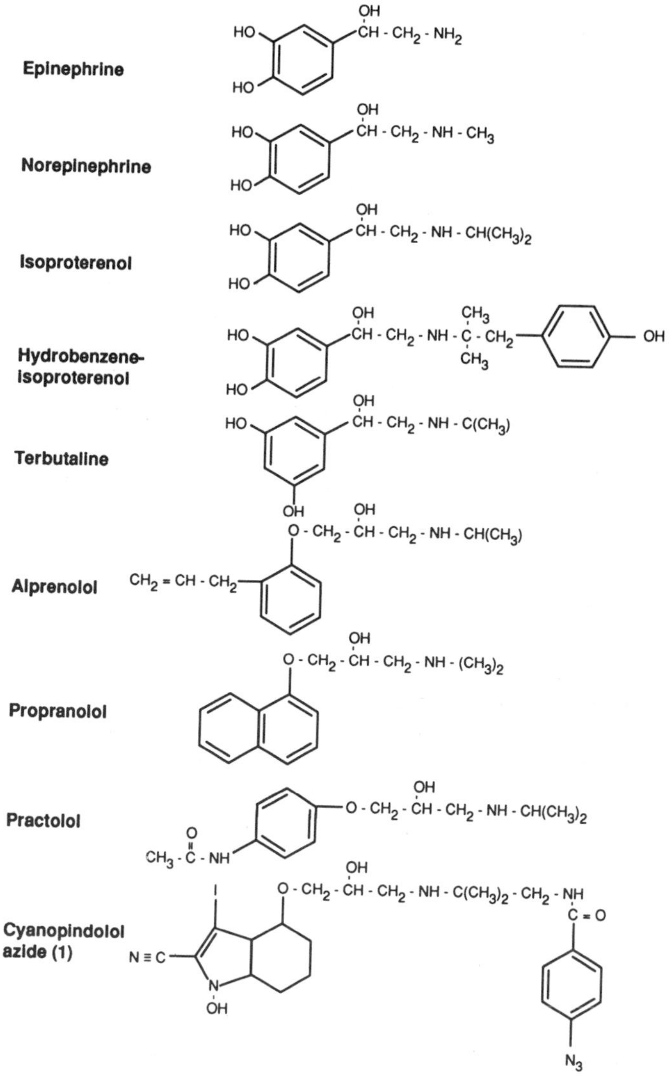

Figure 7.6. Adrenergic agonists and antagonists. Several adrenergic agonists and antagonists are shown.

mones interact with the several types of adrenergic receptors found in tissues. These several receptors in their various tissue environments generate a coordinated response to stress. β-adrenergic receptors of liver and adipose tissues trigger glycogenolysis and triglyceride hydrolysis, respectively. Similar receptors generally cause smooth muscles to relax and the heart to beat faster. Simultaneously, α-adrenergic receptors in blood vessels diminish circulation to peripheral organs such as the intestine and kidney and stimulate the release of fatty acids from arterioles. The β-adrenergic receptor of immune cells causes them to become refractory to stimuli. Non-immediately essential functions are shut down while oxygen and energy sources are rerouted to the large muscle groups needed in a "fight or flight" response.

7.3.2. Structure–Activity Relationships

The membrane pharmacology of β_1-and β_2-adrenergic receptors has been well characterized. Over 75 agonists and antagonists have been reported. This panel of reagents has: (1) identified adrenergic receptor subtypes, (2) characterized the ligand structures necessary for binding and triggering, and (3) provided clinically useful drugs.

7.3.2.1. Structure–Activity Relationships Define Adrenergic Receptor Subtypes

Classically, adrenergic receptors have been classified as α_1, α_2, β_1, and β_2 receptors according to their pharmacological properties. The α- and β-adrenergic receptors were originally operationally defined by the potency of agonists and antagonists (Ahlquist, 1967). Figure 7.6 shows the structures of several adrenergic-active compounds. α-Adrenergic receptors display the rank order potency norepinephrine > epinephrine > phenylephrine $\gg$ isoproterenol. In contrast, β receptors show a rank order potency of isoproterenol > epinephrine > norepinephrine > phenylephrine. The development of specific antagonists for the α (phentolamine and phenoxybenzamine) and β (dichlorisoproterenol and propranolol) receptors has aided in their identification and characterization. The α-adrenergic receptors are categorized as α_1 and α_2 based on their physiological properties. The α_1 receptors stimulate vascular smooth muscle contraction whereas the α_2 receptors are inhibitory (Berthelsen and Pettinger, 1977). The β_1 and β_2 receptors have distinct pharmacological properties. β_1 receptors of cardiac and adipose tissue exhibit the rank order potency isoproterenol > epinephrine = norepinephrine. In contrast, β_2 receptors found in smooth muscle tissues (bronchial passages, blood vessels, and uteri) show a potency profile of isoproterenol > epinephrine > norepinephrine. Selective β_1 and β_2 antagonists have been developed; these include the β_1 blocker practolol and the β_2 antagonist butoxamine.

Although it has been known for many years that the adrenergic receptors are pharmacologically and physiologically distinct, the molecular reasons for these differences were not known since there were no detailed structural data. However, the genes encoding these functionally distinct receptors have now been cloned (Yarden et al., 1986; Cotecchia et al., 1988; Kobilka et al., 1987a,b, 1988). The deduced amino acid sequences show that these different adrenergic receptor activities correspond to distinct gene products; variations in amino acid sequences may account for pharmacological differences. Recently, hybridization techniques have been used to sift through human genomic DNA libraries to search for proteins homologous to the adrenergic receptors. This approach was used to discover the β_3-adrenergic receptor (Emorine et al., 1987). Any remaining members of the adrenergic receptor family could be found using molecular biological techniques.

7.3.2.2. Structure–Activity Relationships Distinguish Binding from Triggering and Illuminate the Receptor's Ligand Binding Site

The wide range of adrenergic agonists and antagonists have been tested for their abilities to bind β receptors and evoke physiological responses. By comparing their molecular structures, K_d values, and physiological activities, several rules emerged which govern the behavior of β-adrenergic ligands (e.g., Mukherjee *et al.*, 1975a).

As Figure 7.6 shows, these ligands possess a phenyl ring and an ethanolamine side chain. The phenyl ring determines functional activity whereas the side chain is responsible for affinity. It should be stressed that these two ligand activities are *independent* events. High affinities do not ensure *any* level of physiological triggering; indeed, many competitive antagonists bind with high affinity to the ligand's binding site but elicit no response.

The catechol ring is essential for physiological activity. Any change or deletion at the *meta* or *para* hydroxyl group dramatically reduces or eliminates triggering. Some antagonists possessing variations in the phenyl ring's structure are shown in Figure 7.6. These structural alterations are not accompanied by reduced affinities. This also suggests that the *meta* and *para* hydroxyls do something at the receptor's binding site to begin transmembrane signaling.

The ethanolamine chain determines binding affinity. A ligand's affinity is dramatically increased by a ($-$) stereochemistry of the β-carbon's hydroxyl group and a large apolar moiety on the amino nitrogen. The positive charge is necessary for both agonist and antagonist binding. The apolar groups of these compounds may interact with hydrophobic protein or lipid domains. We shall return to the properties of the β-adrenergic receptor's ligand binding site after we explore the receptor's identification and physical properties.

7.3.3. Antagonist Binding to β-Adrenergic Receptors

Figure 7.7 shows the binding of [³H]alprenolol, a competitive *antagonist*, to the β_2-adrenergic receptor of erythrocytes (Mukherjee *et al.*, 1975b). The data yielded a K_d of 5 nM and 1500 sites per cell. As expected for a ligand–receptor interaction, binding demonstrated saturability and stereospecificity; in addition, "cold" ligand displaced the label. The physiological relevance of this interaction is supported by the finding that one-half maximal binding leads to one-half maximal inhibition of agonist-induced cAMP production. The apparent dissociation constants of physiological ligands were first measured by studying their ability to compete with antagonists (Mukherjee *et al.*, 1975a,b). Epinephrine and norepinephrine have K_d's of 20 and 250 μM, respectively, when [³H]alprenolol is used as antagonist. One-half maximal stimulation of cAMP production in these cells by epinephrine and norepinephrine occurred at 15 and 150 μM, respectively. In contrast to ACh receptors, the biochemical and physiological measurements of β-adrenergic K_d's are in good agreement.

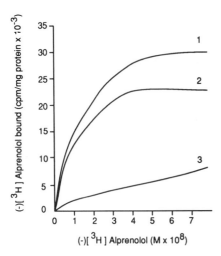

Figure 7.7. Alprenolol binding to erythrocyte membranes. The dose-dependent binding of (−) [³H]alprenolol to frog erythrocyte membranes is shown. The ordinate and abscissa list the amount of alprenolol bound and the amount of alprenolol added, respectively. The three traces show: (1) the total binding, (2) the specific binding, and (3) the nonspecific binding. The total and nonspecific binding were measured in the absence and presence of a large excess of propranolol. The specific binding curve is obtained by subtracting nonspecific from total binding. (Redrawn from Murkherjee *et al.*, 1975, *Proc. Natl. Acad. Sci. USA* **72**:1945.)

7.3.4. Structural Identification of β-Adrenergic Receptors

There are two approaches used to identify the chemical entities, presumably receptors, associated with ligand binding. These are affinity chromatography and photoaffinity labeling, both of which exploit a receptor's ligand binding properties. In the former approach solubilization precedes affinity selection whereas affinity labeling precedes solubilization during the latter.

Alprenolol is a high-affinity β-receptor antagonist. When covalently linked to a solid support such as Sepharose,* alprenolol provides a convenient and high-affinity tool to purify β receptors from cell homogenates. The receptors are isolated from membranes by detergent solubilization followed by affinity chromatography and elution with a β-receptor agonist. Purified receptors bind ligands according to their rank order potency in intact cells or tissues. The apparent molecular mass depends on the tissue of origin, but is usually about 58 kDa for frog erythrocytes (Lefkowitz *et al.*, 1983).

A second approach to identify a membrane component responsible for β-adrenergic binding is photoaffinity labeling. In this approach a photoactivatable moiety is included within a ligand. Figure 7.6 shows the covalent structure of [¹²⁵I]cyanopindolol azide, a representative photoaffinity labeling reagent. The nonphotolyzed probe binds with high affinity to β-adrenergic receptors. The radioactive iodine allows it to be detected by gamma counting or autoradiography. This reagent competes with β-adrenergic ligands according to their rank order potency and requires correct stereochemistry. When the aryl azide is photoactivated, it forms a highly reactive nitrene that inserts into the receptor's binding site. The intact membranes are then solubilized prior to analysis. After SDS-PAGE of photoaffinity-labeled turkey erythrocytes, a single band of 58 kDa corresponding to the β₁ receptor was found. Although photoaffinity labeling and affinity

---

*Sepharose is the registered trademark of Pharmacia, Inc.

chromatography are different techniques, they provide consistent and complementary findings showing that a 58-kDa membrane protein binds to β-adrenergic ligands in a saturable, stereospecific, high-affinity, and pharmacologically appropriate fashion.

These affinity-based methods have satisfied all of the criteria necessary to define a receptor (Section 7.1) except the ability to trigger a response. A true receptor must communicate its binding response into a physiological event; how can we *know* that this is occurring via *this* protein in complex cells? This distinction was not a problem for ACh receptors since the opening of the receptor's channel *was* the physiological event that led to a nerve impulse. RGC receptors must communicate with other cellular proteins to trigger an effector pathway (see below) that ultimately leads to a physiological response. This issue can be addressed by reconstitution of a receptor's function in deficient cells (cellular reconstitution) or in lipid vesicles.

To accomplish a cellular reconstitution, the β-adrenergic receptor was first purified to homogeneity by affinity chromatography. It was then reconstituted into lipid vesicles (Cerione *et al.*, 1983a). These vesicles were mixed with *Xenopus laevis* erythrocytes, which are deficient in β receptors. Polyethylene glycol was then added to promote fusion of β-receptor-containing vesicles with β-receptor-negative cells (Section 8.1.2.1). These hybrids, but not controls, produced cAMP in response to agonists according to their original rank order potency. These results indicate that β-adrenergic receptors can be coupled to the cAMP-generating machinery of foreign cells. Furthermore, both a ligand binding site and signaling unit must be present in the isolated receptor.

Recently, it has become possible to reconstitute the β receptor's transmembrane signaling pathway in lipid vesicles (Section 7.3.6.3). This reconstitution requires the presence of three proteins: the receptor, a guanine nucleotide binding protein, and adenylyl cyclase. We will defer the discussion of this complete reconstitution until we have considered these two additional proteins.

7.3.5. β-Adrenergic Receptor Structure and Identification of Its Functional Domains

The apparent molecular mass of affinity-labeled or -purified β-adrenergic receptors vary with the species and tissue of origin. Reported values extend from about 45 kDa to 64 kDa; much of this heterogeneity may be due to differences in glycosylation. The affinity-purified receptors were cleaved and partial amino acid sequences were then obtained. This allowed the cloning and sequencing of the genes encoding the β_1- and β_2-adrenergic receptors. The receptors' deduced amino acid sequences showed the presence of seven transmembrane domains (Figure 7.8). The presence of one or more transmembrane domains was expected because its solubility properties indicated that it was an integral membrane protein. The β_1, β_2, and β_3 receptors are homologous to one another and to the α_2 receptor, rhodopsin, the muscarinic receptor, and yeast pheromone receptors.

The structure of the β_2 receptor and its several functional regions are illustrated in Figure 7.8. The NH_2-terminal domain at the *trans* membrane face of the β_2 receptor has

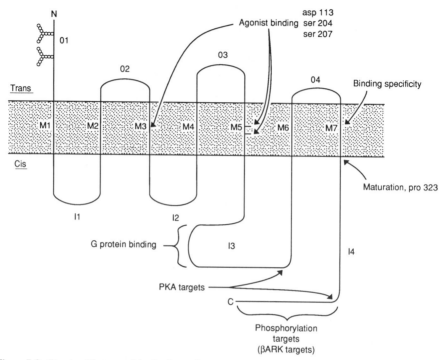

Figure 7.8. Structural features of the β-adrenergic receptor. Several structural properties of the β-adrenergic receptor are illustrated. The seven transmembrane (M1–M7), four outer (O1–O4), and four inner (I1–I4) domains are indicated. Sites participating in glycosylation (small circles), G protein binding, and phosphorylation are shown.

two potential sites for *N*-linked glycosylation. Proteolytic digestion of the receptor in intact membranes does not affect ligand binding, thus indicating that extramembrane sequences play no role in ligand binding. Similar results were obtained when intracellular loops were removed in deletion mutants. In contrast to other classes of receptors, the ligand binding domain of β receptors and other RGC receptors is buried within the polypeptides' membrane structures. This result focused attention on the receptors' intramembrane domains. In particular, investigations examined the potential roles of polar residues within the membrane and those residues conserved within the sequences of all RGC receptors. The receptors' M3 helices were found to govern high-affinity binding. The negative charge of Asp-113 interacts with a ligand's positively charged amine. If Asp-113 is replaced by Asn, the receptor loses ligand binding ability (Strader *et al.*, 1987b, 1988). Ligand binding activity is dramatically reduced by replacing Asp-113 by Glu; this change retains the negative charge but moves it the distance of one methylene group farther from the M3 helix. Apparently, the distance from the receptor to the amino nitrogen is far more important than the presence of bulky groups extending from the nitrogen.

Figure 7.9 shows a schematic drawing of a β-adrenergic binding site as viewed

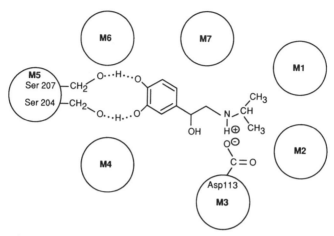

Figure 7.9. Postulated features of the β-adrenergic receptor's ligand binding site. The transmembrane portion of the receptor is shown as viewed perpendicular to the *trans* face. Potential sites of ligand–receptor interaction are indicated. (Redrawn from Strader *et al.*, 1989, *J. Biol. Chem.* **264:**13572, with permission from the American Society for Biochemistry and Molecular Biology.)

from outside a cell (Strader *et al.*, 1988, 1989). Serine residues 204 and 207 of M5 interact with the *meta* and *para* hydroxyls of the ligand, respectively. When the hydroxyls of either Ser-204 or -207 were replaced by Ala using site-directed muta-genesis, stimulation of adenylyl cyclase was dramatically reduced (~10,000-fold) while the affinity was only somewhat reduced (~10-fold). By comparing the selectively altered β-adrenergic receptors and ligands with various phenyl rings, the relationships between these particular hydroxyl groups were established (Strader *et al.*, 1989). These inter-actions are necessary to generate physiological signals and are absent when competitive antagonists, which have altered phenyl rings, bind.

In addition to identifying its receptor, photoaffinity labels are also used to identify the portions of receptors near a label's binding site. After photoaffinity labeling, membranes are solubilized and treated with proteases. The peptide fragment containing the photoaffinity label is compared with a proteolytic map of the receptor. By using this approach, the label was found to become attached to the M2 helix. This clearly shows that M2 participates in forming the β-adrenergic binding site. This may seem odd because we have already mentioned that the helices M3 and M5 interact with ligands. However, this observation makes sense when one considers the structures of β-adren-ergic photoaffinity labels (e.g., Figure 7.6) and binding sites (Figure 7.9). The labels' aryl azide moieties are attached via the amino nitrogen of the ethanolamine chain, away from the phenyl ring. An inspection of Figure 7.9 shows that the aryl azide group should be near M2.

Photoaffinity labels also become covalently attached to neighboring glycolipids. These are found near the bottom of SDS-PAGE gels after labeling and solubilization. This suggests that glycolipids are also near the ligand's binding site. This is consistent with the binding site's intramembrane location. This may neatly explain why ligands with large hydrophobic substituents extending from the amino nitrogen bind more tightly (e.g., K_d = 30 pM for cyanopindolol) than physiological ligands such as epinephrine.

Although the above site-directed mutagenesis experiments explain high-affinity binding and triggering, they fail to account for the pharmacological differences between receptor subtypes. For example, the β_2 receptor's Asp-113, Ser-203, and Ser-207 are conserved in the α_2 receptor. The β_2 receptor's Ser-204 is replaced by a Cys residue, another hydrogen bond donor. To address the question of why β_2 receptors are β_2 and not α_2 in specificity, Kobilka et al. (1988) used gene fusion experiments to combine matching portions of the α_2 and β_2 receptor genes to create chimeric receptors with amino acid sequences derived from each subtype. These genetically engineered chimeric receptors were expressed in cells and then tested for their rank order potencies to adrenergic agonists or antagonists. These studies indicate that the M7 helix was largely responsible for the pharmacological differences between the α_2 and β_2 receptors. However, a partial gene product containing M6, M7, and I4 could not bind ligands, consistent with a role of the other domains in ligand binding. Helix M7 contributes to the ligand binding site or is critically important in determining its conformation. As expected, gene fusion experiments have supported the conclusion that I3 confers α_2 inhibitory or β_2 stimulatory activity upon adenylyl cyclase. Site-directed mutagenesis experiments (Strader et al., 1987b) have shown the Pro-323, at the M7 to I4 interface, is required for proper β-receptor folding and/or sorting during biosynthesis.

After ligand binding, the M5 helix participates in transmitting a signal into the cytoplasm. The first step in this chain of events is the interaction of the β receptor with a G protein (see below). Deletion analysis experiments have localized the G_s binding domain to amino acids 239 to 272 of the I3 region (Figure 7.8). Mutants lacking this region were unable to stimulate adenylyl cyclase and had unusually high affinities for isoproterenol. The M5 helix, which is most closely associated with physiological triggering, immediately precedes the intracellular domain I3. This provides reasonably close proximity between the ligand activating site and the effector site. Presumably agonist binding transmits a conformational change to I3 via M5. The COOH-terminal domain of G_s is bound to the β receptor's I3 domain. This allows the G_s subunit to change conformation leading to adenylyl cyclase stimulation.

7.3.6. The β-Adrenergic Receptors' Transmembrane Signaling: Roles of G Proteins and Adenylyl Cyclase

Receptors must transduce ligand binding (the "first" messenger) into a physiological response. The signal is carried into the cytosol by "second" messengers. The second messenger for β_1 and β_2 receptors is the generation of cAMP. In comparison, α_1-adrenergic receptors stimulate Ca^{2+} fluxes and α_2 receptors inhibit cAMP production. We will now discuss the mechanism responsible for cAMP accumulation and its metabolic consequences.

7.3.6.1. G_s Proteins Stimulate Adenylyl Cyclase

G proteins are a family of homologous membrane-associated proteins that couple many types of membrane receptors to various second messenger systems (Gilman, 1987; Casperson and Bourne, 1987; Simon *et al.*, 1991). G proteins regulate second messenger production by binding guanine nucleotides (GDP and GTP). β receptors are coupled to adenylyl cyclase via a stimulatory G protein, G_s. G_s is composed of three protein subunits: α (42 kDa), β (35 kDa), and γ (10 kDa). The β and γ subunits are hydrophobic integral membrane proteins that anchor the α subunit, a peripheral protein, to a plasma membrane's *cis* face. In addition to binding the βγ heterodimer, the $α_s$ subunit binds either to the β-adrenergic receptor or the adenylyl cyclase, depending on which guanine nucleotide is bound to its active site.

The deduced amino acid sequence of the $α_s$ subunit, in conjunction with biochemical studies, has assigned functional roles to certain sequences within the subunit. There are four putative sites for GTP binding and hydrolysis within the $α_s$ subunit. Adenylyl cyclase is believed to bind between residues 61 and 132. Cholera toxin's modification site is at the COOH-terminal side of this domain. The $α_s$ subunit's βγ binding domain is thought to reside between residues 153 and 204. The protein's COOH-terminal sequence binds to β receptors. A map of these functional domains follows:

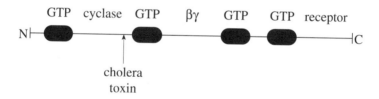

Further functional and molecular biological studies should lead to a detailed picture of the mechanism of action of G_s.

Under nonstimulatory conditions the G_s complex is associated with the *cis* face of a plasma membrane and a β-adrenergic receptor (Figure 7.10). When ligand binding changes the conformation of the β receptor's I3 domain, the neighboring $G_sα$ subunit's conformation is also altered. This allows GTP to replace GDP within the α subunit's nucleotide binding pocket. The $G_sα$·GTP and βγ complexes dissociate from β-adrenergic receptors. The $G_sα$·GTP component then binds to adenylyl cyclase and thus stimulates the production of cAMP. Although some models propose that $G_sα$·GTP dissociates from membranes, recent studies suggest that this may not be responsible for cAMP production (Levitzki, 1988). Therefore, Figure 7.10 shows the simplest scheme that retains the $α_s$·GTP interaction with βγ. It is possible that the dissociated $α_s$·GTP is itself a second messenger in communication with cytoplasmic membranes.

The $G_sα$·GTP and/or $G_sα$·GTP–βγ complex binds to adenylyl cyclase and thus activates the production of cAMP by consumption of ATP. In addition to its guanine nucleotide binding properties, the α subunit possesses an intrinsic GTPase activity. After about 15 sec the α subunit's GTPase activity cleaves the GTP into GDP and P_i. The α subunit reverts to its former GDP-bound conformation which is unable to

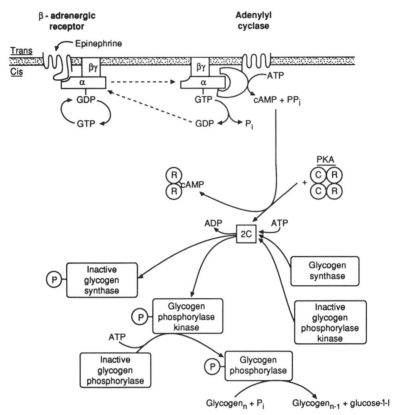

Figure 7.10. A model of β-adrenergic receptor-mediated events during glycogenolysis. The membrane and cytosolic events accompanying β-adrenergic receptor-mediated triggering of glycogenolysis are shown. The terms are defined in the text.

stimulate adenylyl cyclase. This conformation rebinds to β receptors. Therefore, α subunits rapidly shuttle between the receptor and the catalytic enzyme adenylyl cyclase.

When the nonhydrolyzable GTP analogue GMPPCP (**5**)

$$\text{guanine-ribose-}\overset{\overset{\displaystyle O}{\|}}{\underset{\underset{\displaystyle O_-}{|}}{P}}-O-\overset{\overset{\displaystyle O}{\|}}{\underset{\underset{\displaystyle O_-}{|}}{P}}-CH_2-\overset{\overset{\displaystyle O}{\|}}{\underset{\underset{\displaystyle O_-}{|}}{P}}-O^-$$

(**5**)

is added to cell homogenates with agonist, heightened levels of cAMP production occur over a long period of time. Since this analogue cannot be cleaved by the α subunit's GTPase activity, the adenylyl cyclase is constantly stimulated. This supports the role of GTP in triggering cAMP production.

The importance of G_s in homeostasis is illustrated by disease states involving G_s.

Pseudohypoparathyroidism patients are unable to respond appropriately to endogenous or exogenous parathyroid hormone (Spiegel *et al.*, 1985). Many of these patients have been found to have a systemic defect in G_s. The disease is inherited in an autosomal dominant fashion and has been associated with the α_s subunit of G_s. This form of pseudohypoparathyroidism is particularly severe because many tissues are involved and numerous hormones rely on the same G_s pool for their signaling. Consequently, these patients have blunted responses to many hormones resulting in mental retardation, metabolic insufficiencies, ionic imbalances in urine, inability to grow, and failure to sexually mature. Acquired defects in G_s proteins occur during certain infections. For example, cholera toxin interacts with G_s to deregulate intestinal epithelial cell trans-membrane signaling causing electrolyte and water loss and thus leading to diarrhea. Mutant forms of G proteins can deregulate cell growth and thus cause cancer (Section 9.2.5).

7.3.6.2. G Proteins Regulate the β-Adrenergic Receptor's Affinity for Agonists

In Section 7.3.3 we discussed the β receptor's ability to bind *antagonists*. The binding properties of *agonists* were purposely avoided because their affinity for the receptor is determined by both the receptor *and* G_s. In the preceding paragraphs we stressed how events at the ligand binding site are communicated to a cell's cytosol. Similarly, information regarding events in the cytosol is sent to the ligand binding pocket. This second avenue of communication is well illustrated by Figure 7.11. In this experiment the binding of the agonist [³H]hydroxybenzylisoproterenol (HBI; see Figure 7.6) to frog erythrocyte β receptors was measured (Williams and Lefkowitz, 1977). After addition of HBI, the amount of membrane-bound HBI increases during the first 8 min and then stabilizes at a maximal value. At 12 min unlabeled isoproterenol or a nonhydrolyzable GTP analogue was added to separate samples. As expected, iso-

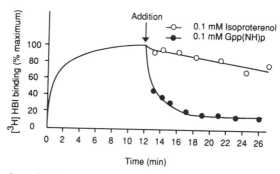

Figure 7.11. Kinetics of agonist binding and dissociation. The β-adrenergic agonist [³H]hydroxybenzyl-isoproterenol rapidly binds to frog erythrocyte membranes. At 12 min isoproterenol (●) or a nonhydrolyzable GTP analogue (○) is added. Isoproterenol competes for the ligand binding site, thus dissociating radiolabel. The GTP analogue causes a very rapid loss of label. (Redrawn from Williams and Lefkowitz, 1977, *J. Biol. Chem.* **252:**7207, with permission from the American Society for Biochemistry and Molecular Biology.)

proterenol competes with HBI for ligand binding sites, causing the radiolabel's slow dissociation. Although GTP does not interact with the ligand's binding site, it causes a dramatic increase in the agonist's rate of dissociation from β receptors. In contrast, GTP and its analogues have little or no effect on the binding of antagonists with β receptors. These data show that β receptors possess two different affinities for agonists but only one for antagonists.

The β receptor's affinity for agonists is regulated by the formation of a ligand–receptor–G_s complex (DeLean *et al.*, 1980). A cartoon illustrating this agonist affinity regulatory system is shown in Figure 7.12. When free receptors are studied they have a low affinity for agonist. This low affinity is also observed when a high concentration of GTP is present. The low-affinity state makes sense because it would be useless to bind ligand in the absence of $G_s \cdot$GDP. When $G_s \cdot$GDP is attached to a receptor's *cis* face, it binds agonist with high affinity. An analysis of HBI binding to membranes has shown that the high- and low-affinity states have K_d's of 0.6 and 45 nM, respectively (DeLean *et al.*, 1980). When agonist and $G_s \cdot$GDP are bound to a β receptor, GTP is exchanged for GDP at the α_s's nucleotide binding site. As discussed above, the G_s complex then dissociates from a receptor leading to cAMP production. The binary agonist–receptor complex then binds another $G_s \cdot$GDP complex or its ligand dissociates.

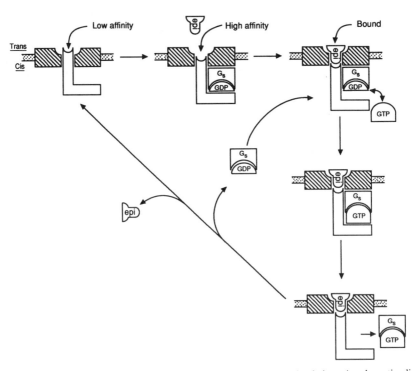

Figure 7.12. An illustration of events occurring during β-receptor stimulation. A schematic diagram illustrating receptor binding and $G_s \cdot$GTP activation is shown.

7.3.6.3. Adenylyl Cyclase

The $\alpha_s \cdot$GTP complex interacts with the enzyme adenylyl cyclase to stimulate the formation of cAMP (**6**)

(**6**)

from ATP. Adenylyl cyclase is a 120-kDa integral membrane protein. Its activity is triggered by $\alpha_s \cdot$GTP or forskolin, a diterpene produced by *Coleus forskohilii* (Seamon and Daly, 1981). Forskolin's reversible stimulation is apparently due to binding to adenylyl cyclase's catalytic domain. In addition to providing a useful exogenous means of regulating cAMP levels, forskolin provided the key to isolating and characterizing the cyclase. Adenylyl cyclase is purified by affinity chromatography using forskolin coupled to an insoluble matrix. Partial amino acid sequencing of this protein allowed its gene to be selected from a bovine brain cDNA library (Krupinski *et al.*, 1989).

The deduced amino acid sequence of adenylyl cyclase revealed 12 apparent transmembrane segments and one potential *N*-linked glycosylation site. The number of transmembrane domains was unexpected since adenylyl cyclase is an enzyme exposed at a membrane's *cis* face. The protein's overall structure is most similar to a transporter. The NH_2-terminal sequence at the *cis* face has a region (positions 16–21) that is identical to the guanine nucleotide binding region of G proteins. Sites for potential PKA and PKC phosphorylation are also present. There is complete confidence that this sequence represents an adenylyl cyclase since the gene product can be reconstituted in target cells using transfection and the purified gene product produces cAMP when co-reconstituted in lipid vesicles with β receptor and G_s. The functional dissection of this signaling membrane protein is just beginning.

It is now certain that the combination of β-adrenergic receptor, G_s protein, and adenylyl cyclase constitutes the necessary and sufficient proteins required for hormone-stimulated cAMP production. May *et al.* (1985) were the first to describe the complete reconstitution of the β-adrenergic signaling pathway from purified proteins. The purified β receptors, G_s proteins, and adenylyl cyclase were reconstituted at a mole ratio of 1:10:1 in lipid vesicles. The lipid composition of 3:2 PE:PS seems to be important in forming functional systems (May *et al.*, 1985; Feder *et al.*, 1986). This amino lipid composition resembles that usually found at a membrane's *cis* face. In the presence of agonist, GTP, Mg^{2+}, and ATP, these lipid vesicles synthesize cAMP. This receptor-triggered enzyme activity requires proper agonist stereochemistry and displays the appropriate receptor specificity. If any of these three proteins are missing, the system will not generate cAMP.

7.3.6.4. Physiological Changes Triggered by β Receptors Are Mediated by cAMP

Many cell surface receptors transmit their recognition events into a cell's cytosol via G proteins and adenylyl cyclase. All responses triggered by β-adrenergic receptors are mediated by this signaling pathway. The G_s protein and adenylyl cyclase components of the signaling pathway are not the "property" of any specific receptor, but are shared among many different receptors. For example, in tissues that express both β_1 and β_2 receptors, their simultaneous ligation activates a common pool of G_s and adenylyl cyclase leading to a combined response. Distinct receptors such as β-adrenergic and glucagon receptors combine in a similar way. The cAMP produced by these interactions can modify enzyme reactions, cytoskeletal assembly, membrane transport, and gene expression.

During resting conditions a typical cell's cAMP concentration is <1 μM. This concentration increases in response to activation of adenylyl cyclase or decrease due to the activation of cAMP phosphodiesterase. cAMP phosphodiesterase is a cytosolic enzyme that catalyzes the reaction:

$$cAMP + H_2O \rightarrow 5'\text{-AMP} \tag{7.10}$$

This enzyme is regulated by Ca^{2+} via calmodulin. These simultaneous opposing enzyme reactions control cAMP concentrations. When agonists bind to β receptors, the balance is tipped toward higher cAMP levels.

The second messenger cAMP simultaneously activates or inhibits several biochemical pathways. The key step in this process is the binding of cAMP to a cAMP-dependent protein kinase (PKA). Figure 7.10 illustrates the PKA signaling mechanism brought about by adenylyl cyclase activation. When cAMP binds to PKA, its regulatory subunits dissociate, thus activating its two catalytic subunits. The catalytic subunits transfer a phosphate from ATP to a serine or threonine residue of a target protein. A target protein can be activated or inactivated by phosphorylation. Figure 7.10 shows one protein that is inactivated by phosphorylation. All of these enzymes participate in the regulation of glycogen metabolism.

When glycogen synthase is phosphorylated by PKA, its ability to make glycogen is lost. The concomitant phosphorylation of glycogen phosphorylase kinase causes its activation (Figure 7.10). The activated glycogen phosphorylase *kinase* phosphorylates its substrate glycogen phosphorylase which, in turn, breaks down glycogen into glucose-1-phosphate. The glucose obtained from glucose-1-phosphate is secreted into the blood by liver cells, thus mobilizing energy reserves, as mentioned in Section 7.3.1; this completes just one arm of the physiological web outlined above. *In vivo*, analogous processes are believed to lead to triglyceride hydrolysis, changes in heart rate, and blood pressure elevation.

7.3.7. Physiological Regulation and Pharmacological Control

The physiological regulation of the β receptor's activity is an active area of research. Although several likely regulatory pathways have been proposed, their details have not yet been sorted out. Potential regulatory mechanisms include endocytosis, phosphorylation, and lipid environment.

Within 2 to 3 min of adding β agonist to target cells, the cells become unresponsive to agonists. This desensitization is specific for the β receptor since other RGC receptors remain functional. After longer incubation times, the signaling pathway also becomes refractory to agonists and exogenous compounds such as GTP analogues. Therefore, there are at least two forms of receptor desensitization.

7.3.7.1. Endocytotic Regulation

In addition to their localization on plasma membranes, β-adrenergic receptors migrate into endocytotic vesicle membranes and, ultimately, to an endolysosomal compartment (Stadel *et al.*, 1983; Waldo *et al.*, 1983). These two compartments can be distinguished from one another using hydrophilic ligands that only bind to plasma membrane receptors and hydrophobic ligands that bind to both plasma membrane-bound and intracellular receptors. They can also be physically separated from one another by ultracentrifugation. Figure 7.13 shows the specific binding of the antagonist [^{125}I]cyano-pindolol (ICyP) to plasma membrane and vesicle fractions obtained by centrifugation. During resting conditions ICyP binds to plasma membranes (panel a) but not to vesicles (panel b). However, if the agonist isoproterenol is added to cells for 3 hr at 37°C followed by extensive washing to remove both free and bound agonist, the binding of ICyP to these cell compartments is dramatically changed. The number of plasma membrane-associated receptors is decreased by about 60% (panel a), whereas the binding to the vesicle fraction dramatically increases (panel b). The vesicular β receptors are completely unresponsive to ligand because they cannot communicate with adenylyl cyclase. When the agonist is washed away from living cells, these receptors are recycled to the plasma membrane. This translocation of β receptors is one regulatory control mechanism. After prolonged incubation (11 hr) the total number of receptors in both compartments is decreased due to lysosomal destruction (Waldo *et al.*, 1983).

7.3.7.2. Regulatory Phosphorylation

In Section 7.2.5 we mentioned the role of phosphorylation in the desensitization of ACh receptors. Similarly, evidence has also been presented for the phosphorylation-dependent desensitization of β-adrenergic receptors (Bouvier *et al.*, 1988; Levitzki, 1988). This is consistent with the receptor's deduced amino acid sequence since the I3 and I4 domains contain amino acids and consensus amino acid sequences associated with protein kinase substrates (Figure 7.8). There are at least three types of protein

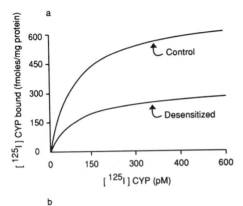

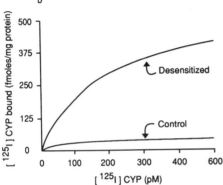

Figure 7.13. Specific antagonist binding to plasma and endosomal membranes of frog erythrocytes. Antagonist binding to plasma membranes (panel a) and endosomal membranes (panel b) are shown. During resting (control) conditions, only plasma membranes exhibit binding. However, after desensitization the number of internal receptors dramatically increases. (Redrawn from Stadel *et al.*, 1983, *J. Biol. Chem.* **258**:3032, with permission from the American Society for Biochemistry and Molecular Biology.)

kinases that interact with β receptors: PKA, PKC, and a specific β-adrenergic receptor kinase (βARK).

The β_2 receptor's PKA sites are located at residues 259–262 and 343–348 of the I3 and I4 domains, respectively. These sites have the sequences Lys/Arg-Arg-X-(X)-Ser. The PKA site within I3 is particularly interesting since this domain interacts with G_s. Phosphorylation of the I3 domain might be responsible for blocking its interaction with G_s. A role for I3 in endocytotic regulation is suggested by experiments showing that deletion of residues 239–272 inhibits receptor internalization (Strader *et al.*, 1987a). Endocytotic regulation seems unique to I3 since deletion of I4 does not affect endocytosis. The phenomenological coupling between G_s and endocytosis should not be surprising since G proteins participate in endomembrane trafficking.

The β_2 receptor's I4 domain possesses one PKA site of unknown significance and 11 potential sites for βARK. Removal of these phosphorylation sites within I4 by deletion or site-directed mutagenesis leads to greatly delayed desensitization (Bouvier *et al.*, 1988). Normal agonist-induced desensitization requires phosphorylation of I3. After 3 hr the desensitization of both I4 domain mutants and wild-type receptors are similar.

Therefore, phosphorylation of I4 acts as a catalyst for rapid desensitization, perhaps by accelerating the phosphorylation of I3.

Dose–response studies suggest that βARK's principal physiological activity is the regulation of β-adrenergic receptor function at sympathetic synapses. βARK's protein kinase catalytic domain is very similar to those of PKA and PKC (Benovic *et al.*, 1989). When the β receptor is occupied by agonist, it can be recognized and phosphorylated by βARK. βARK's desensitization activity is likely carried out when it is bound to membranes via the G protein βγ subunits (Pitcher *et al.*, 1992). In other words, the βγ complex serves as a locus for both the stimulation *and* inactivation of transmembrane signaling. However, phosphorylation *per se* is insufficient to trigger desensitization. A phosphorylated adrenergic receptor then binds to β-arrestin, a 47-kDa cytosolic protein, which promotes the inhibition of receptor function (Lohse *et al.*, 1990a). Interestingly, β-arrestin's deduced amino acid sequence possesses five regions homologous to α subunits of G proteins; these similarities include the COOH-terminal region, which is thought to mediate α subunit binding to receptors. When β-arrestin is bound to the receptor, the G_s stimulatory cycle cannot begin, thus causing desensitization.

There are multiple pathways of regulatory phosphorylation in intact cells. In addition to the PKA- and βARK-mediated phosphorylation events described above, PKC is also believed to participate in receptor phosphorylation. PKC's effects on β-receptor function are mediated by other receptors. For example, ligation of a PKC-dependent receptor would trigger a cascade of phosphorylation reactions which might include the β receptor as a substrate. Similarly, triggering the β receptor may influence the functions of other receptors.

7.3.7.3. Lipid Regulation

Several lines of evidence suggest that membrane lipid composition affects receptor-mediated production of cAMP (e.g., Hirata *et al.*, 1979; Hanski *et al.*, 1979; Houslay *et al.*, 1976; Puchwein *et al.*, 1974). Both receptor-mediated and exogenous changes in lipid composition may affect β-adrenergic stimulation of adenylyl cyclase. The ability of lipids to influence hormone action is also supported by the fact that lipid vesicle reconstitution of cAMP production is highly sensitive to lipid composition (May *et al.*, 1985; Feder *et al.*, 1986). Unfortunately, no systematic study of lipid effects on reconstituted β receptor, G_s, and adenylyl cyclase function has been reported. Studies using intact cells are complicated by the heterogeneity of structural components and the presence of multiple protein and lipid domains. In one internally consistent experiment, Hanski *et al.* (1979) used a fluidizing *cis* unsaturated fatty acid and DPH, a fluid domain fluorescence polarization probe, to study β-adrenergic-mediated cAMP production. This study suggests that an increase in bilayer fluidity leads to enhanced receptor coupling due to the components' greater lateral diffusion. Adenylyl cyclase activation might therefore be a diffusion-controlled collisionally coupled phenomenon, as implicitly suggested by Figure 7.10. Although the formation of diffusion-controlled encounter complexes is an efficient mechanism and a simple and attractive hypothesis analogous to that of electron transport in mitochondrial membranes (Chapter 5), it remains unproven.

7.3.7.4. Pharmacological Regulation

Modern patient treatment for numerous conditions relies on the use of selective adrenergic agonists and antagonists. Some of the compounds described above have been used in clinical medicine. For example, the β_1-specific antagonist practolol (Figure 7.6) slows heart rate and decreases its contraction strength. Practolol is used to treat patients with angina and cardiac arrhythmias. The β_2-selective agonist terbutaline is used in treating asthma. However, as we learned above, adrenergic ligands bind simultaneously to many tissues. By defining all adrenergic receptor subtypes and developing new drugs targeted to these subtypes, scientists hope to eliminate the side effects of adrenergic drug treatment.

7.4. POLYPEPTIDE HORMONE RECEPTORS: THE INSULIN RECEPTOR

Polypeptide hormone receptors constitute an unusually broad class of membrane receptors that participate in intercellular communication. In contrast to channel, RGC, and molecular transport receptors, polypeptide hormone receptors possess a single transmembrane domain in their monomeric state. Their function is to convey information, not ions, sugars, or electrons, across a bilayer. When its putative signal reaches a membrane's *cis* face, the receptor's catalytic domain, often a tyrosyl kinase, amplifies the signal by distributing a second messenger(s). The identity of a relevant second messenger(s) and its interactions are now under active investigation.

The insulin receptor is the most thoroughly studied polypeptide hormone receptor, although its mechanism of action remains poorly understood. The insulin receptor possesses endogenous tyrosyl kinase activity. Also included among the tyrosyl kinase receptors are: the epidermal growth factor receptor, the platelet-derived growth factor receptor, the type 1 insulin-like growth factor (IGF-1) receptor, colony-stimulating factor-1 receptor, and certain oncogene products, many of which represent mutant forms of growth factor receptors (Section 9.2.3). Tyrosyl kinase activity is believed to trigger the many biochemical pathways needed to initiate the physiological responses of these receptors. This section will explore how insulin receptors alter cell physiology.

7.4.1. Insulin and Its Physiological Setting

Insulin is a pluripotent hormone that participates in a wide variety of physiological responses. Indeed, the uncertainties in our understanding of how insulin works are due to its numerous effects on a cell. In this section we will sketch some background information on insulin and its physiological effects.

7.4.1.1. Insulin

Insulin is a small globular protein found in vertebrates and invertebrates. It is one member of a family of structurally and functionally similar polypeptide hormones. This family also includes IGF-1, IGF-2, and relaxin. In humans insulin's gene is located on chromosome 11. Several alleles coding for both normal and pathological forms of insulin have been identified (e.g., Bell *et al.*, 1982; Tager *et al.*, 1979; Haneda *et al.*, 1984). An individual's two insulin genes are co-dominantly expressed. Insulin is synthesized as a single precursor protein, proinsulin. As the protein matures, amino acid sequences are removed, resulting in an active ligand composed of two chains, A and B, linked by two disulfide bridges. The physiological concentration range of insulin is 0.01 to 1 nM. The mature protein has two fairly hydrophobic surfaces, at least one of which is involved in receptor interactions. At unphysiologically high concentrations (1 μM) insulin dimerizes between these interfaces. Although dimerization is not important in its physiological signaling, it is likely important for its storage in pancreatic β cells.

7.4.1.2. Insulin's Functions, Especially Blood Glucose Regulation

Under resting conditions the blood's glucose level is established by the hormones insulin and glucagon. When glucose levels are low, A cells of the pancreas secrete glucagon into the bloodstream. Its receptors are primarily found on liver cells. These cells respond by producing cAMP which in turn stimulates glycogen breakdown and glucose release, just as described in Section 7.3.6.4 for epinephrine. The physiological action of insulin opposes that of glucagon. When glucose levels are high, such as after a meal, insulin is secreted by the islets of Langerhans's β cells. Insulin receptors are found on most cell types at concentrations of 100 to 100,000 per cell. Its binding stimulates the entry and storage of glucose in target cells.

In addition to its role in blood glucose regulation, insulin leads to pleiomorphic effects on cell physiology, especially anabolic pathways. Table 7.2 outlines insulin's effects on a target cell. These effects have been categorized according to their ability to influence membrane transport, protein dephosphorylation, or protein seryl group

Table 7.2.
Effects of Insulin Binding

Membrane transport	Sites affected by seryl phosphorylation
Glucose	Ribosome S6 component
Amino acids	Acetyl coenzyme A carboxylase
Nucleotides	Gene expression
Ions	Sites affected by tyrosyl phosphorylation
Sites affected by dephosphorylation	Insulin receptor
Glycogen synthase	pp185
Hormone-sensitive kinase	ERK seryl/threonyl kinases
Pyruvate dehydrogenase	

phosphorylation. Some of these changes such as glucose transport occur within seconds of ligand binding whereas others like cell growth require hours of exposure to affect a cell's biosynthetic machinery.

7.4.2. Structure–Activity Relationships Define Insulin's Receptor Binding Domain

We have previously discussed the development of structure–activity relationships for the ACh and adrenergic receptors. It is not possible to formulate such precise relationships for polypeptide hormone receptors. ACh and epinephrine have molecular weights of 146 and 184, respectively. In comparison, insulin has a molecular weight of 5734. Therefore, due to these size differences, it has not been possible to refine insulin's structure–activity relationships to an atomic level.

To establish structure–activity relationships for any molecule or macromolecule, one must have a panel of homologous ligands with slight structural variations to test. For a molecule such as epinephrine, which is about the size of an amino acid, a small change in a single functional group can have a major impact on its physiological response. For a protein such as insulin, one examines the effects of altered amino acid sequences on activity. A wide variety of insulin analogues are listed in Table 7.3. Natural analogues of insulin with variable amino acid sequences are found in insulins from animals, patients, immature insulins, and homologous polypeptide hormones. A variety of chemically altered or synthesized insulins can be experimentally produced. The biochemical and physiological properties of insulin analogues are then compared with their structural differences.

Insulin's receptor-binding domain includes portions of both its A and B chains (Pullen *et al.*, 1976; Gammeltoft, 1984; Inouye *et al.*, 1981). This region has been highly conserved throughout evolution. The participating residues are believed to include Gly-A1, Gln-A5, Tyr-A19, Asn-A21, Val-B12, Tyr-B16, Phe-B24, Phe-B25, and Tyr-B26, although some of these residues are more important than others. These residues are not scattered about the protein; they form a surface on insulin as shown in Figure 7.14.

Table 7.3.
Insulin Analogues Used in Formulating
Structure–Activity Relationships

Naturally occurring forms
   Insulin from animals
   Proinsulin and intermediate immature forms of insulin
   Mutant forms of insulin from patients
   Polypeptide hormones homologous to insulin (e.g., relaxin and IGF-1)
Synthetic forms
   Cross-linked insulin dimers
   Chemical or enzymatic modifications
   Synthetic insulin or semisynthetic insulin

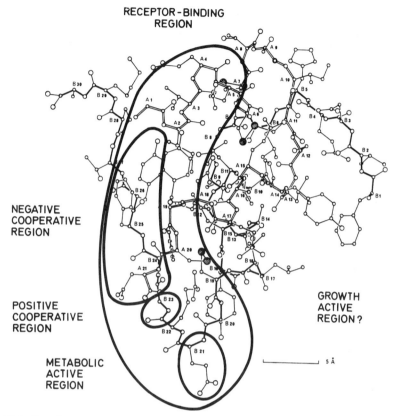

Figure 7.14. Crystal structure of insulin and its physiologically relevant domains. The three-dimensional structure of insulin and proposed functional domains are shown. The A and B chains with residue numbers are listed along the polypeptide backbones. (From Gammeltoft, 1984, *Physiol. Rev.* **64:**1321.)

The three-dimensional structure of insulin shown in this figure was derived from x-ray diffraction studies. The surface formed by these amino acids is believed to create a tight bond with an insulin receptor by forming an intermolecular β-sheet between the peptide strands of the ligand and receptor and by hydrophobic and ionic interactions between the side chains (including the A chain's NH_2-terminal amino).

Several lines of evidence support this consensus view of insulin's interaction with its receptor. Deletion studies have shown that removal of residues B28 to B30 has no effect on insulin activity. However, as additional residues are removed, activity is progressively lost until no activity remains when B23 to B30 are removed. The importance of this region in receptor binding and triggering is supported by the discovery that certain rare forms of diabetes mellitus are due to mutant forms of insulin (Tager *et al.*, 1979; Haneda *et al.*, 1984). Both receptor binding and triggering are dramatically reduced in patients with a Phe-B25-to-Leu-B25 mutation or a Phe-B24 to Ser-B24 mutation. The binding and functional sensitivity of this domain were further explored using semisynthetic insulin analogues (Inouye *et al.*, 1981). These experiments showed that replacement of

Phe-B24 or Phe-B25 by Leu or Ala dramatically decreased binding and activity. However, the switch from Tyr-B26 to Ala-B26 had no effect, suggesting that Phe-B25 and Phe-B24 play a special role in ligand–receptor interactions. A residue with an aromatic ring at position B25 is likely important in triggering a conformational change in insulin when it binds its receptor (Nakagawa and Tager, 1987).

The A chain also participates in receptor binding. The NH_2-terminus is especially important. If the positive charge of the NH_2-terminus is removed by chemical treatment, activity falls to 10% to 40% of normal. When the NH_2-terminus is blocked, as in proinsulin or its intermediate fragmented forms, activity is 0% to 10% of normal. Tyr-A19 has also been closely linked to receptor function. When Tyr-A19 is replaced by Phe-A19, activity falls to 22% of normal. Moreover, selective iodination of Tyr-A19 decreases activity by 50%. In contrast, iodination of Tyr-A14 or its replacement by Phe-A14 has no effect on activity. This indicates that Gly-A1 and Tyr-A19, which are close together in the crystal structure (Figure 7.14), play important roles in receptor binding.

In addition, insulin analogues possessing higher affinities and greater physiological potency have been synthesized (Schwartz et al., 1989). Enhanced activity is observed when His-B10 is modified to Asp-B10 and when B25 to B30 is replaced by Tyr-NH_2. Since these changes make insulin more hydrophobic, insulin's conformation is likely altered—perhaps to a form more closely resembling its receptor-bound conformation. Further details regarding the nature of insulin–receptor interactions must await a detailed analysis of the receptor's tertiary structure.

Similar analyses have localized the regions of insulin responsible for metabolic triggering and cooperative interactions (Gammeltoft, 1984; De Meyts et al., 1978). These regions have been associated with particular areas within the receptor binding domain. Negative cooperativity has been linked to amino acids B24 to B26 and A21. Tentative analyses suggest that the carboxylic acid moiety of Glu-B21 is particularly important in triggering, although the entire receptor binding domain participates.

7.4.3. Molecular Structures and Macromolecular Assemblies of Insulin Receptors

Structural analysis of insulin receptors has lagged behind that of ACh receptors. One of the early reasons for this difference is that a typical cell possesses 10^4 insulin receptors whereas a single synapse from the *Torpedo* electric organ contains 10^{11} ACh receptors, a 10 million-fold difference in receptor availability. Recently, this gap has been narrowed by increasingly sensitive microchemical techniques and the application of molecular biological techniques. However, the structural analysis of insulin receptors remains largely at a descriptive biochemical level.

7.4.3.1. Identification and Characterization of Insulin Receptors

Insulin receptors retain their ability to bind insulin when solubilized in detergent. When passed through an affinity chromatography column containing insulin linked to agarose, the insulin receptors become bound to the matrix. After washing the matrix, receptors are eluted by brief exposure to acidic urea. This procedure leads to a 250,000-fold purification of the receptor. Two subunits of 135 and 90 kDa termed α and β, respectively, are observed after SDS-PAGE under reducing conditions. However, under nonreducing SDS-PAGE, gel chromatography, or sucrose density centrifugation a mass of about 450 kDa is observed (e.g., Helmerhorst *et al.*, 1986). Therefore, the native receptor is composed of two α and two β subunits.

Chemical and photochemical cross-linking studies have indicated that insulin binds to the α subunit. Chemical cross-linking experiments were performed by adding radioactive insulin to purified plasma membranes followed by the addition of a bifunctional cross-linking reagent. After electrophoresis under reducing conditions, a single band of 135 kDa was observed. Similar experiments using radioactive insulin linked to an aryl azide labeled a 135-kDa protein when photoactivated. These experiments indicate that insulin binds to its receptor's α chain.

How can we be absolutely certain that these two proteins are the true insulin receptor, not just insulin binding proteins that regulate the availability of insulin? Our earlier discussions will serve as a guide. We knew that the ACh receptor identification was correct because, when reconstituted in lipid bilayers, it acted as a ligand-gated ion channel; this protein complex was a necessary and sufficient condition to depolarize a membrane, the receptor's physiological role. The β-adrenergic receptor was more complicated. The receptor and its signaling apparatus was reconstituted in liposome membranes. When triggered by agonist, this system behaved in a manner analogous to cellular signaling. The insulin receptor is still more complex. Unfortunately, it has not been possible to reconstitute the insulin receptor's signaling system because we do not understand it. Incontrovertible evidence regarding the receptor's identity is therefore not in hand.

Although foolproof data are not available, there is no doubt that the 135- and 90-kDa proteins are the insulin receptor. First, this putative receptor binds insulin with the same specificity and affinity as found for cell membranes. Moreover, the purified putative receptor has an associated tyrosyl kinase activity which is believed to be the first step in intracellular signal processing (see below). Receptor specificity, antibody binding, and gene sequencing studies have distinguished this putative insulin receptor from homologous receptors, such as the IGF-1 receptor. The best evidence for the identity of the insulin receptor comes from transfection experiments. These experiments are a type of cellular reconstitution of receptor function. CHO cells respond to insulin by a 2.5-fold increase in 2-deoxyglucose transport with a K_d of 0.3 nM. However, when these same cells are transfected with the putative wild-type insulin receptor, their sensitivity to insulin increases by 30-fold (Ebina *et al.*, 1987). This result indicates that this gene product is capable of mediating the insulin receptor's physiological effects.

Although SDS-PAGE experiments have shown the masses of the α and β subunits to be 135 and 90 kDa, respectively, the receptor's gene codes for protein subunits of 84,214 and 69,703 Da. These apparent differences are accounted for by receptor glycosylation. The α chain has 13 potential N-linked glycosylation sites whereas the β chain has 4. Receptor glycosylation is consistent with the ability of glycosidases to interfere with receptor function. Furthermore, the lectins Con A and wheat germ agglutinin bind to insulin receptors. Metabolic labeling studies have identified the presence of glucosamine, galactose, and fucose. Functional receptors cannot be assembled in the presence of tunicamycin, an inhibitor of N-linked glycosylation (Ronnett *et al.*, 1984). The abilities of tunicamycin and glycosidases to inhibit receptor activities suggest that carbohydrates stabilize the receptor in an active conformation.

7.4.3.2. Macromolecular Structure and Functional Domains

The human insulin receptor's gene is located on a 45-kb stretch of chromosome 19. Insulin receptor mRNA encodes a precursor protein ($M_r = 153{,}917$) that includes both the α and β subunits. Alternatively spliced forms of receptor mRNA and receptor isoforms have been observed (Yamaguci *et al.*, 1991). Each mature protein is composed of an α and β subunit held together by a disulfide bridge.

Figure 7.15 shows a schematic model of the insulin receptor. The deduced amino acid sequence revealed several of the receptor's important structural features (Ebina *et al.*, 1985). The α subunit has a Cys-rich domain; this structural attribute is shared

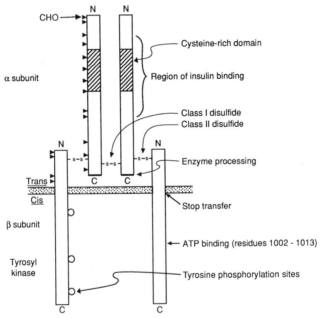

Figure 7.15. Schematic model of the heterotetrameric insulin receptor. A structural model of the heterotetrameric insulin receptor is shown. The cysteine-rich regions are striped. Carbohydrate attachment sites (arrowheads) and tyrosine phosphorylation sites (open circles) are indicated along the left-hand side. Disulfides link the receptor's subunits together. However, the amino acid residues participating in disulfide formation are not certain. The locations of class 1 and 2 disulfides are noted. A possible site of insulin binding is indicated.

among several receptors including the LDL and epidermal growth factor receptors. The Cys-rich and NH_2-terminal regions of the α subunit have been implicated in ligand binding. To more precisely map the insulin binding domain, experiments utilizing chemical cross-linking followed by microsequencing and chimeric receptor constructs have been performed (e.g., Kjeldsen *et al.*, 1991). Although both the Cys-rich and NH_2-terminal regions contribute to the ligand binding pocket, the NH_2-terminal domain plays a greater role in determining ligand binding affinity.

The β subunit extends across a plasma membrane with most of its mass at the *cis* face. The β chain's short extracellular sequence is linked to the α subunit's COOH-terminal region via a disulfide(s). The disulfide bridges linking the α and β chains together are referred to as class 2 disulfides. As we have encountered with other integral membrane proteins, a transmembrane sequence of 23 hydrophobic amino acids is followed by a tripeptide of basic residues at a membrane's *cis* face. The receptor's large COOH-terminal domain possesses an ATP-binding region and sites for Tyr phosphorylation. This observation is consistent with previous studies showing an endogenous tyrosyl kinase activity. Under physiological conditions, the receptor is a heterotetramer $(\alpha\beta)_2$ held together by class 1 disulfides.

7.4.3.3. Insulin Receptors Can Form Heteroreceptors with IGF-1 Receptors

In addition to the monomeric (αβ) and dimeric $(\alpha\beta)_2$ insulin receptor structures described in the preceding paragraphs, the receptor also exists in other forms. Recently, Soos and Siddle (1989) have suggested that the monomeric insulin receptor (αβ) forms a heteroreceptor $(\alpha\beta\beta'\alpha')$ with monomeric $(\alpha'\beta')$ IGF-1 receptor. The IGF-1 receptor is structurally and functionally similar to the insulin receptor. This structural observation may explain the parallel expression and downregulation of insulin receptors, IGF-1 receptors, and perhaps other receptors (e.g., Davis and Czech, 1986).

7.4.3.4. Insulin Binding Sites Are Expressed Individually and Collectively at Cell Surfaces

The next higher level of structural organization is the manner in which insulin receptors are expressed at cell surfaces. For example, the receptors could be found polarized into one large cap, as multiple receptor clusters, or as individual $(\alpha\beta)_2$ complexes. To address this question a technique sensitive to small distances such as electron microscopy must be employed. Jarett and Smith (1977) covalently coupled insulin to the electron-dense protein ferritin. Using a combination of gel filtration and affinity chromatography, a 1:1 conjugate of ferritin–insulin was prepared. This conjugate bound to cell surfaces with appropriate specificity and stimulated insulin's physiological activity. This suggests that the observations with the conjugates are relevant to insulin's biological activity, although one can never know which insulin receptor(s) at a *particular* cell's surface may have triggered a physiological response. The native distribution of receptors can be determined by fixing cells or purified plasma membranes with glutaraldehyde prior to addition of the conjugate. When this experiment is performed with rat adipocytes, microclusters of two to six insulin binding sites are observed. Furthermore, these binding sites are only found on one membrane face (*trans*); this approach provided early evidence for the asymmetry of ligand binding sites and, by inference, their membrane receptors. Other cell types, such as liver cells and transformed murine adipocytes, have been found to possess primarily monomeric binding sites under resting conditions (Jarett *et al.*, 1980). In addition to structural heterogeneity, regulatory differences between cell types may also contribute to differences in binding site expression.

7.4.4. Ligand–Receptor Binding

The binding of insulin to its receptor is fraught with many subtleties which are only now beginning to be sorted out. Since ligands are often radiolabeled with the isotope ^{125}I, one must ensure that insulin is monoiodinated at Tyr-14. If Tyr-19 is labeled, insulin behaves as a less potent analogue (Section 7.4.2). Furthermore, living cells internalize and degrade insulin (Section 7.4.6.1); these possibilities must be accounted for to obtain meaningful data. We will now consider some recent studies of insulin–receptor binding in simple systems.

Insulin's binding affinity to its receptor is potentially affected by the presence of receptor heterodimers ($\alpha\beta$), heterotetramers ($\alpha\beta$)$_2$, and heteroreceptors ($\alpha\beta\beta'\alpha'$). The receptor's affinity is dependent on the association of α subunits whereas its ability to trigger tyrosyl kinase is affected by the association of β subunits (Boni-Schnetzler *et al.*, 1987, 1988; Sweet *et al.*, 1987). Heterotetramers, or native receptors, bind one insulin molecule with high affinity. Heterodimers also bind one insulin molecule but with low affinity.

Affinity chromatography isolates native heterotetrameric receptors because of their high affinity; low-affinity forms are necessarily missed. Heterodimeric receptors are isolated by gel filtration chromatography after a brief exposure of heterotetramers to alkaline dithiothreitol. The ligand binding affinities of heterotetramers and heterodimers can then be measured separately. Figure 7.16 shows Scatchard plots of ligand binding to solubilized receptors (panel a) and intact membranes (panel b). Let us first consider insulin binding to solubilized receptors. Heterotetramers yield a curvilinear Scatchard plot (Boni-Schnetzler *et al.*, 1987; Sweet *et al.*, 1987). The value of K_d is somewhat dependent on the associative model chosen to represent the reaction. For example, the curvilinear Scatchard plot could be due to negative cooperativity or to the presence of two noninteracting sites differing in K_d (Sweet *et al.*, 1987). In either case the K_d is roughly 0.1–0.5 nM. However, the solubilized heterodimeric receptor yields a linear Scatchard plot with a tenfold reduction in affinity. Both of these K_d's are within insulin's physiological range.

Figure 7.16, panel b shows Scatchard plots of insulin binding to intact membranes. In this experiment, untreated membranes are compared with membranes reduced by exposure to alkaline dithiothreitol. Indistinguishable curvilinear Scatchard plots are obtained. The linear region of the plot indicates a K_d of about 1 nM. Although the

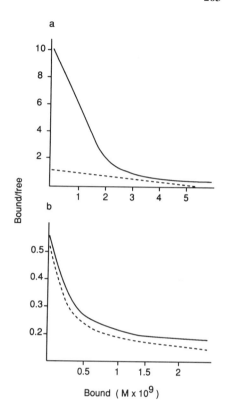

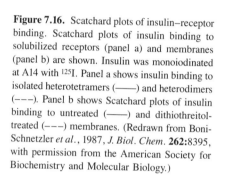

Figure 7.16. Scatchard plots of insulin–receptor binding. Scatchard plots of insulin binding to solubilized receptors (panel a) and membranes (panel b) are shown. Insulin was monoiodinated at A14 with [125]I. Panel a shows insulin binding to isolated heterotetramers (——) and heterodimers (– – –). Panel b shows Scatchard plots of insulin binding to untreated (——) and dithiothreitol-treated (– – –) membranes. (Redrawn from Boni-Schnetzler *et al.*, 1987, *J. Biol. Chem.* **262**:8395, with permission from the American Society for Biochemistry and Molecular Biology.)

disulfide link between the two heterodimers has been broken, they remain noncovalently associated in membranes. This accounts for the retention of the high-affinity binding site.

The transmission of a physiological signal into a cell depends on the receptor's β subunit. The first step in this process is believed to be the activation of a tyrosyl kinase. The receptor's enzyme activity is also dependent on its associative state (Boni-Schnetzler *et al.*, 1988). Insulin stimulates its autophosphorylation in intact cell membranes or when solubilized as heterotetramers. The tyrosyl kinase activity is also triggered in dithiothreitol-treated membranes, indicating that class 1 disulfides are not important in enzyme activation. In contrast, solubilized heterodimers do not trigger tyrosyl kinase activity when they are physically separated from one another. This suggests that the association of heterodimers to form heterotetramers is important in signaling. If heterodimers and insulin are mixed in solution, they spontaneously form insulin-bound heterotetramers. This is particularly interesting because it indicates that, at least *in vitro*, insulin regulates the affinity of the half (or monomeric) receptors for one another and, necessarily, their tyrosyl kinase activity.

How might all these important clues be assembled for students of the subject? Figure 7.17 shows a *purely speculative* model of *in vitro* insulin–receptor interactions

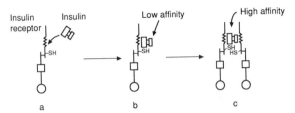

Figure 7.17. A speculative model of insulin–receptor interaction. A model to account for the properties of solubilized insulin receptors is shown. (a) The heterodimer is first solubilized by detergent. (b) When insulin binds to a solubilized receptor, it exposes a domain which has low affinity for receptors. (c) The combined low affinity for self-association and the second low affinity of the ligand for a second α subunit triggers heterotetramer formation. Therefore, the heterotetramer necessarily expresses a single site with higher affinity for ligand.

that illustrates some of the properties discussed above. The heterodimer has one insulin binding site with a K_d of 1 nM. This site binds to insulin's receptor binding domain, as shown in Figure 7.17. The heterodimer has a low affinity for itself; it can only self-associate in the reduced dimensionality of a membrane. When insulin is receptor-bound, its second hydrophobic face is exposed. We suggest that this second face has a weak affinity for insulin receptors, just as insulin has a weak affinity for the IGF-1 receptor. The combined affinity of insulin's second face and the monomeric receptor's native affinity for self-association promotes heterotetramer formation. The apparent binding affinity of insulin for its receptor necessarily increases because it is the combination of the heterodimer's affinity *and* the ligand's weak secondary affinity for another α subunit. This also accounts for single ligand binding to both heterodimeric and heterotetrameric receptors. This is consistent with the apparent negative cooperativity between the receptors (Gammeltoft, 1984). The model suggests that insulin bound to one α subunit interferes with insulin binding to the second α subunit of a heterotetramer. Furthermore, when cells are labeled with radioactive insulin, washed, and then incubated with excess "cold" insulin, the cold ligand may interact with the second α subunit's high-affinity binding site. Thus, labeled insulin would dissociate faster because of "cold" insulin molecules interacting at the second α subunit.

7.4.5. Chasing the Signal: Participants in a Complex Process

The first event in insulin's signal transduction process is its receptor binding step, which was described in the preceding paragraphs. This section will explore the receptor-mediated events that lead to a physiological change within a cell or tissue.

7.4.5.1. Reciprocal Conformational Changes Get Things Started

Studies with a variety of insulin analogues have suggested that insulin undergoes a conformational change when binding to its receptor (Nakagawa and Tager, 1987). One site of conformational change has been associated with the β chain's COOH-terminal region of B26 to B30. This conformational change is believed to be triggered by an interaction between Phe-B25 and the receptor.

The receptor's conformation changes when it binds ligand (Pilch and Czech, 1980). Insulin receptors were subjected to mild proteolytic treatment in the absence or presence of insulin. The α subunit's degradation was substantially increased in the presence of insulin. Apparently, proteolytic cleavage sites within the receptor become more accessible to solvent when insulin is bound. The receptor's conformational change must then be relayed to its tyrosyl kinase domain at a membrane's *cis* face.

7.4.5.2. The Mechanism of Transmembrane Signaling May Involve a Propagated Conformational Change

After insulin binds to the receptor's α subunit, a signal is transmitted across a membrane to the β subunit's tyrosyl kinase. Unfortunately, this signaling mechanism is not understood. Conformational changes in the receptor and/or receptor clustering brought about by ligand binding have been proposed as signaling mechanisms.

Conformational changes are generally invoked by biochemists when they run out of explanations. The insulin receptor system is in the enviable position of having data supporting conformational changes of both the ligand and receptor upon binding. An α-helical transmembrane domain provides an excellent signal conduit. In contrast to β-sheets, α-helices are close-packed structures that move as rigid bodies (Chothia *et al.*, 1983). Conformational changes of more than 0.15 nm (the radius of a lithium atom) cannot be accommodated locally by changes in peptide bond angles. Consequently, a conformational change is propagated along a helix.

It is easy to understand how the β-adrenergic receptor's presumed conformational change is transmitted to the cytosol. The *trans* face of M5 is restricted in motion by the short O3 domain (Figure 7.8). The O3 domain is likely relatively immobile, in analogy with previous NMR studies of the "honorary" RGC protein bacteriorhodopsin (Herzfeld *et al.*, 1987). When a conformational change occurs at the M5 helix, its movement can only be propagated toward the cytosol. In contrast, insulin receptor heterodimers are tethered to membranes by just one domain. Therefore, a conformational change in a heterodimer's α subunit would be propagated via the path of least resistance toward the extracellular environment. This could, for example, explain why the heterodimer is nonfunctional. The functional insulin receptor is a heterotetramer which contains *two* transmembrane domains. Therefore, one membrane domain could anchor the complex while the second propagates the ligand's disturbance to the kinase domain. This is equivalent to saying that insulin induces a relative asymmetry in a heterotetramer's structure that activates a signal.

The idea of a propagated conformational change is supported by studies of heterodimers, heterotetramers, and heteroreceptors. Insulin stimulates the tyrosyl kinase activity of solubilized heterotetrameric receptors (Boni-Schnetzler *et al.*, 1988). This seems to indicate that: (1) membranes are not required to transmit a signal and (2) the heterotetramer is a sufficient condition to generate a response. Recent studies of mutation-bearing insulin/IGF-1 heteroreceptors also support these ideas (Treadway *et al.*, 1991). Therefore, a conformational change transmitted through the receptor's transmembrane domain seems to be a likely possibility.

The transmembrane domain's role in signaling could be directly tested. The insertion of a positive charge in this domain may disturb the α-helical structure (see Section 9.2.1) and physiological signaling. The transmembrane α-helix could also be shortened. This might force the domain into a random coil conformation, thus muting transmembrane conformational changes. Conformational changes in transmembrane domains could be detected by site-directed mutagenesis and spin-labeling of amino acid residues near the membrane–aqueous interface of the insulin receptor (or the adrenergic receptor's M5 helix) coupled with ESR spectroscopy in the presence and absence of ligand. Within the next few years we should have direct evidence regarding the role of transmembrane domains in signaling events.

Receptor clustering has also been proposed to induce the activation of the tyrosyl kinase domain. The formation of large clusters, such as those formed prior to endocytosis, in untenable as a tyrosyl kinase signaling strategy since triggering occurs prior to clustering and endocytosis. Microclusters such as those discussed in Section 7.4.3.4 could participate. However, the ability of solubilized heterotetramers to trigger tyrosyl kinase activity weighs heavily against this hypothesis. Therefore, a propagated conformational change seems to be the most likely alternative.

7.4.5.3. The Mechanism of Transmembrane Signaling Is a Primitive and Ubiquitous Event

The insulin receptor's mechanism of transmembrane signaling can be explored using chimeric receptors, as previously described for adrenergic receptors. Kashles *et al.* (1988) genetically fused the insulin receptor's ligand binding domain with the transmembrane and kinase domains of the epidermal growth factor receptor. When expressed in transfected cells, these chimeric receptors activated autophosphorylation in an insulin-dependent manner. This suggests that the signal originating at the receptor's ligand binding site was correctly relayed through a foreign transmembrane domain to a foreign tyrosyl kinase. This suggests that the signal, perhaps a conformational change, is shared among several receptors.

The ubiquity of the signal is underscored by the observations of Moe *et al.* (1989). These investigators fused the extracellular and transmembrane domains of the aspartate receptor of *E. coli* to the intracellular portion of the insulin receptor. When aspartate binds to this chimeric receptor *in vitro*, it triggers kinase phosphorylation. Although some differences between chimeric receptor and the native insulin receptor were noted, a *prokaryotic* ligand–receptor interaction could trigger a *eukaryotic* intracellular signaling apparatus. This suggests that at least one mechanism of intra-receptor transmembrane signaling is quite primitive.

7.4.5.4. Tyrosyl Kinase Activity Is Required for Signaling

The next step in insulin's mechanistic chain of events is known to involve tyrosyl kinase activation. Early studies indicated that the insulin receptor possesses tyrosyl kinase activity and that the receptor's β subunit becomes phosphorylated by its kinase activity (Kasuga *et al.*, 1982). The two β subunits within each heterotetramer phosphorylate each other. In general, cytoplasmic tyrosyl kinase activity is low since few Tyr residues are phosphorylated. Insulin binding increases the kinase's V_{max} about 400-fold without affecting its affinity for ATP (Czech, 1985). But is this change in the insulin receptor's enzyme activity important in transmembrane signaling?

Several lines of evidence now indicate that tyrosyl kinase activation is important in every aspect of insulin's physiological action. These data have come from antibody binding and site-directed mutagenesis/transfection experiments. Morgan *et al.* (1986) prepared an extensive panel of monoclonal antibodies that recognized the insulin receptor's β subunit. One of these antibodies was found to inhibit the insulin-dependent autophosphorylation of solubilized insulin receptors. When this antibody was microinjected into the cytosol of living *Xenopus* oocytes, it was found to specifically block the cells' insulin-stimulated maturation. This result clearly links the β subunit's tyrosyl kinase activity with cell maturation.

In an important study, Ebina *et al.* (1987) abolished the insulin receptor's kinase activity by site-directed mutagenesis. A single amino acid, Lys-1030, of the ATP-binding region was changed to a Met, Ala, or Arg residue. The wild-type and mutant receptors were separately transfected into CHO cells. Cells transfected with the wild-type receptor had a greatly enhanced autophosphorylation and hexose transport response to insulin in comparison with control CHO cells. The wild-type and mutant receptors were processed normally by the cells and possessed the same affinity for insulin. However, the mutant β subunits did not autophosphorylate or stimulate glucose transport. In fact, glucose transport was actually depressed in cells transfected with a defective receptor. This may be due to defective heterodimers pairing with endogenous heterodimers and thus perturbing a CHO cell's normal response to insulin. It is always possible that the change in amino acid sequence altered the β subunit's conformation, thus interfering with triggering independent of kinase activity. However, the mutant insulin receptor's processing, turnover, and binding were normal. The substitution of a positively charged Arg for a positively charged Lys still abolished activity. Furthermore, monoclonal antibodies sensitive to the β-subunit's conformation could not distinguish between the wild-type and mutant receptors. Therefore, the loss of receptor function is likely due to a loss in kinase activity.

The kinase activity is required for insulin's multifaceted response (Rosen, 1987). In addition to hexose transport, mutant receptors are deficient in activating glycogen synthesis, DNA synthesis, ribosome S6 kinase, and seryl/threonyl phosphorylation. Kinase activation seems to be the last step common to all of insulin's physiological events.

7.4.5.5. β-Subunit Autophosphorylation Distinguishes Glucose Transport from Mitogenic Stimulation

We have learned that a receptor's *trans* face provides ligand specificity by the formation of specific intermolecular interactions. In contrast, a transmembrane signal such as the insulin receptor's putative conformational change, seems to be fairly widespread in nature. However, if numerous receptors signaled via similar conformational changes, how could intracellular metabolic specificity be achieved? Specificity is provided to a large degree by a receptor's cytosolic domain.

Let us begin by considering receptor autophosphorylation. Tyr residues are found in three groups on the β chain: (1) residues 1162 and 1163, (2) residues 1328 and 1334, and (3) residues 949, 965, and 972. The first two groups participate in signaling whereas the significance of the third is unknown. Using site-directed mutagenesis, Ellis *et al.* (1986) replaced Tyr-1162 and Tyr-1163 with phenylalanine. This mutant receptor did not display tyrosyl kinase activity toward exogenous substrates nor did it stimulate hexose transport or glycogen synthesis. However, cells transfected with this mutant receptor *did* exhibit an insulin-stimulated hypersensitivity for DNA synthesis (Debant *et al.*, 1988). Some autophosphorylation of the β subunit was also noted in these latter studies. Therefore, insulin's glucose transport and DNA synthesis stimulating pathways bifurcate prior to autophosphorylation of residues 1162 and 1163. This suggests that at least two different signals are carried away from the receptor into the cytosol. There seems to be no shortage of potential mediators of these signals.

7.4.5.6. The β-Subunit's COOH-Terminus Collaborates with Autophosphorylation to Stimulate Glucose Transport

Although autophosphorylation is required to stimulate glucose transport, additional conditions must also be met. The primary structure of the tyrosyl kinase domain affects the ability of the autophosphorylated receptor to propagate an intracellular signal. For example, a chimeric receptor was formed from the extracellular portion of the insulin receptor and the tyrosyl kinase from the viral oncogene *v-ros*. Insulin could stimulate autophosphorylation but was unable to trigger hexose uptake in transfected cells. The importance of the tyrosyl kinase's sequence is also illustrated by a deletion mutant of the insulin receptor. The 43 COOH-terminal amino acids were removed from cDNA encoding the insulin receptor. When expressed in transfected cells, these mutant receptors exhibited normal endocytosis, recycling, insulin binding, and autophosphorylation. However, insulin was unable to stimulate glucose uptake or glycogen synthesis. Potential sites for serine and threonine phosphorylation are lost in this deletion mutant. This region of the β subunit is functionally important in communicating with the cytosol.

7.4.5.7. Receptor-Mediated Tyrosyl Phosphorylation of pp185 and/or ERK Seryl/ Threonyl Kinases May Be Early Events in Signaling

As described in the preceding paragraphs, the best characterized substrate for the insulin receptor's tyrosyl kinase activity is the receptor itself. However, recent studies have shown that insulin binding leads to tyrosyl phosphorylation of a seryl/threonyl kinase and a 185-kDa phosphoprotein (pp185) which is a specific cellular substrate (While *et al.*, 1985; Sun *et al.*, 1991). The gene encoding pp185 (*IRS-1*) has been cloned (Sun *et al.*, 1991). The deduced amino acid sequence of this cytoplasmic protein shows numerous potential phosphorylation sites: 10 Tyr phosphorylation sites and 35 potential Ser/Thr phosphorylation sites. An apparent ATP binding site was also revealed by the sequence. When insulin binds to its receptor, pp185 undergoes Tyr phosphorylation and binds the enzyme PI_3-kinase, which catalyzes the formation of the rare lipid PI_3-phosphate. Although the role of PI_3-phosphate in transmembrane signaling is not appreciated, these data suggest that pp185 may link the insulin receptor's tyrosyl kinase to cytoplasmic signaling machinery.

Extracellular signal-related kinases (ERK) may also link insulin receptors to their signal transduction apparatus (Boulton et al., 1991). ERKs become Tyr phosphorylated in response to insulin, possibly via insulin receptors. Future studies should establish where ERKs participate in key phosphorylation pathways that regulate metabolism and the physiological roles of the various potential molecular intermediaries (Sun et al., 1991; Boulton et al., 1991; Chen et al., 1992).

7.4.5.8. Activation of Seryl/Threonyl Kinases Leads to Altered Protein Phosphorylation Patterns

Insulin binding leads to both protein phosphorylation and dephosphorylation reactions. Table 7.2 lists some of the proteins altered by insulin's transmembrane signaling. For example, glycogen synthase is dephosphorylated during signaling, thus leading to its activation and the storage of glucose as glycogen. On the other hand, some enzymes such as acetyl CoA carboxylase are phosphorylated due to exposure to insulin treatment. However, the physiological roles of these phosphorylations are not yet completely clear. The way in which these dephosphorylations and phosphorylations are controlled is unknown.

Figure 7.18 shows two hypothetical models of seryl kinase activation by insulin (Czech et al., 1988). The kinase cascade hypothesis is shown on the left and the mediator hypothesis on the right. The cascade hypothesis proposes that a seryl kinase becomes phosphorylated by the insulin receptor's tyrosyl kinase activity. An insulin receptor binding seryl kinase has been identified (Czech et al., 1988; Smith et al., 1988). As mentioned above, a family of seryl/threonyl kinases which become activated and phosphorylated on Tyr residues in response to ligands such as insulin has been identified (Boulton et al., 1991). Tyr phosphorylation is believed to activate the seryl kinase(s) which then trigger appropriate phosphorylation reactions leading to the observed phosphorylation and dephosphorylation events. Since the autophosphorylated receptor remains active *after* insulin has dissociated from the α subunit, the β subunit could turn over many seryl kinases, thus amplifying a cell's response. The seryl kinase activation is thought to be propagated within a cell ultimately leading to phosphorylation of appropriate substrates. However, the components of the phosphorylation pathway and their physiological roles have not yet been characterized.

Seryl kinase activation has also been proposed to be mediated by small cytoplasmic molecules. One of the likely cytoplasmic messengers is the phosphoinositol glycan head group (Figure 7.18). Insulin has been shown to stimulate the release of glycan head groups from PI-glycans at a membrane's *cis* face (Saltiel et al., 1986; Saltiel and

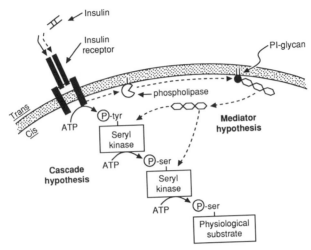

Figure 7.18. Models for the activation of seryl kinases. The cascade hypothesis (solid arrows) suggests that the insulin receptor's tyrosyl kinase activity triggers a series of phosphorylation reactions that leads to alterations in cellular physiology. The mediator hypothesis proposes that the tyrosyl kinase activates a phospholipase C. The phospholipase C, in turn, triggers the release of glycans from PI-glycans. These glycans are thought to modulate the activities of seryl kinases. These two hypotheses are not mutually exclusive. (Redrawn from Czech *et al.*, 1988, *J. Biol. Chem.* **263**:11017, with permission from the American Society for Biochemistry and Molecular Biology.)

Cuatrecasas, 1986). This hypothesis proposes that an autophosphorylated insulin receptor undergoes a change that allows it to activate a membrane-associated phospholipase C, which is enzymatically distinct from the PI_3-kinase discussed above. Recently, Downing *et al.* (1989) have discussed a phospholipase C activity that is phosphorylated in response to ligation of the platelet-derived growth factor receptor, another member of the tyrosyl kinase-containing receptor family. A similar phospholipase C activity may participate in insulin's signaling apparatus. The phospholipase releases the glycan moiety from the membrane while the diacylglycerol remains associated with a membrane. The diacylglycerol can then stimulate PKC activity. When added to cell-free extracts, this glycan stimulated the phosphorylation of insulin-sensitive proteins in a fashion indistinguishable from that observed for insulin stimulation of intact cells. Furthermore, this glycan enhanced phosphodiesterase activity, thus accounting for reduced cAMP levels during insulin stimulation. These studies suggest that either or both pathways may contribute to insulin action.

7.4.6. Physiological Regulation

The insulin receptor's physiological activity is regulated by several mechanisms. However, these mechanisms remain rather sketchy and concrete physiologically relevant regulatory pathways are unknown. These pathways likely involve protein phosphorylation, dephosphorylation, and endocytosis.

7.4.6.1. Adrenergic Receptor Binding Diminishes the Insulin Receptor's Transmembrane Signaling

Cells must interact with their environment in a controlled fashion to yield a coordinated physiological response. In the complex environment *in vivo*, cells are often presented with mutually exclusive signals. However, cells can only display a single sort of response at any given time. For example, it makes no sense to simultaneously stimulate glycogen storage and glycogenolysis. One mechanism that cells use to avoid conflicting cytosolic signals is interreceptor signaling.

After exposure to adrenergic agonists, adipocytes become refractive to insulin stimulation. Catecholamines dramatically reduce insulin receptor tyrosyl kinase activity and eliminate insulin-sensitive glucose transport (Haring *et al.*, 1986). As we have previously described, β-adrenergic receptors trigger the production of cAMP in response to agonist binding. In turn, cAMP activates PKA which then phosphorylates regulatory and catalytic enzymes leading to a cell's response (Figure 7.10). In addition to these proteins, PKA is believed to phosphorylate Ser residues of the insulin receptor's β subunit. The insulin receptor's activity is influenced in at least two fashions by PKA-mediated seryl phosphorylation. First, the α subunit's insulin binding activity is reduced. This reduction, however, is insufficient to account for the insulin resistance observed after catecholamine treatment. Furthermore, an adrenergic agonist was found to increase the β subunit's K_d for ATP binding by 400% (Haring *et al.*, 1986). These two effects inhibit insulin's transmembrane signaling. Therefore, a uniform *in vivo* response to catecholamines is generated by the β-adrenergic receptor's effector pathway and its inhibitory action on insulin receptor function.

7.4.6.2. Dephosphorylation Reactions Control Phosphoprotein Activity

We have previously encountered tyrosyl phosphorylation reactions that activate insulin's signaling and seryl phosphorylations that inhibit its activity. The phosphorylation state of insulin's receptor and its cytosolic signaling apparatus is not constant; it continually changes in response to environmental signals. To allow these changes to occur, there must also be enzymes that remove phosphate groups from proteins. These enzymes are called phosphatases. Equation (7.11) illustrates the reciprocal roles of kinases and phosphatases. The roles of phosphatases in regulating receptor-mediated signaling are just beginning to be appreciated (Cohen, 1989; Hunter, 1989).

As suggested by Eq. (7.11), protein phosphatases reverse the effects of protein kinases in a controlled manner. Although the regulatory pathways are not known, they are likely to be just as complicated as the protein kinase regulatory pathways. There are two independent families of protein phosphatases: the protein–serine phosphatases (PSPase) and protein–tyrosine phosphatases (PTPase). As their names imply, these enzymes remove phosphate from seryl and tyrosyl residues, respectively. The multiple

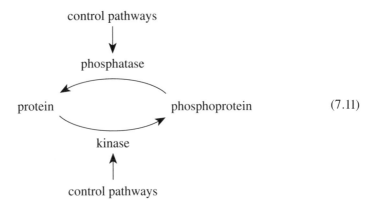

$$\text{(7.11)}$$

classes of PSPases and PTPases suggest that they have distinct substrate specificities. Certain specificities of the PSPases have been established (Cohen, 1989). The PSPases have regulatory subunits, in analogy with PKA.

Both soluble and membrane-bound protein phosphatases have been identified. PTPases associated with either the cytosol or membrane may regulate transmembrane signaling. For example, it appears as though some cell surface receptors express PTPase activity at their cytoplasmic surfaces (Hunter, 1989). This suggests that some receptors may function via tyrosyl dephosphorylation. It is attractive to speculate that PTPases downregulate receptor function whereas PSPases may upregulate receptor function (Figure 7.19). Similarly, numerous intracellular protein targets are likely regulated by these phosphatases.

7.4.6.3. Endocytotic Regulation

The endocytosis of insulin is not required for its transmembrane signaling. This is illustrated by the fact that insulin covalently linked to a large immobilized surface is capable of stimulating the physiological response of native insulin. Furthermore, it has been observed that triggering occurs more rapidly than endocytosis. Although internalization is not required for cytoplasmic signaling, it does participate in receptor downregulation. The number of cell surface insulin receptors is regulated by their internalization, as we have previously seen for β-adrenergic receptors (Section 7.3.7.1). Fluorescence microscopy and electron microscopy of ferritin–insulin conjugates and [125I]insulin show that insulin ligand–receptor complexes are usually distributed uniformly about a cell's perimeter (Jarett and Smith, 1977; Schlessinger et al., 1978; Carpentier et al., 1979). Initially, most insulin receptors exhibit rapid lateral diffusion. However, as time passes lateral mobility disappears. Concomitantly, the ligand–receptor complexes accumulate as clusters in coated pits. Each of the receptor's β subunits possesses two tyrosine-containing β-turns (Section 6.2.1.3) which contribute to adaptor binding at coated pits (Rajagopalan et al., 1991; Backer et al., 1992). However, these coated pits are not specific for insulin receptors since insulin and epidermal growth factor can simultaneously collect at the same coated pit. Insulin then migrates from these

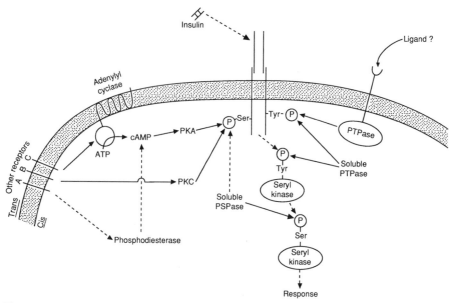

Figure 7.19. Potential interactions among phosphorylation and dephosphorylation reactions during signal transduction. Several of the activating (– – –) and inhibitory (——) pathways of cell signaling during insulin binding are shown.

membrane invaginations into the endosomal and endolysosomal compartments where it is destroyed (Carpentier *et al.*, 1979).

7.5. ADHERENCE RECEPTORS

The life cycles of many prokaryotes and eukaryotes revolve about cell adherence to other cells and to extracellular materials. Bacteria can adhere to one another to form clumps. This occurs, for example, during infections such as cholera. Cell-to-cell attachment also takes place during bacterial conjugation and in the formation of fruiting bodies by amoebae. Metazoans possess a bewildering array of adherence events; the scope of this problem is just beginning to be appreciated. For example, cells must interact to form tissues and tissues must assemble to form organisms. Sperm and egg cells adhere to create a new organism. During development, cells must sort to form specific patterns; these interactions require highly specific and informational adhesive events. Adhesive events represent bona fide receptor interactions since they trigger physiologically relevant processes such as cell patterning, cell shape and cytoskeletal assembly, host defense, cell maintenance or death, and a vast array of phenomena which generally require the participation of second messengers. This section will explore the receptor-mediated recognition of certain peptide and carbohydrate structures.

7.5.1. RGD Receptors

RGD receptors constitute a broad and newly discovered class of adherence receptors. They all recognize the amino acid sequence Arg-Gly-Asp (in the single-letter system, R-G-D), although they can vary with regard to *how* they recognize this tripeptide. Table 7.4 shows a summary of the RGD receptor family. In addition to these broad functions of RGD receptors in vertebrates, adherence reactions that mimic these receptors' RGD specificity have been found in certain prokaryotic and eukaryotic microorganisms.

The first RGD receptor to be recognized was the fibronectin receptor. Fibronectin is a large ($\sim$250 kDa) extracellular protein that binds to cell surface fibronectin receptors. Fibronectin is composed of multiple domains that separately bind to heparin, collagen, fibrin, or the fibronectin receptor. After a brief proteolytic treatment, each of fibronectin's domains can be isolated. The cell binding (or fibronectin receptor binding) domain was purified and then cleaved into small fragments. Both the intact cell binding domain and its proteolytic fragments inhibit cell adhesion to fibronectin-coated surfaces. When these molecules are in the supernate, they saturate a cell's fibronectin receptors, thus inhibiting adhesion.

The importance of the RGD sequence was recognized during studies of cell adherence (Pierschbacher and Ruoslahti, 1984; Yamada *et al.*, 1985). Certain proteolytic fragments or synthetic peptides mimicking portions of fibronectin's amino acid sequence inhibit fibronectin-mediated cell adherence in a competitive and reversible fashion. These same peptides promote cell attachment when linked to a solid support. By testing progressively smaller peptides, the amino acid sequence Gly-Arg-Gly-Asp-Ser (GRGDS) was found to possess nearly full receptor binding activity. Soluble RGD tripeptide was found to be a necessary and sufficient condition to inhibit attachment, although it is slightly less potent than the GRGDS sequence.

Several structure–activity relationships of the RGD tripeptide have been explored. The addition of just one methyl group abolishes activity. Furthermore, similarly charged

Table 7.4.
RDG Receptors and Functions

Receptor	Ligand	Function
Fibronectin receptor	Fibronectin	Cell attachment and phagocytosis
Vitronectin receptor	Vitronectin	Cell attachment
gpIIb/IIIa	Fibronectin	Blood clotting
	Fibrinogen	
	Vitronectin	
	Von Willebrand factor	
iC3b receptor	Complement fragment iC3b	Immune adhesion and phagocytosis
LFA-1	ICAM-1	Cell adhesion
p150,95		Cell adhesion

amino acids cannot be substituted for Arg and Asp. With one exception, any alteration in amino acid order eliminates its ability to affect adhesion. When the order of amino acids is exactly inverted (SDGR), its dose-dependent inhibition of attachment is almost identical to that of RGDS (Yamada and Kennedy, 1985). These data indicate that a highly restrictive spatial arrangement of a positive and a negative charge is required for recognition.

As mentioned above, the RGD sequence is used by many different ligand–receptor systems within the same organism (Table 7.4). This would, of course, lead to chaos if many ligands could be recognized by different receptors. The amino acids neighboring an RGD sequence likely confer receptor specificity upon biological ligands. Furthermore, the neighboring residues likely adopt different three-dimensional conformations, thus creating different shapes at the surface of a ligand. For example, fibronectin's shape in the vicinity of the RGD sequence probably determines its receptor specificity. Figure 7.20, panel a schematically illustrates how neighboring ligand and receptor structures might sterically regulate RGD-dependent ligand–receptor interactions.

Native fibronectin binds to fibroblasts in a specific and saturable fashion with a $K_d = 8 \times 10^{-7}$ M and $R_{tot} \cong 5 \times 10^5$ per cell. Fibronectin's receptor is a member of the integrin family of adherence receptors. As previously discussed (Figure 3.12), integrins link extracellular components to a cell's cytoskeleton at focal contacts.

The fibronectin receptor has been purified by both affinity and immuno-affinity chromatography. The affinity chromatography matrix is prepared by covalently linking

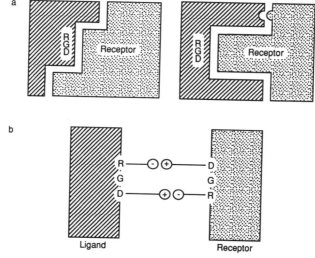

Figure 7.20. Possible principles involved in RGD ligand–receptor interactions. Several different ligand–receptor interactions display RGD specificity. Panel a illustrates how different ligand–receptor systems could utilize this tripeptide. The three-dimensional structures of the ligand and receptor allow only certain lock-and-key interactions, thus allowing specificity. The reverse sequence DGR has been observed in the ligand binding region of fibronectin's receptor. It is possible that charge–charge interactions, as shown in panel b, take place. (The ligands and receptors are crosshatched and stipled, respectively.)

fibronectin to large beads. When radiolabeled membrane extracts are passed through this column, fibronectin's receptor is retained on the matrix. RGD peptides are then added at a high concentration to elute receptors. This receptor has been found to be a heterodimer composed of an α (160 kDa) and a β (140 kDa) subunit (Figure 7.21). These same proteins were immunoprecipitated when antiadherence antibodies were added to membrane extracts. The α and β chains are held together in a noncovalent fashion.

The deduced amino acid sequences of both subunits of fibronectin's receptor have been determined (Argraves et al., 1987). These studies revealed that both subunits are transmembrane proteins with large extracellular domains (Figure 7.21). The cDNA sequences have also identified several additional structural features of this receptor. The α chain, which recognizes extracellular ligands, has an external domain that is homologous to the Ca^{2+} binding domain of calmodulin. The presence of a Ca^{2+} binding domain within this receptor was not unexpected since Ca^{2+} is required for ligand-to-receptor interaction. The β subunit possesses an extracellular Cys-rich domain and an intracellular region that can become phosphorylated. The Cys-rich domain contains over 20% Cys residues. Although this type of domain is found in many receptors, its function is unknown. The β subunit's COOH-terminus includes Ser, Thr, and Tyr phosphorylation sites. Furthermore, fibronectin receptors can become phosphorylated upon cell transformation (Hirst et al., 1986). Fibronectin receptors have affinity for the cytoskeletal proteins α-actinin and talin (e.g., Section 3.2.2.2). The talin binding site is contained within the β subunit's COOH-terminus. It seems likely that β-subunit phosphorylation

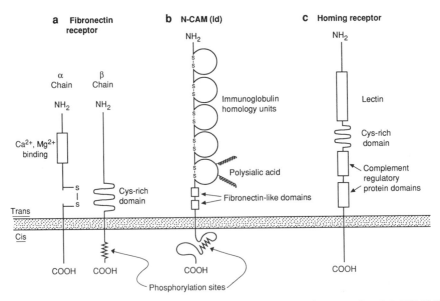

Figure 7.21. Structural features of three adherence receptors. The fibronectin receptor (panel a), NCAM (ld) (panel b), and lymphocyte homing receptor (panel c) are shown. Several important structural features are indicated.

affects its ability to interact with cytoskeletal proteins. The β chain's cytoplasmic domain also includes two sequences containing a Tyr residue within a β-turn (Reszka et al., 1992), as we have previously seen for endocytotic receptors (Section 6.2.1.3). These sequences participate in integrin clustering at adherence sites.

As mentioned above, the fibronectin receptor's ligand binding specificity is contained within the α subunit. The deduced amino acid sequence of the fibronectin receptor has shown the presence of a DGR tripeptide within the α subunit's Ca^{2+} binding region. This raises the possibility that antiparallel charge pairing (Figure 7.20, panel b) might contribute to ligand–receptor interactions.

The lateral mobility of fibronectin receptors reflects a cell's physiological state (Duband et al., 1988). Most fibronectin receptors are mobile on migrating cells. However, when cells become stationary and develop contacts, only 16% of the receptors remain mobile. As cells adhere, the fibronectin receptors form immobile clusters linked to the extracellular matrix and cytoskeleton.

RGD-dependent adherence reactions are of broad significance in biology. They participate in the aggregation of sea urchin cells and Dictyostelium. This type of adherence event also participates in the development of Drosophila and Xenopus embryos. In humans, chronic bleeding and infections result from genetic deficiencies of the adherence proteins gpIIa/IIIb and the iC3b receptor, respectively. As we will discuss in Section 9.4, RGD peptides may be useful in inhibiting the metastatic spread of cancer cells.

7.5.2. Cell Adherence Molecules of the Immunoglobulin Supergene Family

A second class of cell surface adherence receptors is homologous to domains within immunoglobulin molecules. The immunoglobulin supergene family includes protein products that contain an immunoglobulin homology unit. The immunoglobulin homology unit is just under 100 amino acids in length with strategically placed Cys residues. Disulfides formed from these residues fold the domain into a characteristic structure possessing antiparallel β sheets. Many cell surface recognition structures belong to the immunoglobulin supergene family. For example, HLA class I and II antigens, Thy-1 antigens, and the poly-Ig receptor are members. Several nonimmunological adhesion receptors are also in this family. They include the neural cell adhesion molecule (NCAM), myelin-associated glycoprotein (MAG), fasciclin II, and neuron–glial cell adhesion molecule (NgCAM). We shall now focus on the properties of NCAM, one of the more thoroughly studied cell adhesion molecules.

One important property of NCAM is its homophilicity. That is, NCAM simultaneously acts as both a ligand and a receptor. Homophilic interactions were first suggested by the ability of NCAM to spontaneously self-aggregate in solution. This property of NCAM is likely responsible for cell-to-cell adhesion. NCAM at the surface of a neural cell binds to NCAM on a second cell's surface. However, the many details of this binding interaction are presently unknown.

A structural model of NCAM is shown in Figure 7.21. Three forms of NCAM can result from the expression of the single NCAM gene (Cunningham *et al.*, 1987). Figure 7.21 shows the large cytoplasmic domain or ld structure. In addition, membrane proteins with a small cytoplasmic domain (sd) and a glycophospholipid-linked form without a cytoplasmic domain (ssd) are produced. One striking physical feature of NCAM is the presence of five immunoglobulin homology units at the *trans* membrane face. The three outermost immunoglobulin-like domains contain the binding site. Two segments of amino acids have also been found to be similar to fibronectin (Harrelson and Goodman, 1988).

NCAM undergoes several covalent modifications. The cytoplasmic COOH-terminal domain contains several Ser and Thr residues that can become phosphorylated. This protein is fatty acid acylated prior to expression. Four potential Asn-linked oligosaccharide side chains are found on NCAM. About one-third of embryonic NCAM's mass is carbohydrate. Of this carbohydrate, 80% is Asn-linked NANA in the form of an α-2,8-linked polysialic acid. NANA is rarely found in this form in vertebrates. Most of the polysialic acid is found in the vicinity of the fifth immunoglobulin homology unit.

The factors contributing to NCAM's regulation of cell adhesion are being studied in many laboratories. Although polysialic acid does not directly participate in molecular binding, it likely regulates the binding ability of NCAM (Rutishauser *et al.*, 1988). Due to charge–charge interactions, polysialic acid forms extended rodlike structures. These highly charged domains may prevent two NCAM molecules on the same cell's surface from interacting. Embryonic NCAM has 13 moles of NANA per 100 moles of amino acids. As cells develop, the total amount of NCAM-associated NANA decreases. Embryonic NCAM is 30% NANA whereas adult NCAM is only 10%. The higher NCAM charge density may minimize NCAM–NCAM binding, thus inhibiting cell attachment. Adult cells, which have formed specific patterns, retain specific interactions; the decreased NANA content may be required for these long-term interactions of NCAM.

Several physicochemical parameters, in addition to glycosylation, regulate NCAM's adherence function. Small changes in the cell surface NCAM density lead to large changes in cell adherence. The ability of the single NCAM gene to code for three membrane proteins suggests that these gene products vary in their function and/or regulation. For example, NCAM's ld form may be regulated by phosphorylation of its COOH-terminal domain. On the other hand, the ssd polypeptide could not be regulated in such a fashion. Furthermore, the ssd form likely displays very high lateral mobility in neural membranes and therefore it may affect adherence kinetics.

7.5.3. Carbohydrate Recognition Mechanisms in Membrane Adhesion

In addition to the protein–protein interactions described above, protein–carbohydrate binding reactions also contribute to receptor-mediated cell adherence. We have already encountered the protein-to-carbohydrate recognition events that occur during lectin binding (Section 3.1.5). Plant lectins, such as Con A, may play a role in stabilizing cell walls, host resistance against pathogens, and/or the developmental regulation of cell patterning in plants (Etzler, 1985). Membrane-bound lectins, which often act as receptors for another cell's oligosaccharide chains, participate in adhesive recognition. This section discusses the lectin-like recognition systems of sponge cells and human lymphocytes.

7.5.3.1. Adherence of Sponge Cells

Marine sponges are the oldest metazoans. These first confederations of cells needed adherence mechanisms to stick together and to discriminate self from nonself. Although sponges differ in membrane recognition mechanisms, they all employ lectin-like binding interactions during adhesion. In *Geodia cydonium*, different lectins participate in cell–cell and cell–substrate (or extracellular matrix) adherence. Other mechanisms of sponge cell adherence, such as collagen binding, will not be discussed.

Sponge cells can adhere to each other and to components of the extracellular matrix. When individual sponge cells are reassociated *in vitro*, aggregation proceeds in two distinct phases called primary and secondary aggregation. Primary aggregates are small ($\sim$60 μm) clumps that form in the absence of aggregation factors. This collagen-dependent adherence mechanism does not provide species selectivity. After the formation of primary aggregates, large secondary aggregates several millimeters in size form. The species-specific secondary aggregates are formed when two cells are bridged by aggregation factors.

Adhesive lectins play a major role in secondary aggregation. They are both membrane-bound and soluble proteins. The adhesion receptor of *Microciona prolifera* is a peripheral membrane glycoprotein expressing a lectin-like activity. This receptor binds to glucuronic acid attached to an extracellular glycoprotein called aggregation factor (Weinbaum and Burger, 1973). The interaction between the aggregation factor and its lectin-like receptor is a sufficient condition to promote aggregation. In the presence of Ca^{2+}, aggregation factors bound to different cells promote adherence by bridging two aggregation factor receptors.

The membrane adherence events of cells from *G. cydonium* have been studied in greater detail (Muller, 1985). Cell adherence is mediated by a very large (10^4 to 10^5 kDa) extracellular aggregation factor that bridges two cells. This complex glycoprotein contains a functionally diverse set of subunits including: (1) Ca^{2+} binding sites, (2) glycosyltransferase enzyme activity, (3) lectin binding sites, and (4) a lectin-like cell binding domain. Calcium is required for both primary and secondary sponge cell aggregation. During secondary aggregation calcium stabilizes the aggregation factor to promote binding. As described below, the glycosyltransferase activity regulates the aggregation factor receptor's function. The carbohydrate lectin binding sites of the aggregation factor interact with galactose-specific polylectins of the extracellular matrix to promote cell–substrate binding. The aggregation factor also possesses a cell binding domain that recognizes glucuronic acid of the aggregation factor receptor.

The aggregation factor's cell binding domain has been localized to a 47-kDa glycoprotein. About 12 copies of the cell binding domain are present in each copy of the aggregation factor. This suggests that the aggregation factor extensively cross-links the surfaces of adjoining cells. As expected, monoclonal antibodies directed against this domain bind in the extracellular region between cells (Gramzow *et al.*, 1986). It binds to a cell's aggregation factor receptors with high affinity ($K_a = 7 \times 10^8$ M^{-1}). Competition experiments indicate that 4×10^6 receptors are present on each cell. Although cell aggregation requires Ca^{2+}, the binding of the *G. cydonium* aggregation factor to its receptor is Ca^{2+}-independent. The aggregation factor receptor ($M_r = 16,000$) is over three-quarters carbohydrate. A terminal D-glucuronic acid moiety controls its ability to bind the aggregation factor. Figure 7.22 illustrates the bridging of two sponge cells by the aggregation factor.

The functional activity of the aggregation factor receptors is regulated at the level of the cell surface. The ability of the aggregation factor receptor to bind its ligand is regulated by its terminal glucuronic acid residue. The presence of this glucuronic acid moiety is in turn regulated by the presence of nearby enzymes. Sponge cells possess a cell surface β-glucuronidase activity (25 kDa). This enzyme cleaves terminal glucuronic

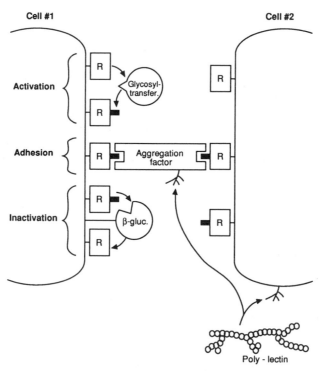

Figure 7.22. Mechanisms of sponge cell adherence. Two adherent sponge cells are shown. The steps involved in activation, adhesion, and inactivation are indicated. The aggregation factor receptor is activated by the addition of glucuronic acid by a glycosyltransferase (glycosyltransfer.). The aggregation factor joins two cells together by binding to glucuronic acid at each cell's surface. The receptor is inactivated by the removal of glucuronic acid by a β-glucuronidase (β-gluc.). The cell surface and aggregation factors can also bind to an extracellular poly-lectin, a component of the extracellular matrix.

acid residues from aggregation factor receptors; in doing so the receptor becomes inactivated (Figure 7.22). On the other hand, the receptor is activated by the action of glycosyltransferases. Depending on the species, glycosyltransferases are found as free enzymes in the interstitial space and as a component of aggregation factors (e.g., *G. cydonium*). This enzyme adds glucuronic acid to the receptor's oligosaccharide chains. Thus, the receptor becomes functionally activated and recognized by the lectin-like activity of the aggregation factor's cell binding domain. This provides sponge cells with convenient up- and down-regulatory adherence mechanisms.

After the aggregation factor is bound by its receptor, it is not internalized or chemically changed. However, the receptor transmits a proliferative signal to a cell's interior. Recent studies have shown that these signals may be accounted for by changes in PI levels (Muller *et al.*, 1987). The presence of PI signaling mechanisms in sponges demonstrates their evolutionary significance.

7.5.3.2. Lymphocyte Homing

After production in the bone marrow, lymphocytes become distributed throughout an organism via the cardiovascular circulatory system. Lymphocytes must cross an epithelial cell layer to reach the lymphatic system. During their life span, lymphocytes exhibit sessile phases within lymphoid organs, such as the thymus, and motile phases of recirculation throughout the lymphatic and circulatory systems. The ability of lymphocytes to nonrandomly migrate between anatomical sites is called lymphocyte homing. Lymphocyte homing is controlled by lectin-like interactions between lymphocytes and certain epithelial cells.

When leaving the cardiovascular circulation, lymphocytes adhere to cube-shaped endothelial cells of high endothelial venules (HEV). Specific cell surface structures of endothelial cells and lymphocytes mediate this interaction. Early studies showed that D-mannose and L-fucose could inhibit binding. Mannose-6-phosphate and fructose-1-phosphate, a structural analogue, are 20 to 50 times more potent binding inhibitors (Yednock *et al.*, 1987). Other phosphorylated or nonphosphorylated carbohydrates were without effect. These experiments provide a strong indication that mannose-6-phosphate is required for this cellular interaction. To determine which cell recognizes mannose-6-phosphate, lymphocytes and epithelial cells were preincubated with phosphomonoester mannan fragment (PPME), an oligosaccharide containing numerous mannose-6-phosphate residues. Epithelial cell-to-lymphocyte binding could only be inhibited when PPME was first added to lymphocytes. This indicates that lymphocytes express a mannose-6-phosphate receptor that recognizes epithelial cells. Subsequent studies showed that PPME-derivatized beads avidly bound to lymphocytes. These data show that lymphocytes possess a receptor for mannose-6-phosphate.

In a separate line of investigation, an anti-lymphocyte monoclonal antibody, MEL-14, was developed that inhibited lymphocyte-to-cuboidal HEV cell adhesion. MEL-14 binds only to a subset of lymphocytes that home to lymph nodes. These and other observations indicate that the antigen recognized by MEL-14 is the lymphocyte homing receptor* (Imai *et al.*, 1990). After immunoprecipitation this membrane antigen was found to be a 90-kDa glycoprotein by SDS-PAGE. Experiments using glycosidases and inhibitors of glycosylation suggested that the MEL-14 antigen is extensively glycosylated. However, glycosylation is solely via *N*-linked sites. Biochemical analysis showed that ubiquitin is covalently linked to MEL-14.

A cDNA encoding the putative lymphocyte homing receptor has been characterized (Siegelman *et al.*, 1989). This cDNA encodes a transmembrane protein with a short 18-amino-acid cytoplasmic tail. The mass calculated from the deduced amino acid sequence is 37 kDa. The difference between the deduced mass and the 90 kDa indicated by SDS-PAGE is accounted for by the addition of one or more ubiquitin chains (8.5 kDa) and glycosylation. The extracellular portion of the homing receptor is composed of four domains, as illustrated in Figure 7.21. The domain protruding farthest into the extracellular space bears a high degree of homology to other animal lectins. This domain doubtlessly binds to mannose-6-phosphate. A short Cys-rich sequence makes up an EGF-like domain within the receptor. This region is followed by two domains that exhibit homology to complement regulatory proteins, such as factor H. Although the roles of the homing receptor's EGF-like and complement regulatory protein-like domains are unknown, their homologues participate in molecular recognition at cell surfaces. These domains may also participate in adherence and/or accessory functions during lymphocyte homing. Although considerable progress has been made in the molecular characterization of the homing receptor, very little is known regarding its regulation at a molecular level.

7.6. A SURVEY OF SIGNALING STRATEGIES AND THEIR CYTOSKELETAL COMMUNICATION

In this chapter we have explored several types of ligand–receptor interactions that participate in neural signaling, endocrine signaling, and direct cell–cell or cell–substrate interactions. Certain properties of the receptors discussed in this and the preceding chapter are summarized in Table 7.5. These data illustrate the broad variety of physicochemical and biochemical properties associated with receptors. All bona fide receptors, as defined in Section 7.1, possess the ability to relay a message across a membrane. Their extracellular domains confer *ligand specificity* on a receptor. There are now reasonable grounds for believing that transmembrane domains provide a useful, perhaps bidirectional, transmitter for communicating physiological information across a membrane. The cytosolic domains provide *signaling specificity* for directing a cell's metabolic, genetic, and cytoskeletal machinery. This includes both the type of signal and the enzyme(s) that recognize the signal.

Apart from transport signaling, we have encountered three major types of intracellular signaling. These intracellular signals include: (1) transmembrane potentials, (2) G protein-linked adenylyl cyclase activity, and (3) tyrosyl kinase activity. All of these

---

*This is also now known as L-selectin.

Table 7.5.
Survey of Receptor Properties

Class	Receptor	Transmembrane domains/monomer	K_d (μM)	Transmembrane effect	Messengers	Regulates phosphorylation
Channel	ACh	20	0.008 0.025	Ions	Voltage	Yes
RGC	β-Adrenergic	7	20	Conformation	G protein, cAMP	Yes
Polypeptide hormone	Insulin	1	0.0001	Conformation	Tyrosyl kinase, seryl kinase	Yes
Adherence	RGD	2	0.8	?	Ca^{2+}, DAG ?, cytoskeleton	Yes
Macromolecular transport	LDL	1			Endocytosis	?
Molecular transport	Lac permease	12	5	Conformation	Internalized ligand	?

forms of signaling are regulated by phosphorylation/dephosphorylation reactions. PKC is one important seryl kinase that participates in the regulation of these receptors. Moreover, PKC also couples certain receptors to their physiological response, although its activities are poorly understood. We briefly mentioned the activity of PKC in Section 7.4.5.8. We will now summarize the possible actions and roles of PKC in signal transduction and its communication with other receptor systems.

7.6.1. More on PKC and Its Role as an Activating Signal

Earlier in this chapter we learned that the seryl kinase PKA downregulates the functional activity of the ACh and insulin receptors. On the other hand, PKA was found to be responsible for epinephrine's metabolic responses. In addition to PKA, PKC was also found to downregulate the functional activity of ACh and insulin receptors. In addition to its inhibitory role, PKC acts as a cytosolic messenger for several receptor systems.

PKC is known to stimulate a variety of physiological processes. This family of responses involves the breakdown of phosphoinositides and calcium fluxes. For example, muscarinic ACh receptors are found on pancreatic β cells and smooth muscle cells. When ligated, these receptors stimulate insulin release from β cells and contraction of smooth muscle cells. Similarly, histamine, a mediator of allergic reactions, is released from mast cells after antibody molecules activate PKC. The molecular events responsible for these responses are just beginning to be sorted out.

At least seven types of PKC ($M_r \simeq 77,000$) have been identified in gene cloning experiments (Kikkawa *et al.*, 1989). In general, a given cell type will have several different PKC types which are generated by alternative RNA splicing or separate genes. PKC's NH_2-terminal domain provides regulatory functions while its COOH-terminal region is a phosphokinase. DAG and Ca^{2+} presumably bind to the NH_2-terminal domain. The NH_2-terminal domain contains a Cys-rich region that is homologous to the Cys–zinc DNA-binding fingers found in gene regulatory proteins. This presents the exciting possibility that these cytosolic messenger proteins are also nuclear gene regulatory proteins. Such a possibility would provide a convenient link between transmembrane signaling and gene expression. The catalytic COOH-terminal possesses an ATP binding site ($K_d = 4$ μM) and sequence homology to other protein kinases.

Although PKC's mechanism of action is not completely clear, Figure 7.23 outlines some of the elements of PKC-mediated signaling. After a ligand binds to a PKC-linked receptor, it stimulates the activity of a G protein, G_p. This protein is called G_p because it binds guanine nucleotides and requires phospholipids for activation. If pertussis toxin is added to most cells, it inactivates G_p. A stimulated G_p protein activates a membrane-associated PI-specific phospholipase C (see Section 7.4.5.8). The phospholipase C then cleaves PI-bisphosphate (PIP_2), which is located at a plasma membrane's *cis* face, into DAG and inositol trisphosphate (IP_3; **7**). The signal is then propagated along two paths, as Figure 7.23 shows.

When IP_3 is released from a membrane, it diffuses into the cytosol. Cytosolic IP_3,

$$\begin{array}{ccc}
O^- & & O^- \\
| & & | \\
^-O-P=O & & O=P-O^- \\
| & & | \\
O & & O
\end{array}$$

OH

OH HO

$$\begin{array}{c}
O \\
| \\
O=P-O^- \\
| \\
O_-
\end{array}$$

(7)

but not other inositol phosphates, binds to an IP_3 receptor on endoplasmic reticulum membranes with high affinity (K_d = 5 nM). The IP_3 receptor is a 260-kDa integral membrane glycoprotein that exists as a tetramer *in situ* and after careful solubilization (Supattapone *et al.*, 1988; Ross *et al.*, 1989; Chadwick *et al.*, 1990). Reconstitution studies have shown that the receptor is also a Ca^{2+} channel. Its deduced amino acid sequence (Mignery *et al.*, 1990) shows a channel-like area composed of eight transmembrane domains in the COOH-terminal region. Its NH_2-terminal region is believed to participate in IP_3 binding. When IP_3 is bound to this receptor, it undergoes a conformational change (Mignery and Sudhof, 1990) which is thought to trigger Ca^{2+} release into the cytosol. The ability of IP_3 receptors to release Ca^{2+} from the endoplasmic reticulum is regulated by PKA-dependent phosphorylation in at least some cell types. This

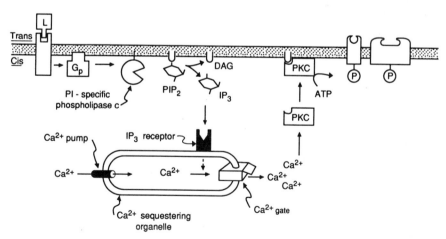

Figure 7.23. PKC-mediated signaling. A current model of protein kinase C-mediated (PKC) transmembrane signaling is shown. A ligand (L) binds to a membrane-bound receptor which in turn activates a G_p complex and a PI-specific phospholipase C. This phospholipase C cleaves PIP_2 into DAG and IP_3. DAG and cytosolic Ca^{2+} interact to trigger PKC activity.

provides another interactive pathway between signaling systems to regulate physiological responses.

DAG and Ca^{2+} collaborate to activate PKC at a membrane's *cis* face. The activated enzyme then begins phosphorylation of Ser and Thr residues of target proteins. These phosphorylation targets may include other plasma membrane-bound or intracellular receptors and a Na^+/H^+ membrane-bound translocator. Phosphorylation of the Na^+/H^+ exchanger by PKC is thought to alter cell pH and, consequently, a cell's biochemical pathways. In contrast to the opening and closing of a membrane channel, these numerous membrane interactions in PKC-mediated signaling require some time. Therefore, it should not be surprising that muscarinic responses are considerably slower than nicotinic cholinergic responses. Much current research activity is focused on sorting out the molecular details of PKC's mechanism of action and its roles in physiological responses.

7.6.2. Signaling Systems Interact with the Cytoskeleton

In the preceding discussions we have emphasized both how receptors accomplish their physiological assignments and how their behavior is regulated. We have not yet discussed the relationship between intracellular signals and the cytoskeleton. The cytoskeleton plays a key role in regulating a cell's shape and activities.

In Section 3.2.2.6 we briefly discussed cytoskeletal changes mediated by regulatory pathways. The number of potential regulatory pathways between the complex signaling mechanisms and the equally complex cytoskeleton are astronomical. The several interactions that have been analyzed do not approach the quantitative molecular and cellular physiology discussed earlier in this chapter.

Microtubules are influenced by free Ca^{2+} and MAPs (microtubule-associated proteins). As previously mentioned (Section 3.2.2.6), MAP2 binds to PKA and calmodulin, a calcium-binding protein. This provides two ways to regulate MAP2, which in turn regulates microtubules. Tau, another microtubule-associated protein, is regulated by both calmodulin and phosphorylation sites. The state of cellular microtubules can be acutely regulated by intracellular messengers produced by ligand–receptor interactions.

Cellular microfilaments are also regulated by intracellular signaling molecules. Their assembly is affected by calcium and MAP2 (Section 3.2.2.6). In erythrocytes PIP_2 participates in linking band 4.1 to glycophorin (Section 3.2.1.1). Its hydrolysis by phospholipase C alters the mobility of cell surface proteins and reorganizes membrane linkages. Recent studies have linked receptor ligation, PKA, and especially PKC activity with microfilament organization (Lassing and Lindberg, 1988; Burn, 1988; Petty and Martin, 1989; Zhou *et al.*, 1991). Synthesis of PIP_2 leads to dissociation of profilin, an actin-binding protein (Section 3.2.2.3), from actin, thus increasing a cell's functional actin pool. Moreover, profilin may regulate the PI signaling pathway (Goldschmidt-Clermont *et al.*, 1990). The formation of DAG by phospholipase C during cell triggering, which may or may not affect the steady-state level of PIP_2, potentiates membrane–microfilament interactions. In addition to PKC, DAG directly interacts with α-actinin.

DAG and α-actinin trigger a lateral phase separation in membranes. These DAG/α-actinin foci serve as nucleation and attachment sites for microfilaments. These molecular events may participate in the assembly of microfilaments observed in cells treated with ligands that signal via the PKC system. The protein tensin may also couple phosphorylation events into changes of cytoskeletal structure (Davis *et al.*, 1991). Although no clear picture of cytoskeletal signaling has emerged, our understanding of its regulation should improve as the molecular and cellular studies of receptor–cytoskeleton interactions merge. Plasma membrane receptor-mediated control is therefore very broad, extending from cytoskeletal architecture and cell metabolism to gene expression.

REFERENCES AND FURTHER READING

Receptors: An Introduction

Berg, H. C., and Purcell, E. M. 1977. Physics of chemoreception. *Biophys. J.* **20**:193–219.

DeLisi, C. 1981. The magnitude of signal amplification by ligand-induced clustering. *Nature* **289**: 322–323.

Hill, A. V. 1913. The combination of hemoglobin with oxygen and with carbon monoxide. *Biochem J.* **7**:471–480.

Leckband, D. E., *et al.* 1992. Long-range attraction and molecular rearrangements in receptor–ligand interactions. *Science* **255**:1419–1421.

Levitzki, A. 1984. *Receptors: A Quantitative Approach.* Benjamin–Cummings, Menlo Park, Calif.

Racker, E. 1985. *Reconstitutions of Transporters, Receptors, and Pathological States.* Academic Press, New York.

Scatchard, G. 1949. The attraction of proteins for small molecules and ions. *Ann. N.Y. Acad. Sci.* **51**:660–672.

Channel Receptors: The Nicotinic ACh Receptor

Barkas, T., *et al.* 1987. Mapping the main immunogenic region and toxin-binding site of the nicotinic acetylcholine receptor. *Science* **235**:77–80.

Changeux, J.-P., and Podleski, T. R. 1968. On the excitation and cooperativity of the electroplax membrane. *Proc. Natl. Acad. Sci. USA* **59**:944–950.

Changeux, J.-P., *et al.* 1981. Acetylcholine receptor: An allosteric protein. *Science* **225**:1335–1345.

Chothia, C., and Pauling, P. 1970. The conformation of cholinergic molecules at nicotinic nerve receptors. *Proc. Natl. Acad. Sci. USA* **65**:477–482.

Claudio, T., *et al.* 1987. Genetic reconstitution of functional acetylcholine receptor channels in mouse fibroblasts. *Science* **238**:1688–1694.

Cohen, J. B., and Changeux, J.-P. 1975. The cholinergic receptor protein in its membrane environment. *Annu. Rev. Pharmacol.* **15**:83–103.

Fels, G., *et al.* 1982. Equilibrium binding of acetylcholine to the membrane-bound acetylcholine receptor. *Eur. J. Biochem.* **127**:31–38.

Fong, T. M., and McNamee, M. G. 1986. Correlation between acetylcholine receptor function and structural properties of membranes. *Biochemistry* **25**:830–840.

Gershoni, J. M. 1987. Expression of the α-bungarotoxin binding site of the nicotinic acetylcholine receptor by *Escherichia coli* transformants. *Proc. Natl. Acad. Sci. USA* **84**:4318–4321.

Gonzalez-Ros, J. M., et al. 1983. Ligand-induced effects at regions of acetylcholine receptor accessible to membrane lipids. *Biochemistry* **22**:3807–3811.

Heidmann, O., et al. 1986. Chromosomal localization of muscle nicotinic acetylcholine receptor genes in the mouse. *Science* **234**:866–868.

Hemmings, H. C., et al. 1989. Role of protein phosphorylation in neuronal signal transduction. *FASEB J.* **3**:1583–1592.

Hopfield, J. F., et al. 1988. Functional modulation of the nicotinic acetylcholine receptor by tyrosine phosphorylation. *Nature* **336**:677–680.

Huganir, R. L., and Greengard, P. 1983. cAMP-dependent protein kinase phosphorylates the nicotinic acetylcholine receptor. *Proc. Natl. Acad. Sci. USA* **80**:1130–1134.

Huganir, R. L., et al. 1986. Phosphorylation of the nicotinic acetylcholine receptor regulates its rate of desensitization. *Nature* **321**:774–776.

Imoto, K., et al. 1988. Rings of negatively charged amino acids determine the acetylcholine receptor channel conductance. *Nature* **335**:645–648.

Leonard, R. J., et al. 1988. Evidence that the M2 membrane-spanning region lines the ion channel pore of the nicotinic receptor. *Science* **242**:1578–1581.

Miles, K., et al. 1989. Calcitonin gene-related peptide regulates phosphorylation of the nicotinic acetylcholine receptor in rat myotubes. *Neuron* **2**:1517–1524.

Nelson, N., et al. 1980. Reconstitution of purified acetylcholine receptors with functional ion channels in planar lipid bilayers. *Proc. Natl. Acad. Sci. USA* **77**:3057–3061.

Neumann, D., et al. 1986. Analysis of ligand binding to the synthetic dodecapeptide 185–196 of the acetylcholine receptor subunit. *Proc. Natl. Acad. Sci. USA* **83**:9250–9253.

Noda, M., et al. 1982. Primary structure of α-subunit precursor of *Torpedo californica* acetylcholine receptor deduced from cDNA sequence. *Nature* **299**:793–797.

Oiki, S., et al. 1988. M2δ, a candidate for the structure lining the ionic channel of the nicotinic cholinergic receptor. *Proc. Natl. Acad. Sci. USA* **85**:8703–8707.

Okamoto, M. 1980. Cholinergic receptors. In: *Advanced Cell Biology* (L. M. Schwartz and M. M. Azar, eds.) Van Nostrand–Reinhold, Princeton, N. J., pp. 288–303.

Phillips, W. D., et al. 1991a. ACh receptor-rich membrane domains organized in fibroblasts by recombinant 43-kilodalton protein. *Science* **251**:568–570.

Phillips, W. D., et al. 1991b. Mutagenesis of the 43-kD postsynaptic protein defines domains involved in plasma membrane targeting and AChR clustering. *J. Cell Biol.* **115**:1713–1723.

Poo, M.-M. 1982. Rapid lateral diffusion of functional ACh receptors in embryonic muscle cell membrane. *Nature* **295**:332–334.

Poulter, L., et al. 1989. Structure, oligosaccharide structures, and posttranslationally modified sites of the nicotinic acetylcholine receptor. *Proc. Natl. Acad. Sci. USA* **86**:6615–6649.

Qu, T., et al. 1990. Regulation of tyrosine phosphorylation of the nicotinic acetylcholine receptor at the rat neuromuscular junction. *Neuron* **2**:367–378.

RGC Receptors: The Adrenergic Receptor

Ahlquist, R. P. 1967. Development of the concept of alpha and beta adrenotropic receptors. *Ann. N.Y. Acad. Sci.* **139**:549–552.

Benovic, J. L., *et al.* 1989. β-adrenergic receptor kinase: Primary structure delineates a multigene family. *Science* **246**:235–240.

Berthelsen, S., and Pettinger, W. A. 1977. A functional basis for classification of α-adrenergic receptors. *Life Sci.* **21**:595–606.

Bouvier, M., *et al.* 1988. Removal of phosphorylation sites from the β_2-adrenergic receptor delays onset of agonist-promoted desensitization. *Nature* **333**:370–373.

Casperson, G. F., and Bourne, H. R. 1987. Cytochemical and molecular genetic analysis of hormone-sensitive adenylyl cyclase. *Annu. Rev. Pharmacol. Toxicol.* **27**:371–384.

Cerione, R. A., *et al.* 1983a. Reconstitution of β-adrenergic receptors in lipid vesicles: Affinity chromatography-purified receptors confer catecholamine responsiveness on a heterologous adenylate cyclase system. *Proc. Natl. Acad. Sci. USA* **80**:4899–4903.

Cerione, R. A., *et al.* 1983b. Pure β-adrenergic receptor: The single polypeptide confers catecholamine responsiveness to adenylate cyclase. *Nature* **306**:562–566.

Cotecchia, S., *et al.* 1988. Molecular cloning and expression of the cDNA for the hamster α_1-adrenergic receptor. *Proc. Natl. Acad. Sci. USA* **85**:7159–7163.

DeLean, A., *et al.* 1980. A ternary complex model explains the agonist-specific binding properties of the adenylate cyclase-coupled β-adrenergic receptor. *J. Biol. Chem.* **255**:7108–7117.

Dixon, R. A. F., *et al.* 1987. Ligand binding to the β-adrenergic receptor involves its rhodopsin-like core. *Nature* **326**:73–77.

Eimerl, S., *et al.* 1980. Functional implantation of solubilized β-adrenergic receptor in the membrane of a cell. *Proc. Natl. Acad. Sci. USA* **77**:760–764.

Emorine, L. J., *et al.* 1987. Structure of the gene for human β_2-adrenergic receptor: Expression and promoter characterization. *Proc. Natl. Acad. Sci. USA* **84**:6995–6999.

Feder, D., *et al.* 1986. Reconstitution of β_1-adrenoceptor-dependent adenylate cyclase from purified components. *EMBO J.* **5**:1509–1514.

Gilman, A. G. 1987. G proteins: Transducers of receptor-generated signals. *Annu. Rev. Biochem.* **56**:615–649.

Hanski, E., *et al.* 1979. Adenylate cyclase activation by the β-adrenergic receptors as a diffusion-controlled process. *Biochemistry* **18**:846–853.

Hirata, F., *et al.* 1979. β-adrenergic receptor agonists increase phospholipid methylation, membrane fluidity, and β-adrenergic receptor–adenylate cyclase coupling. *Proc. Natl. Acad. Sci. USA* **76**:368–372.

Houslay, M. D., *et al.* 1976. The lipid environment of the glucagon receptor regulates adenylate cyclase activity. *Biochim. Biophys. Acta* **436**:495–504.

Kobilka, B. K., *et al.* 1987a. cDNA for the human β_2-adrenergic receptor: A protein with multiple membrane-spanning domains and encoded by a gene whose chromosomal location is shared with that of the receptor for platelet-derived growth factor. *Proc. Natl. Acad. Sci. USA* **84**:46–50.

Kobilka, B. K., *et al.* 1987b. Cloning, sequencing, and expression of the gene coding for the human platelet α_2-adrenergic receptor. *Science* **238**:650–656.

Kobilka, B. K., *et al.* 1988. Chimeric α_2-,β_2-adrenergic receptors: Delineation of domains involved in effector coupling and ligand binding specificity. *Science* **240**:1310–1316.

Krupinski, J., *et al.* 1989. Adenylyl cyclase amino acid sequence: Possible channel- or transmitter-like structure. *Science* **244**:1558–1564.

Lefkowitz, R. J., *et al.* 1983. Adenylate cyclase-coupled beta-adrenergic receptors. *Annu. Rev. Biochem.* **52**:159–186.

Levitzki, A. 1988. From epinephrine to cyclic AMP. *Science* **241**:800–806.

Lohse, M. J., *et al.* 1990a. β-arrestin: A protein that regulates β-adrenergic receptor function. *Science* **248**:1547–1550.

Lohse, M. J., *et al.* 1990b. Multiple pathways of rapid β_2-adrenergic receptor desensitization. *J. Biol. Chem.* **265:**3202–3209.

May, D. C., *et al.* 1985. Reconstitution of catecholamine-stimulated adenylate cyclase activity using three purified proteins. *J. Biol. Chem.* **260:**15829–15833.

Mukherjee, C., *et al.* 1975a. Identification of adenylate cyclase coupled β-adrenergic receptors in frog erythrocytes with $(-)$ [³H]alprenolol. *J. Biol. Chem.* **250:**4869–4876.

Mukherjee, C., *et al.* 1975b. Catecholamine-induced subsensitivity of adenylate cyclase associated with loss of β-adrenergic receptor binding sites. *Proc. Natl. Acad. Sci. USA* **72:**1945–1949.

Pitcher, J. A., *et al.* 1992. Role of $\beta\gamma$ subunits of G proteins in targeting the β-adrenergic receptor kinase to membrane-bound receptors. *Science* **257:**1264–1267.

Puchwein, G., *et al.* 1974. Uncoupling of catecholamine activation of pigeon erythrocyte membrane adenylate cyclase by filipin. *J. Biol. Chem* **249:**3232–3240.

Ross, E. M., and Gilman, A. G. 1980. Biochemical properties of hormone-sensitive adenylate cyclase. *Annu. Rev. Biochem.* **49:**533–564.

Seamon, K. B., and Daly, J. W. 1981. Forskolin: A unique diterpene activator of cyclic AMP-generating systems. *J. Cyclic Nucleotide Res.* **7:**201–224.

Sibley, D. R., and Lefkowitz, R. J. 1985. Molecular mechanisms of receptor desensitization using the β-adrenergic receptor-coupled adenylate cyclase system as a model. *Nature* **317:**124–129.

Simon, M. I., *et al.* 1991. Diversity of G proteins in signal transduction. *Science* **252:**802–808.

Spiegel, A. M., *et al.* 1985. Clinical implications of guanine nucleotide-binding proteins as receptor–effector couplers. *N. Engl. J. Med.* **312:**26–33.

Stadel, J. M., *et al.* 1980. A high affinity agonist·β-adrenergic receptor complex is an intermediate for catecholamine stimulation of adenylate cyclase in turkey and frog erythrocyte membranes. *J. Biol. Chem.* **255:**1436–1441.

Stadel, J. M., *et al.* 1983. Desensitization of the β-adrenergic receptor of frog erythrocytes: Recovery and characterization of the down-regulated receptors in sequestered vesicles. *J. Biol. Chem.* **258:**3032–3038.

Strader, C. D., *et al.* 1987a. The carboxyl terminus of the hamster β-adrenergic receptor expressed in mouse L cells is not required for receptor sequestration. *Cell* **49:**855–863.

Strader, C. D., *et al.* 1987b. Identification of residues required for ligand binding to the β-adrenergic receptor. *Proc. Natl. Acad. Sci. USA* **84:**4384–4388.

Strader, C. D., *et al.* 1988. Conserved aspartic acid residues 79 and 113 of the β-adrenergic receptor have different roles in receptor function. *J. Biol. Chem.* **263:**10267–10271.

Strader, C. D., *et al.* 1989. Identification of two serine residues involved in agonist activation of the β-adrenergic receptor. *J. Biol. Chem.* **264:**13572–13578.

Strulovici, B., *et al.* 1984. Direct demonstration of impaired functionality of a purified desensitized β-adrenergic receptor in a reconstituted system. *Science* **227:**837–840.

Valiquette, M., *et al.* 1990. Involvement of tyrosine residues located in the carboxyl tail of the human β_2-adrenergic receptor in agonist-induced down regulation of the receptor. *Proc. Natl. Acad. Sci. USA* **87:**5089–5093.

Waldo, G. L., *et al.* 1983. Characterization of an altered membrane form of the β-adrenergic receptor produced during agonist-induced desensitization. *J. Biol. Chem.* **258:**13900–13908.

Williams, L. T., and Lefkowitz, R. J. 1977. Slowly reversible binding of catecholamine to a nucleotide-sensitive state of the β-adrenergic receptor. *J. Biol. Chem.* **252:**7207–7213.

Yarden, Y., *et al.* 1986. The avian β-adrenergic receptor: Primary structure and membrane topology. *Proc. Natl. Acad. Sci. USA* **83:**6795–6799.

Polypeptide Hormone Receptors: The Insulin Receptor

Auberger, P., *et al.* 1989. Characterization of a natural inhibitor of the insulin receptor tyrosine kinase: cDNA cloning, purification, and anti-mitogenic activity. *Cell* **58**:631–640.

Backer, J. M., *et al.* 1992. The insulin receptor juxtamembrane region contains two independent tyrosine/β-turn internalization signals. *J. Cell Biol.* **118**:831–839.

Bell, G. I., *et al.* 1982. The highly polymorphic region near the human insulin gene is composed of simple tandemly repeating sequences. *Nature* **295**:31–35.

Boni-Schnetzler, M., *et al.* 1987. The insulin receptor: Structural basis for high affinity ligand binding. *J. Biol. Chem.* **262**:8395–8401.

Boni-Schnetzler, M., *et al.* 1988. Ligand-dependent intersubunit association within the insulin receptor complex activates its intrinsic kinase activity. *J. Biol. Chem.* **263**:6822–6828.

Boulton, T. G., *et al.* 1991. ERKs: A family of protein-serine/threonine kinases that are activated and tyrosine phosphorylated in response to insulin and NGF. *Cell* **65**:663–675.

Carpentier, J. L., *et al.* 1979. Lysosomal association of internalized 125-I-insulin in isolated rat hepatocytes. Direct demonstration by quantitative electron microscopic autoradiography. *J. Clin. Invest.* **63**:1249–1261.

Carpentier, J. L., *et al.* 1992. Insulin-induced surface redistribution regulates internalization of the insulin receptor and requires its autophosphorylation. *Proc. Natl. Acad. Sci. USA* **89**:162–166.

Cautrecasas, P. 1974. Membrane receptors. *Annu. Rev. Biochem.* **43**:169–214.

Chan, B. L., *et al.* 1988. Insulin-stimulated release of lipoprotein lipase by metabolism of its phosphatidylinositol anchor. *Science* **241**:1670–1672.

Chen, J., *et al.* 1992. Regulation of protein serine-threonine phosphatases type-2A by tyrosine phosphorylation. *Science* **257**:1261–1264.

Cheng, K., and Larner, J. 1985. Intracellular mediators of insulin action. *Annu. Rev. Physiol.* **47**:405–424.

Chothia, C., *et al.* 1983. Transmission of a conformational change in insulin. *Nature* **302**:500–505.

Christiansen, K., *et al.* 1991. A model for the quaternary structure of human placental insulin receptor deduced from electron microscopy. *Proc. Natl. Acad. Sci. USA* **88**:249–252.

Clark, S., and Harrison, L. C. 1983. Disulfide exchange between insulin and its receptor. *J. Biol. Chem.* **258**:11434–11437.

Cohen, P. 1989. The structure and regulation of protein phosphatases. *Annu. Rev. Biochem.* **58**:453–508.

Czech, M. P. 1985. The nature and regulation of the insulin receptor: Structure and function. *Annu. Rev. Physiol.* **47**:357–381.

Czech, M. P., *et al.* 1988. Insulin receptor signaling: Activation of multiple serine kinases. *J. Biol. Chem.* **263**:11017–11020.

Davis, R. J., and Czech, M. P. 1986. Regulation of transferrin receptor expression at the cell surface by insulin-like growth factors, epidermal growth factor and platelet-derived growth factor. *EMBO J.* **5**:653–658.

Debant, A., *et al.* 1988. Replacement of insulin receptor tyrosine residues 1162 and 1163 does not alter the mitogenic effect of the hormone. *Proc. Natl. Acad. Sci. USA* **85**:8032–8036.

De Meyts, P., *et al.* 1978. Mapping of the residues responsible for the negative cooperativity of the receptor-binding region of insulin. *Nature* **273**:504–509.

Downing, J. R., *et al.* 1989. Phospholipase C-γ, a substrate for PDGF receptor kinase, is not phosphorylated on tyrosine during the mitogenic response to CSF-1. *EMBO J.* **8**:3345–3350.

Ebina, Y., et al. 1985. The human insulin receptor cDNA: The structural basis for hormone-activated transmembrane signalling. Cell **40**:747–758.

Ebina, Y., et al. 1987. Replacement of lysine residue 1030 in the putative ATP-binding region of the insulin receptor abolishes insulin- and antibody-stimulated glucose uptake and receptor kinase activity. Proc. Natl. Acad. Sci. USA **84**:704–708.

Ellis, L., et al. 1986. Replacement of insulin receptor tyrosine residues 1162 and 1163 compromises insulin-stimulated kinase activity and uptake of 2-deoxyglucose. Cell **45**:721–732.

Fan, J. Y., et al. 1982. Receptor-mediated endocytosis of insulin: Role of microvilli, coated pits, and coated vesicles. Proc. Natl. Acad. Sci. USA **79**:7788–7791.

Fehlmann, M., et al. 1982. Internalized insulin receptors are recycled to the cell surface in rat hepatocytes. Proc. Natl. Acad. Sci. USA **79**:5921–5925.

Freychet, P., et al. 1971. Insulin receptors in the liver: Specific binding of 125-I insulin to the plasma membrane and its relation to insulin bioactivity. Proc. Natl. Acad. Sci. USA **68**:1833–1837.

Gammeltoft, S. 1984. Insulin receptors: Binding kinetics and structure–function relationship of insulin. Physiol. Rev. **64**:1321–1378.

Grunfeld, C. 1984. Antibody against the insulin receptor causes disappearance of insulin receptors in 3T3-L1 cells: A possible explanation of antibody-induced insulin resistance. Proc. Natl. Acad. Sci. USA **81**:2508–2511.

Haneda, M., et al. 1984. Familial hyperinsulinemia due to a structurally abnormal insulin. N. Engl. J. Med **310**:1288–1294.

Haring, H., et al. 1986. Decreased tyrosine kinase activity of insulin receptor isolated from rat adipocytes rendered insulin-resistant by catecholamine treatment in vivo. Biochem. J. **234**:59–66.

Helmerhorst, E., et al. 1986. High molecular weight forms of the insulin receptor. Biochemistry **25**:2060–2065.

Herzfeld, J., et al. 1987. Contrasting molecular dynamics in red and purple membrane fractions of Halobacterium halobium. Biophys. J. **52**:855–858.

Hunter, T. 1989. Protein-tyrosine phosphatases: The other side of the coin. Cell **58**:1013–1016.

Inouye, K., et al. 1981. Semisynthesis and properties of some insulin analogs. Biopolymers **20**:1845–1858.

Jarett, L., and Smith, R. M. 1977. The natural occurrence of insulin receptors in groups on adipocyte plasma membranes as demonstrated with monomeric ferritin–insulin. J. Supramol. Struct. **6**:45–59.

Jarett, L., et al. 1980. Insulin receptors: Differences in structural organization on adipocyte and liver plasma membranes. Science **210**:1127–1128.

Kashles, O., et al. 1988. Ligand-induced stimulation of epidermal growth factor receptor mutants with altered transmembrane regions. Proc. Natl. Acad. Sci. USA **85**:9567–9571.

Kasuga, M., et al. 1982. Insulin stimulates the phosphorylation of the 95,000-dalton subunit of its own receptor. Science **215**:185–187.

Kjeldsen, T., et al. 1991. The ligand specificities of the insulin receptor and the insulin-like growth factor I receptor reside in different regions of a common binding site. Proc. Natl. Acad. Sci. USA **88**:4404–4408.

Luly, P., and Shinitzky, M. 1979. Gross structural changes in isolated liver cell plasma membrane upon binding of insulin. Biochemistry **18**:445–450.

Maxfield, F. R, et al. 1978. Collection of insulin, EGF, and α_2 macroglobulin in the same patches on the surface of cultured fibroblasts and common internalization. Cell **14**:805–810.

Moe, G. R., et al. 1989. Transmembrane signaling by a chimera of the Escherichia coli aspartate receptor and the human insulin receptor. Proc. Natl. Acad. Sci. USA **86**:5683–5687.

Morgan, D. O., *et al*. 1986. Insulin action is blocked by a monoclonal antibody that inhibits the insulin receptor kinase. *Proc. Natl. Acad. Sci. USA* **83**:328–332.

Nakagawa, S. H., and Tager, H. S. 1987. Role of the COOH-terminal β-chain domain in insulin–receptor interactions. *J. Biol. Chem.* **262**:12054–12058.

Pang, D. T., and Shager, J. A. 1984. Evidence that insulin receptor from human placenta has a high affinity for only one molecule of insulin. *J. Biol. Chem.* **259**:8589–8596.

Pilch, P. F., and Czech, M. P. 1979. Interaction of cross-linking agents with the insulin effector system of isolated fat cells. Direct linkage of 125-I insulin to plasma membrane receptor protein of 140,000 daltons. *J. Biol. Chem.* **254**:3375–3381.

Pilch, P. F., and Czech, M. P. 1980. Hormone binding alters the conformation of the insulin receptor. *Science* **210**:1152–1153.

Pullen, R. A., *et al*. 1976. Receptor-binding region of insulin. *Nature* **259**:369–373.

Rajagopalan, M., *et al*. 1991. Amino acid sequence gly-pro-leu-tyr and asn-pro-glu-tyr in the submembranous domain of the insulin receptor are required for normal endocytosis. *J. Biol. Chem.* **266**:23068–23073.

Ronnett, G. V., *et al*. 1984. Role of glycosylation in the processing of newly translated insulin proreceptor in 3T3-L1 adipocytes. *J. Biol. Chem.* **259**:4566–4575.

Rosen, O. M. 1987. After insulin binds. *Science* **237**:1452–1458.

Roth, R. A., and Cassell, D. J. 1983. Insulin receptor: Evidence that it is a protein kinase. *Science* **219**:299–301.

Saltiel, A. R., and Cuatrecasas, P. 1986. Insulin stimulates the generation from hepatic plasma membranes of modulators derived from an inositol glycolipid. *Proc. Natl. Acad. Sci. USA* **83**:5793–5797.

Saltiel, A. R., *et al*. 1986. Insulin-stimulated hydrolysis of a novel glycolipid generates modulators of cAMP phosphodiesterase. *Science* **233**:967–972.

Schlessinger, J., *et al*. 1978. Quantitative determination of the lateral diffusion coefficients of the hormone-receptor complexes of insulin and epidermal growth factor on the plasma membrane of cultured fibroblasts. *Proc. Natl. Acad. Sci. USA* **75**:5353–5357.

Schwartz, G. P., *et al*. 1989. A highly potent insulin: Des-(B26-B30)-[asp^{B10},tyr^{B25}-NH$_2$]insulin-(human). *Proc. Natl. Acad. Sci. USA* **86**:458–461.

Smith, C. J., *et al*. 1979. Insulin-stimulated protein phosphorylation in 3T3-L1 preadipocytes. *Proc. Natl. Acad. Sci. USA* **76**:2725–2729.

Smith, D. M., *et al*. 1988. Two systems *in vitro* that show insulin-stimulated serine kinase activity towards the insulin receptor. *Biochem. J.* **250**:509–519.

Soos, M. A., and Siddle, K. 1989. Immunological relationships between receptors for insulin and insulin-like growth factor I: Evidence for structural heterogeneity of insulin-like growth factor I receptors involving hybrids with insulin receptors. *Biochem. J.* **263**:553–563.

Sun, X. J., *et al*. 1991. Structure of the insulin receptor substrate IRS-1 defines a unique signal transduction protein. *Nature* **352**:73–77.

Sweet, L. J., *et al*. 1987. Isolation of functional αβ heterodimers from the purified human placental α$_2$β$_2$ heterotetrameric insulin receptor complex. *J. Biol. Chem.* **262**:6939–6942.

Tager, H., *et al*. 1979. A structurally abnormal insulin causing human diabetes. *Nature* **281**:122–125.

Treadway, J. L., *et al*. 1991. Transdominant inhibition of tyrosine kinase activity in mutant insulin/insulin-like growth factor I hybrid receptors. *Proc. Natl. Acad. Sci. USA* **88**:214–218.

White, M. F., *et al*. 1985. Insulin rapidly stimulates tyrosine phosphorylation of a M_r-185,000 protein in intact cells. *Nature* **318**:183–186.

Yamaguchi, Y., *et al*. 1991. Functional properties of two naturally occurring isoforms of the human insulin receptor in Chinese hamster ovary cells. *Endocrinology* **129**:2058–2066.

Adherence Receptors

Argraves, W. S, *et al*. 1987. Amino acid sequence of the human fibronectin receptor. *J. Cell Biol.* **105**:1183–1190.

Cunningham, B. A., *et al*. 1987. Neural cell adhesion molecule: Structure, immunoglobulin-like domains, cell surface modulation, and alternative RNA splicing. *Science* **236**:799–806.

Diehl-Seifert, B., *et al*. 1985. Physicochemical and functional characterization of the polymerization process of the *Geodia cydonium* lectin. *Eur. J. Biochem.* **147**:517–523.

Duband, J.-L., *et al*. 1988. Fibronectin receptor exhibits high lateral mobility in embryonic locomoting cells but is immobile in focal contacts and fibrillar streaks in stationary cells. *J. Cell Biol.* **107**:1385–1396.

Edelman, G. M. 1988. *Topobiology: An Introduction to Molecular Embryology*. Basic Books, New York.

Etzler, M. E. 1985. Plant lectins: Molecular and biological aspects. *Annu. Rev. Plant Physiol.* **36**:209–234.

Gramzow, M., *et al*. 1986. Identification and further characterization of the specific cell binding fragment from sponge aggregation factor. *J. Cell Biol.* **102**:1344–1349.

Harrelson, A. L., and Goodman, C. S. 1988. Growth cone guidance in insects: Fasciclin II is a member of the immunoglobulin superfamily. *Science* **242**:700–708.

Hirst, R., *et al*. 1986. Phosphorylation of the fibronectin receptor complex in cells transformed by oncogenes that encode tyrosine kinases. *Proc. Natl. Acad. Sci. USA* **83**:6470–6474.

Imai, Y., *et al*. 1990. Direct demonstration of the lectin activity of gp90[MEL], a lymphocyte homing receptor. *J. Cell Biol.* **111**:1225–1232.

Muller, W. E. G. 1985. Cell membranes in sponges. *Int. Rev. Cytol.* **77**:129–181.

Muller, W. E. G., *et al*. 1987. Role of the aggregation factor in the regulation of phosphoinositide metabolism in sponges. Possible consequences on calcium efflux and on mitogenesis. *J. Biol. Chem.* **262**:9850–9858.

Naidet, C., *et al*. 1987. Peptides containing the cell-attachment recognition signal Arg-Gly-Asp prevent gastrulation in *Drosophila* embryos. *Nature* **325**:348–350.

Ouaissi, M. A., *et al*. 1986. *Trypanosoma cruzi* infection inhibited by peptides modeled from a fibronectin cell attachment domain. *Science* **234**:603–607.

Pierschbacher, M. D., and Ruoslahti, E. 1984. Cell attachment activity of fibronectin can be duplicated by small synthetic fragments of the molecule. *Nature* **309**:30–33.

Pytela, R, *et al*. 1985. Identification and isolation of a 140 kd cell surface glycoprotein with properties expected of a fibronectin receptor. *Cell* **40**:191–198.

Pytela, R., *et al*. 1986. Platelet membrane glycoprotein IIb/IIa: Member of a family of Arg-Gly-Asp-specific adhesion receptors. *Science* **231**:1559–1562.

Reszka, A. A., *et al*. 1992. Identification of amino acid sequences in the integrin β1 cytoplasmic domain implicated in cytoskeletal association. *J. Cell Biol.* **117**:1321–1330.

Ruoslahti, E. 1989. Proteoglycans in cell regulation. *J. Biol. Chem.* **264**:13368–13372.

Ruoslahti, E., and Pierschbacher, M. D. 1987. New perspectives in cell adhesion: RGD and integrins. *Science* **238**:491–497.

Rutishauser, U., *et al*. 1988. The neural cell adhesion molecule (NCAM) as a regulator of cell–cell interactions. *Science* **240**:53–57.

Schaller, M. D., *et al*. 1992. pp125[FAK], a structurally distinctive protein-tyrosine kinase associated with focal adhesions. *Proc. Natl. Acad. Sci. USA* **89**:5192–5196.

Siegelman, M. H., *et al*. 1989. Mouse lymph node homing receptor cDNA clone encodes a glycoprotein revealing tandem interaction domains. *Science* **243**:1165–1172.

Weinbaum, G., and Burger, M. M. 1973. Two component system for surface guided reassociation of animal cells. *Nature* **244**:510–512.

Yamada, K. M., and Kennedy, D. W. 1985. Amino acid sequence specificities of an adhesive recognition signal. *J. Cell. Biochem.* **28**:99–104.

Yamada, K. M., *et al.* 1985. Recent advances in research on fibronectin and other cell attachment proteins. *J. Cell. Biochem.* **28**:79–97.

Yednock, T. A., *et al.* 1987. Phosphomannosyl-derivatized beads detect a receptor involved in lymphocyte homing. *J. Cell Biol.* **104**:713–723.

Signaling Strategies and Their Cytoskeletal Communication

Berridge, M. J. 1987. Inositol trisphosphate and diacylglycerol: Two interacting second messengers. *Annu. Rev. Biochem.* **56**:159–193.

Burn, P. 1988. Phosphatidylinositol cycle and its possible involvement in the regulation of cytoskeleton–membrane interactions. *J. Cell. Biochem.* **36**:15–24.

Chadwick, C. C., *et al.* 1990. Isolation and characterization of the inositol triphosphate receptor from smooth muscle. *Proc. Natl. Acad. Sci. USA* **87**:2132–2136.

Cheung, W. Y. 1979. Calmodulin plays a pivotal role in cellular regulation. *Science* **207**:19–27.

Davis, S., *et al.* 1991. Presence of an SH2 domain in the actin-binding protein tensin. *Science* **252**:712–715.

Edelman, A. M., *et al.* 1987. Protein serine/threonine kinases. *Annu. Rev. Biochem.* **56**:567–613.

Exton, J. H. 1988. Mechanisms of action of calcium-mobilizing agonists: Some variations on a young theme. *FASEB J.* **2**:2670–2676.

Gille, H., *et al.* 1992. Phosphorylation of transcription factor p62[TCF] by MAP kinase stimulates ternary complex formation at c-fos promoter. *Nature* **358**:414–417.

Goldschmidt-Clermont, P. J., *et al.* 1990. The actin-binding protein profilin binds to PIP_2 and inhibits its hydrolysis by phospholipase C. *Science* **247**:1575–1578.

Hanks, S. K., *et al.* 1988. The protein kinase family: Conserved features and deduced phylogeny of the catalytic domains. *Science* **241**:42–52.

Khan, A. A., *et al.* 1992. IP3 receptor: Localization to plasma membrane of T cells and co-capping with the T cell receptor. *Science* **257**:815–818.

Kikkawa, U., and Nishizuka, Y. 1986. The role of protein kinase C in transmembrane signalling. *Annu. Rev. Cell Biol.* **2**:149–178.

Kikkawa, U., *et al.* 1989. The protein kinase C family: Heterogeneity and its implications. *Annu. Rev. Biochem.* **58**:31–44.

Knighton, D. R., *et al.* 1991. Crystal structure of the catalytic subunit of cyclic adenosine monophosphate-dependent protein kinase. *Science* **253**:407–414.

Kyriakis, J. M., *et al.* 1992. Raf-1 activates MAP kinase-kinase. *Nature* **358**:417–421.

Lassing, I., and Lindberg, U. 1988. Specificity of the interaction between phosphatidylinositol 4,5-bisphosphate and profilin:actin complex. *J. Cell. Biochem.* **37**:255–267.

Majerus, P. W., *et al.* 1986. The metabolism of phosphoinositide derived messenger molecules. *Science* **234**:1519–1526.

Mignery, G. A., and Sudhof, T. C. 1990. The ligand binding site and transduction mechanism in the inositol-1,4,5-triphosphate receptor. *EMBO J.* **9**:3893–3898.

Mignery, G. A., *et al.* 1990. Structure and expression of the rat inositol 1,4,5-triphosphate receptor. *J. Biol. Chem.* **265**:12679–12685.

Otsu, M., *et al.* 1991. Characterization of two 85kd proteins that associate with receptor tyrosine kinases, middle-T/pp60[c-src] complexes, and PI3-kinase. *Cell* **65**:91–104.

Parker, P. J., *et al.* 1986. The complete primary structure of protein kinase C—the major phorbol ester receptor. *Science* **233**:853–859.

Petty, H. R., and Martin, S. 1989. Combinative ligand–receptor interactions: Effects of cAMP, epinephrine, and met-enkephalin of RAW264 macrophage morphology, spreading, adherence, and microfilaments. *J. Cell. Physiol.* **138**:247–256.

Ross, C. A., *et al.* 1989. Inositol 1,4,5-triphosphate receptor localized to endoplasmic reticulum in cerebellar Purkinje neurons. *Nature* **339**:468–470.

Supattapone, S., *et al.* 1988. Solubilization, purification, and characterization of an inositol triphosphate receptor. *J. Biol. Chem.* **263**:1530–1534.

Taylor, S. S., *et al.* 1988. cAMP-dependent protein kinase: Prototype for a family of enzymes. *FASEB J.* **2**:2677–2685.

Yoshimasa, T., *et al.* 1987. Cross-talk between cellular signalling pathways suggested by phorbol ester-induced adenylate cyclase phosphorylation. *Nature* **327**:67–70.

Zhou, M.-J., *et al.* 1991. Detection of transmembrane linkages between immunoglobulin or complement receptors and the neutrophil's cortical microfilaments by resonance energy transfer microscopy. *J. Mol. Biol.* **218**:263–268.

Chapter 8

Membrane Fusion, Formation and Flow

A membrane's composition undergoes constant change due to the insertion and removal of its components. For example, bacteria alter their membranes' fatty acyl composition in response to temperature shifts. Another example of membrane restructuring is the turnover of membrane proteins as they become worn out. Changes in membrane composition are achieved by the direct insertion of a new component into a membrane or a fusion event between two membranes. This chapter will begin by considering simple fusion events *in vitro*. We will then discuss direct membrane insertion and a few complex fusion events observed *in vivo*.

8.1. FUSION

Membranes are said to fuse when two separate bilayers become one. Many *in vivo* and *in vitro* examples of membrane fusion are discussed in the following paragraphs. Membrane fission also takes place. During fission one membrane becomes two; pinocytosis is a good example of membrane fission. Figure 8.1 illustrates the processes of fusion and fission. Since membrane fusion, but not fission, is readily accomplished *in vitro*, most research has focused on fusion reactions. This section describes studies aimed at understanding and using membrane fusion.

8.1.1. An Overview of Biological Fusion

Fusion is a primary means of changing a membrane's composition both *in vitro* and *in vivo*. For example, *in vivo* myoblasts fuse to become skeletal muscle fibers, macrophages fuse to become giant multinucleated cells, some viruses fuse with cells to cause infections, and a sperm cell fuses with an egg during fertilization. In addition to fusion events at plasma membranes, eukaryotic cells possess endomembranes that also undergo

Membrane fusion

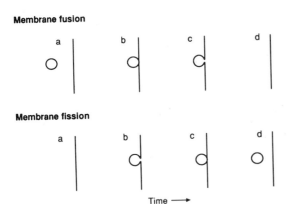

Membrane fission

Time ⟶

Figure 8.1. Membrane fusion and fission. The processes of membrane fusion and fission are illustrated. The time-dependent events of fusion and fission proceed from steps a to d.

fusion. The assembly of a nuclear envelope after mitosis and phagosome-to-lysosome fusion are two examples of endomembrane fusion. Fusion also occurs between certain endomembranes and plasma membranes. The visualization of synaptic vesicles fusing at a presynaptic membrane (Figure 4.24) is a particularly good example of a secretory fusion event. Although there are many examples of biological fusion, one must remember that it is a highly regulated and comparatively rare event. It is the *inability* of membranes to fuse that allows the formation of distinct organelles and cells. Although the mechanism of *in vivo* fusion is poorly understood, the events accompanying *in vitro* fusion are becoming well defined.

8.1.2. Fusing Membranes

The development of new molecular and cellular techniques allows us to artificially change the composition of a living cell's plasma membrane. In doing so, one can tailor a membrane to a particular experimental question. In this section we will explore some of the methods and applications of *in vitro* membrane fusion. *In vitro* fusion can occur between two plasma membranes, two lipid vesicle membranes, or a plasma membrane and a vesicle membrane. Membrane fusion is promoted by four methods: (1) chemical treatment, (2) vesicle phospholipid composition, (3) the presence of viral fusion proteins, and (4) electrofusion techniques.

8.1.2.1. Chemical Fusion

Kao and Michayluk (1974) introduced the use of polyethylene glycol (PEG) as a membrane fusion-promoting agent or fusogen for higher plant protoplasts. It has since been used to fuse many cells including yeast protoplasts, erythrocytes, and mammalian cells. When protoplasts of two higher plant species are fused, the hybrid cells develop into complete hybrid plants. PEG is also routinely used to create hybridomas, cell lines derived from the *in vitro* fusion of lymphocytes and tumor cells, that produce monoclonal antibodies.

Chemical fusogens such as PEG, glycerol, dimethyl sulfoxide (DMSO), and polyhistidine decrease the surface potential of membranes (Maggio *et al.*, 1976). The reduction in surface charge density decreases the electrostatic contribution to cell–cell repulsion. In general, chemical fusogens are also capable of hydrogen bonding. This property of fusogens likely disorganizes ordered water layers near membranes to diminish hydration pressure and potentiate fusion between bilayers. Since DMSO and glycerol-mediated fusion require calcium, fusion will be enhanced due to the additional reduction in hydration pressure and the presence of calcium ion-correlation effects (Rand and Parsegian, 1986). Therefore, the physicochemical properties of fusogens diminish repulsive intermembrane forces and augment attractive forces.

In some cases PEG is poorly fusogenic (Schramm, 1979). The fusogenic activity of PEG is influenced by the composition of a recipient cell's plasma membrane (Roos and Choppin, 1985). The variability in fusogenic potential is apparently due to the ability of certain head groups and fatty acyl chains to form inverted micellar intermediates (IMI) (Section 8.1.2.4). The inclusion of lyso-PC, which destabilizes bilayers, potentiates PEG-mediated fusion of resistant recipient cells (Schramm, 1979). Freeze-fracture electron microscopy has shown that intramembrane particles aggregate when cell membranes are treated with DMSO or glycerol. There is evidence indicating that fusion often occurs between the intramembrane particle-free zones during *in vitro* but not necessarily *in vivo* membrane fusion. However, we do not know if bringing two particle-free cell membranes into very close proximity is sufficient to induce chemical fusion. For example, chemical fusogens also destabilize a bilayer's structure.

8.1.2.2. Phospholipid-Enhanced Fusion

A lipid vesicle's phospholipid composition regulates its ability to fuse with other vesicles or cells. For example, lipid vesicles composed of PE and PS (3:1) destabilize and fuse on addition of Ca^{2+} (Duzgunes *et al.*, 1981). This property of lipid vesicles can be exploited to enhance vesicle-to-cell fusion. Correa-Freire *et al.* (1984) showed that liposomes composed of PE and PS (1:1) fuse readily with cell membranes in the presence of Ca^{2+}. This system effectively delivers exogenous membrane proteins into the plasma membranes of recipient cells. The fusogenic potential of this lipid composition can be understood based on the physical properties of their head groups. PE is the least hydrated phospholipid head group; consequently, the hydration repulsion of PE bilayers is substantially less than that of other phospholipid compositions. Furthermore, PE molecules easily form the hexagonal (HII) phase, a presumed intermediate step during fusion (Section 8.1.2.4). PS is capable of tight Ca^{2+} binding. Membrane-bound Ca^{2+} reduces net surface charge density and hydration repulsion and potentiates intermembrane van der Waals attraction (Rand and Parsegian, 1986).

8.1.2.3. Viral Fusion

Viruses must gain access to a cell's cytosol to cause an infection. Some viruses accomplish this by membrane fusion. These viruses possess an outer bilayer that contains membrane proteins which recognize and then fuse with a target membrane. The envelope proteins required for influenza, Sendai, Semliki Forest, and vesicular stoma-titis virus (VSV) fusion with cells are hemagglutinin, F protein, spike glycoprotein, and VSV G protein, respectively. This discussion will focus on influenza virus's hemagglu-tinin, which is one of the most fully characterized membrane proteins.

As we learned in Section 4.1.3, the mature hemagglutinin of influenza virus consists of two domains, HA1 and HA2. The HA1 domain contains a recognition site that binds to sialic acid at a target cell's surface. The transmembrane HA2 subunit is bound to the viral envelope via its COOH-terminus. Its NH_2-terminus inserts into a target membrane upon activation by low pH. About 1000 hemagglutinin molecules are found in one influenza virus envelope. Hemagglutinin is a trimeric molecule (Figures 4.4 and 8.2). After binding, influenza viruses accumulate at coated pits followed by internaliza-tion within coated vesicles. The low pH of the vacuolar system triggers an irreversible conformational change in hemagglutinin. This conformational change has been verified by measuring alterations in its antigen exposure, protease sensitivity, and spectral properties. During its conformational change, each of hemagglutinin's monomers spreads apart, as schematized in Figure 8.2, panel b. This exposes the NH_2-terminal domains of the HA2 subunits, which are normally buried within the trimers. The hydrophobic NH_2-terminal domains insert into a target membrane. This brings the viral

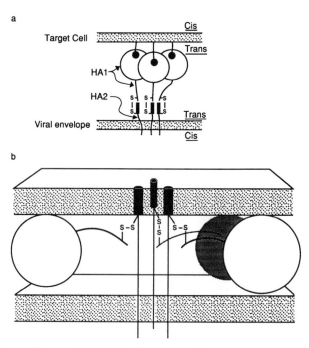

Figure 8.2. Early events in hemagglutinin-induced membrane fusion. As panel a illustrates, the HA1 subunit of influenza virus's hemagglutinin binds to a target's membrane. When exposed to a low pH such as that within an endosome, hemagglutinin's trimer unfolds like the petals of a flower (panel b). The hydrophobic NH_2-terminal domain of HA2 becomes anchored in the target's membrane.

and target membranes to within about 3 nm. This explains how the virus adheres and brings two membranes to within a few hydration layers. However, the events causing membrane destabilization are unknown.

Viruses or viral fusogenic proteins can be used to artificially induce fusion. When reconstituted into lipid vesicles, viral fusogenic proteins mediate lipid vesicle-to-cell fusion. This approach has been employed to insert many kinds of membrane proteins (e.g., band 3, adrenergic receptors, and histocompatibility antigens) into foreign cell types (Volsky *et al.*, 1979; Schramm, 1979; Prujansky-Jakobovits *et al.*, 1980).

8.1.2.4. Fusion Intermediates

In the preceding paragraphs we discussed how chemicals, lipids, and viral proteins potentiate membrane fusion. We focused on how two membranes adhere and/or approach one another to within 2 or 3 nm. We have not yet touched on how two apposed bilayers become one. Although the mechanism(s) of membrane fusion has not been established with certainty, recent studies have provided insight into its details. After stable short-range binding is achieved, both membranes must become destabilized. Bilayer perturbations have been observed by various techniques, such as freeze-fracture electron microscopy. Model membranes are believed to form IMIs during fusion. Figure 8.3 shows a series of likely events participating in membrane fusion. Two membranes adhere tightly to one another, undergo destabilization (including IMI formation), and then fuse. As two model membranes are brought together, the energy required to remove intermembrane water generates sufficient tension to rupture a bilayer (Rand and Parsegian, 1986).

IMIs (Figure 8.3, panel c) are believed to be an intermediate state during fusion (Rand and Parsegian, 1986). IMIs represent a phase (and gross structural) change from a lamellar to hexagonal (HII) packing. Siegel's analysis shows that the IMIs formed during membrane destabilization could: (1) reform bilayer structures making leaflets 1 and 4 continuous and thereby fusing the membranes or (2) aggregated IMIs would lead

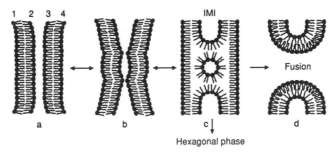

Figure 8.3. Molecular events participating in fusion. The leaflets of the two bilayers are labeled 1 through 4 as shown on the left. Two bilayers approach (a) and then adhere tightly (b). The bilayers become destabilized and then form IMIs (c). The destabilized region can spontaneously form a single bilayer (d). Courtesy of Dr. V. A. Parsegian.

to extensive hexagonal structures that disrupt the bilayers (Siegel, 1984). Destabilization occurs only at points of contact and dehydration. Although it may not be an absolute requirement, the presence of hexagonal phase-forming lipids such as PE facilitate fusion. Alterations in fatty acyl composition may also enhance IMI formation. Regions of membrane contact and IMI formation are likely enriched in PE. At local contact sites, hydration repulsion may be sufficient to force other more hydrated lipids away.

The regulation of biological fusion *in vivo* likely involves both protein and lipid interactions. Although biological fusion is poorly understood, attractive speculations may provide important leads for future study. Rand and Parsegian (1986) have suggested that PI turnover could be physically linked to membrane fusion. Earlier studies have shown that secretory phenomena and myoblast fusion are associated with increases in PI turnover. Diacylglycerol (DAG), a membrane-bound product of PI turnover and intracellular messenger, is uncharged and much more hydrophobic than PI. Therefore, cleavage of PI to DAG simultaneously eliminates electrostatic repulsion and decreases hydration pressure. Furthermore, DAG perturbs bilayer structure. Consequently, sites of PI turnover may greatly accelerate membrane adhesion and destabilization leading to fusion.

8.2. BIOSYNTHESIS OF MEMBRANE LIPIDS, PROTEINS, AND CARBOHYDRATES AND THEIR TRANSLOCATION ACROSS MEMBRANES

The biosynthetic assembly of membranes is an essential feature of life. Furthermore, to remain viable, living membranes must regularly replace worn or lost components. Typically, membrane lipids and proteins of active cells have half-lives of only a few days. Occasionally, cells must alter their membrane composition to compensate for environmental stresses or nutrient supply or to respond to intercellular signals. In this section we shall focus on the biosynthetic machinery of membranes and the biosynthesis and translocation mechanisms of membrane lipids, proteins, and carbohydrates.

8.2.1. Lipid Biosynthesis

In Section 2.1 we learned about the great variety of lipids found in cells. However, these many lipid molecules are made from a rather small number of precursor molecules. Figure 8.4 shows a survey illustration of mammalian phospholipid biosynthetic pathways. In addition to the qualitative structural differences among types of lipids, there are quantitative differences among various membranes in their lipid compositions. For example, plasma membranes are enriched in glycosphingolipids, PS, SM, and cholesterol whereas endoplasmic reticulum membranes contain very little if any of these lipids. In this section we will explore some of the terminal steps in lipid biosynthesis and the origins of transmembrane lipid asymmetry in cells.

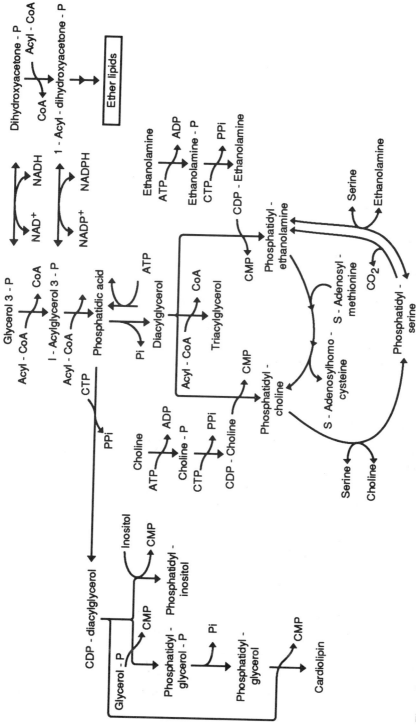

Figure 8.4. Phospholipid biosynthetic pathways of mammalian cells. The biochemical steps participating in phospholipid synthesis are shown. Certain mechanistic details and enzyme activities involved in these pathways are presented later in this chapter. (Redrawn with permission from the *Annual Review of Cell Biology*, Vol. 4, copyright © 1988 by Annual Reviews, Inc.)

8.2.1.1. PC Biosynthesis

The primary synthetic route of PC is the CDP-choline or Kennedy pathway (Vance, 1986). A series of three enzymatic reactions [Eqs. (8.1) to (8.3)] catalyze the transfer of choline to DAG.

$$\text{choline} + \text{ATP} \xrightarrow{\text{choline kinase}} \text{phosphocholine} + \text{ADP} \qquad (8.1)$$

$$\text{phosphocholine} + \text{CTP} \xrightarrow{\text{CTP:phosphocholine cytidylyltransferase}} \text{CDP-choline} + \text{PP}_i \quad (8.2)$$

$$\text{CDP-choline} + \text{DAG} \xrightarrow{\text{CDP-choline:1,2-DAG phosphocholine transferase}} \text{PC} \qquad (8.3)$$

We will now examine these reactions in greater detail.

Choline enters cells via a high-affinity transporter ($K_m = 10 \ \mu M$). The cytosolic enzyme choline kinase catalyzes the formation of phosphocholine [Eq. (8.1)]. It has a mass of 120 to 160 kDa. Its K_m's for choline and ATP are 0.03 and 2 mM, respectively (Vance, 1986). The higher value for ATP is likely a manifestation of ATP's higher cytosolic concentration. The rate-controlling step in PC biosynthesis is catalyzed by CTP:phosphocholine cytidylyltransferase [Eq. (8.2)]. It is a dimer of 45-kDa subunits. Cytidylyltransferase activity is found in both the cytosol and endoplasmic reticulum (ER). The relative amounts of cytosolic and ER-associated enzyme are a reflection of PC synthetic activity. When PC synthesis is activated, the enzyme is translocated from the cytosol to the ER's membrane. It catalyzes the local formation of CDP-choline, which participates in PC synthesis.

CTP:phosphocholine cytidylyltransferase activity is regulated by intracellular signaling molecules. Protein kinases (Sections 7.3 and 7.6) play a major role in regulating cytidylyltransferase activity (Bishop and Bell, 1988). PKA-mediated phosphorylation reactions release cytidylyltransferase from ER membranes. On the other hand, when cells are treated with PKC-activating phorbol esters, cytidylyltransferase is simultaneously activated and translocated to ER membranes. These regulatory events make sense from a physiological perspective. As previously mentioned (Section 7.3.1), one role of β-adrenergic receptor stimulation of PKA is to mobilize energy reserves. Consequently, anabolic pathways such as PC synthesis are shut down. Alternatively, some growth-promoting substances are believed to act through PKC stimulation. Since the assembly of new membrane is required for cell growth, PC synthesis is stimulated by this message.

The last step in PC synthesis is the delivery of a choline group to DAG. This is catalyzed by CDP-choline:1,2-diacylglycerol phosphorylcholine transferase. This enzyme is an integral membrane protein whose active site resides at the ER's *cis* face. Little else is known about this enzyme's structure or mechanism. After their synthesis, PC molecules become incorporated into the ER membrane's *cis* face. Newly synthesized PC molecules have several potential fates. They could be delivered from the ER to the cytosolic faces of other organelles or the plasma membrane. They could also be flip-flopped to the luminal ER membrane face. We shall return to these possibilities in the following section.

In addition to the CDP-choline pathway, yeast cells are known to synthesize PC by methylation of PE. This is accomplished by the sequential transfer of methyl groups from three S-adenosylmethionine (AdoMet) molecules to PE. The reactions are:

$$PE + AdoMet \xrightarrow{\text{PE methyltransferase}} N\text{-methyl-PE} + AdoHCy \qquad (8.4)$$

$$N\text{-methyl-PE} + AdoMet \xrightarrow{\text{phospholipid methyltransferase}} N,N\text{-dimethyl-PE} + AdoHCy \qquad (8.5)$$

$$N,N\text{-dimethyl-PE} + AdoMet \xrightarrow{\text{phospholipid methyltransferase}} PC + AdoHCy \qquad (8.6)$$

where AdoHCy is the reaction product S-adenosylhomocysteine. As Eqs. (8.4) to (8.6) indicate, at least two methyltransferases can participate in PC biosynthesis from PE in yeast. The yeast genes encoding these enzyme activities have been cloned (Kodaki and Yamashita, 1987). These methyltransferases are hydrophobic integral membrane proteins of the ER. The PE methyltransferase ($M_r = 101{,}202$) and phospholipid methyltransferase ($M_r = 23{,}150$) possess five and three putative transmembrane domains, respectively. These two membrane proteins are homologous to one another, suggesting that they evolved by gene duplication. Regions of both methyltransferases are homologous to other methyltransferases that use AdoMet as the methyl donor. These homologous regions are very likely in the AdoMet binding site; therefore, they should be exposed at the ER's cytosolic face. The membrane topology of these methyltransferases is not known. However, it seems likely that the phospholipid methyltransferase's NH$_2$-terminal domain is exposed at the ER's *trans* face since this would position the apparent AdoMet binding site at the *cis* face. Moreover, it is interesting to note that the phospholipid methyltransferase's first hydrophobic domain overlaps the AdoMet binding site by about ten amino acids. Presumably, the intercalation of a portion of the active site into the ER's bilayer enhances both the binding and release of phospholipid intermediates.

Hepatocytes of vertebrates also synthesize PC by methylation of PE. A single enzyme of 18 kDa catalyzes all three steps [Eqs. (8.4)–(8.6)]. Although its deduced amino acid sequence is unavailable, functional studies have shown that its active site is expressed at the ER's cytosolic surface.

8.2.1.2. Membrane Trafficking of PC Molecules

Since phospholipids are exclusively synthesized on the ER's *cis* face, a mechanism must be available to move some newly synthesized lipids to the *trans* bilayer leaflet. This movement is provided by phospholipid transporters or flippases. Although PC does not spontaneously flip from one membrane face to the other (flip-flop) at an appreciable rate on plasma membranes, it rapidly ($t_{1/2} < 2$ min) flip-flops across ER membranes (Bishop and Bell, 1985). This is apparently catalyzed by a membrane protein. In microsomal membranes flip-flop is inhibited by proteases and certain chemical reagents. Although microsomal lipids are unable to support flip-flop when reconstituted in vesicles, reconstituted microsomal proteins can catalyze flip-flop (Backer and Dawidowicz, 1987). Therefore, the ER's membrane is believed to contain proteins that catalyze the rapid transbilayer migration of phospholipids.

As Figure 8.5 illustrates, PC could leave the ER by several pathways. PC at either the *cis* or *trans* faces could meander along the secretory pathway through the Golgi, secretory vesicles, and eventually reach the plasma membrane. This pathway is most important for PC at the *trans* face since there is no other pathway for lipid movement from the ER to a plasma membrane's *trans* face.

PC is also transported from the ER's cytosolic face to cytosolic faces of other organelle membranes and the plasma membrane by translocation proteins. Since PC is insoluble in water, transport through the cytosol in a monomeric form is very, very slow. An attractive possibility that may account for the shuttling of PC and other lipids among membranes are phospholipid transfer proteins.

Although PC accounts for 39% of mitochondrial lipids, it cannot be synthesized by mitochondria. Therefore, mitochondrial membranes are a particularly good system to study PC transport pathways. Newly synthesized PC is transported from ER to mito-chondria with a $t_{1/2} < 30$ min. Cytosolic phospholipid transfer proteins may partici-pate in PC's movement. Transfer proteins have been isolated from numerous cell types and shown to promote the exchange or net transfer of lipids *in vitro*. However, phospholipid transfer proteins have not yet been shown to play a role *in vivo*. After the genes encoding these proteins have been cloned, it may be possible to sort out their functions by gene disruption experiments in yeast.

Phospholipid transfer proteins may also contribute to the transport of PC to the plasma membrane's cytosolic face. Kaplan and Simoni (1985) have shown that PC can be rapidly transported to plasma membranes in a vesicle-independent fashion. As van Meer

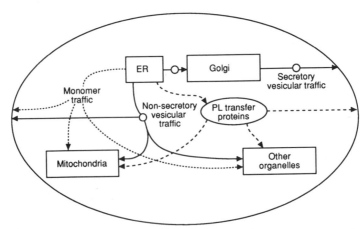

Figure 8.5. Potential phospholipid trafficking pathways among endomembranes. Lipids synthesized in the ER must be translocated to other membranes. Cells handle lipid sorting and transfer by several mechanisms. Secretory vesicular traffic, phospholipid transfer proteins, and monomer traffic are illustrated. The mito-chondria and Golgi also participate in the synthesis of essential lipids. Local lipid synthesis and modification are not shown.

(1989) points out, there are two plasma membrane PC pools corresponding to each of a bilayer's leaflets. The rapidly and slowly exchanging pools could be the *cis* and *trans* membrane faces. The exchange of PC at the cytosolic face may be mediated by transfer proteins whereas the *trans* PC molecules may arrive via the energy- and temperature-dependent secretory pathway.

8.2.1.3. Synthesis and Intracellular Transport of PS and PE

PS is synthesized by two fundamentally distinct pathways in higher eukaryotes versus lower eukaryotes and prokaryotes. In higher eukaryotes, PS is formed by a base exchange reaction catalyzed by PE:serine *O*-phosphatidyltransferase:

$$\text{PE} + \text{serine} \xrightarrow{\text{PE:serine } O\text{-phosphatidyltransferase}} \text{PS} + \text{ethanolamine} \qquad (8.7)$$

This reaction primarily takes place at the ER's *cis* face, although limited activity has been found in other membranes. Although PE is the preferred substrate for this reaction (Vance, 1986), it also synthesizes PS by base exchange with PC (Bishop and Bell, 1988). Experiments with cultured cell lines have shown that base exchange reactions are essential for the growth of higher eukaryotes.

In prokaryotes and lower eukaryotes, PS is directly synthesized from CDP-DAG and serine:

$$\text{CDP-DAG} + \text{serine} \xrightarrow{\text{CDP-DAG:L-serine } O\text{-phosphatidyltransferase}} \text{PS} + \text{CDP} \qquad (8.8)$$

The reaction is catalyzed by the PS synthase (CDP-diacylglycerol:L-serine *O*-phosphatidyltransferase). This enzyme is found in bacteria and yeast, but never in animals or plants (Vance, 1986). It was first purified by affinity chromatography on resins made of DAG covalently linked to Sepharose beads. Early experiments showed that the enzyme was an integral membrane protein of the ER. The PS synthase gene (*chol*) has been cloned from yeast. It encodes a very hydrophobic membrane protein with four potential transmembrane domains (Nikawa *et al.*, 1987a). Although the gene encodes an $M_r = 30,804$ molecule, the mature protein may be somewhat smaller.

In prokaryotes and yeast, PS is a vital precursor of PE and PC:

$$\text{PS} \xrightarrow{\text{decarboxylation}} \text{PE} \xrightarrow{N\text{-methylations}} \text{PC} \qquad (8.9)$$

Therefore, PS production in these cells can regulate the production of other phospholipids (Vance, 1986). One important pathway of PE biosynthesis is via the decarboxylation of PS. This is catalyzed by PS decarboxylase, an enzyme residing in the mitochondrial inner membrane. Due to the locations of PS synthase and PS decarboxylase, PS and PE must be shuttled to and from mitochondria. In contrast to the rapid transport of PC to mitochondria, PS has a transit time of 7 hr (van Meer, 1989). The mechanism of transport is unknown, but it is thought to require the formation of a special class of intracellular transport vesicles. After decarboxylation, the newly formed PE is transported to the ER by a rather obscure energy-dependent mechanism.

Higher eukaryotes synthesize PE by three distinct mechanisms. First, these cells biosynthesize PE by the decarboxylation of PS, as we described in the preceding paragraph. In addition, higher eukaryotes synthesize PE by *de novo* synthesis and base transfer reactions. The steps involved in these processes are analogous to those described above for the CDP-choline pathway of PC synthesis and the base transfer reaction producing PE is just the reverse of Eq. (8.7). These pathways are schematically illustrated in Figure 8.6

8.2.1.4. Genesis and Maintenance of Asymmetric Biological Membranes

As previously described (Section 2.1.7), amino lipids are preferentially located at the plasma membrane's *cis* face whereas choline lipids are primarily found at the *trans* face. This asymmetry might be explained by the asymmetry of phospholipid synthesis. For example, phospholipids are synthesized at the ER's *cis* face whereas SM and glycosphingolipids are assembled at the Golgi's *trans* face. Since PS and PE are not flip-flopped to the *trans* face, biosynthesis may contribute to the *initial* asymmetry of lipids. However, biosynthesis alone cannot account for the maintenance of phospholipid asymmetry. Although ESR spectroscopy has shown that spontaneous PC flip-flop is very slow in erythrocytes, PS and PE spontaneously flip to the *trans* face with $t_{1/2}$'s of 5 min and 1 hr, respectively (Devaux, 1988). Since flip-flop is much faster than the erythrocyte's 120-day lifetime, additional factors must contribute to the *maintenance* of lipid asymmetry in erythrocyte membranes.

A cell surface aminophospholipid flippase is believed to account for the lipid asymmetry of RBC membranes (Devaux, 1988). This flippase activity is mediated by an ATP-driven membrane protein. PS and PE compete for the flippase's binding site. Since PS has the higher flippase binding affinity, it has the faster rate of translocation.

8.2.1.5. Biosynthesis of PI

Although PI accounts for only a small fraction of total membrane lipid, it is a vital regulatory molecule. We previously learned that PI and its phosphorylated derivatives

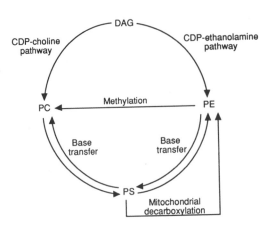

Figure 8.6. Interrelationships among phospholipid biosynthetic routes in eukaryotes. Substrate–product relationships and reaction pathways are shown for PC, PE, DAG, and PS. The complex interactions suggest complex regulatory mechanisms.

PIP and PIP_2 play important roles in transmembrane signaling (Section 7.6). We will now consider the synthesis of PI.

PI is synthesized directly from CDP-DAG and inositol by PI synthase (CDP-diacylglycerol:inositol 3-phosphatidyltransferase):

$$CDP\text{-}DAG + inositol \rightarrow PI + CM \qquad (8.10)$$

$$PI + ATP \rightarrow PIP + ADP \qquad (8.11)$$

$$PIP + ATP \rightarrow PIP_2 + ADP \qquad (8.12)$$

The phosphorylated derivatives PI-4-phosphate (PIP) and PI-4,5-bisphosphate are synthesized by the sequential phosphorylation of PI by ATP.

In analogy with other phospholipids, PI synthesis takes place at the ER's *cis* face. The PI synthase ($M_r = 24,823$) of yeast has been cloned (Nikawa *et al.*, 1987b). Its deduced amino acid sequence revealed two potential transmembrane domains and sequence homology to PS synthase (Section 8.2.1.3). Since both enzymes utilize the substrate CDP-DAG, this homologous region may correspond to a substrate binding site. Several cellular membranes participate in the conversion of PI to PIP and PIP_2.

PI, PIP, and PIP_2 are rapidly transported and turned over within cells. For example, PI is transported to mitochondria within minutes of synthesis (van Meer, 1989). This is likely mediated by a PI/PC transfer protein. The transfer protein is probably responsible for replenishing plasma membrane PI molecules spent during transmembrane signaling.

8.2.1.6. Synthesis of Sphingolipids

Sphingosine is believed to be synthesized at the ER membrane's luminal face. SM, one of its derivatives, is assembled at the Golgi's luminal surface (Pagano and Sleight, 1985). The following base exchange reaction takes place:

$$PC + Cer \xrightarrow{\text{PC:Cer phosphorylcholine transferase}} SM + DAG \qquad (8.13)$$

This enzyme-catalyzed reaction does not require energy. SM's membrane asymmetry is established by its biosynthesis since it is unable to flip-flop across bilayers. It is apparently transferred to other membranes by vesicles since it can only be found in membranes associated with the Golgi, cell surface, secretory vesicles, and endosomes. The assembly of glycosphingolipids will be presented in Section 8.2.3.3.

8.2.2. Cotranslational Insertion of Proteins into the ER's Membrane

All nuclear DNA-directed protein synthesis begins in the cytosol. However, many of a cell's roughly 2000 proteins must be directed to their appropriate organelle. This targeting is accomplished during (cotranslational) or after (posttranslational) protein synthesis. We will now explore the cotranslational biosynthetic delivery of membrane and secretory proteins to the ER; Section 8.2.4 will consider posttranslational movement across membranes. In Section 8.2.5 we will discuss how bacteria utilize both co- and posttranslational membrane insertion.

8.2.2.1. Sequences That Signal: Their Recognition and Targeting

Early *in vitro* translational studies of secretory, lysosomal, and integral membrane proteins found that these proteins possessed 20 to 40 unexpected amino acids at their NH$_2$-termini. These residues are not found after *in vivo* biosynthesis or *in vitro* synthesis in the presence of unmodified microsomes. This string of about 25 amino acids, or signal sequences, specifies the delivery of a nascent polypeptide and its ribosome to the ER's membrane. If these residues are removed or substantively altered by experimental tampering, these proteins accumulate in the cytosol. All signal sequences display considerable hydrophobicity. However, no unambiguous consensus sequence of amino acids that represent an ER signal has been found.

The signal sequence is recognized by the signal recognition particle (SRP). When SRP binds to a signal sequence emerging from a free ribosome, translation is abruptly halted or at least slowed dramatically (Walter and Lingappa, 1986). This allows time for the ribosome–SRP complex to find a binding site on the ER's membrane (Figure 8.7).

SRP is a rod-shaped cytosolic ribonucleoprotein. It consists of a 7 S RNA molecule and six proteins (Figure 8.7). The 300-nucleotide RNA strand possesses considerable secondary structure. Proteins of 9, 14, 19, 54, 68, and 72 kDa are associated with each RNA molecule. The 19- and 54-kDa proteins are monomers whereas the other four proteins exist as two dimers (9 + 14 kDa and 68 + 72 kDa). Purified SRPs can be disassembled and reassembled *in vitro*. This allows the functions of individual proteins to be tested in reconstitution assays. These experiments have defined three regions within SRPs: (1) signal sequence binding, (2) inhibition of protein synthesis (elongation arrest), and (3) membrane docking.

SRP binds very weakly to resting ribosomes. This affinity increases thousands of fold when a signal sequence emerges from a ribosome's active site. SRP's 54-kDa subunit binds to signal sequences. Direct photochemical cross-linking experiments have shown the molecular proximity of the 54-kDa subunit and a signal sequence. The deduced amino acid sequence of the 54-kDa SRP protein has recently been obtained (Romisch *et al.*, 1989; Bernstein *et al.*, 1989). These studies revealed that the 54-kDa protein is a G protein. GTP is apparently used to switch the protein between two conformations. It also contains a Met-rich domain. The Met-rich domain is apparently comprised of three 25-residue amphipathic α-helices. The hydrophobic faces of these helices are largely made up of Met, Ile, and Leu residues (Bernstein *et al.*, 1989). The helices may form a groove or cavity at the protein's surface. When the hydrophobic signal sequence leaves a ribosome's active site, it may fit into this cavity in a lock-and-key fashion. Since Met residues are quite flexible, they may be able to accommodate the many structurally heterogeneous yet hydrophobic signal sequences found in nature.

The SRP's 9- and 14-kDa dimer halts protein synthesis. If the 9- and 14-kDa proteins are deleted in SRP reconstitution experiments, the signal sequence is still recognized and some of the new proteins are targeted to the ER. However, since protein synthesis continues, some proteins mature too quickly to be delivered to the ER. Earlier studies have shown that SRP is only able to functionally bind to certain lengths of nascent polypeptides. Therefore, SRP's elongation arrest function increases the efficiency of protein delivery to the *cis* face of the ER's membrane.

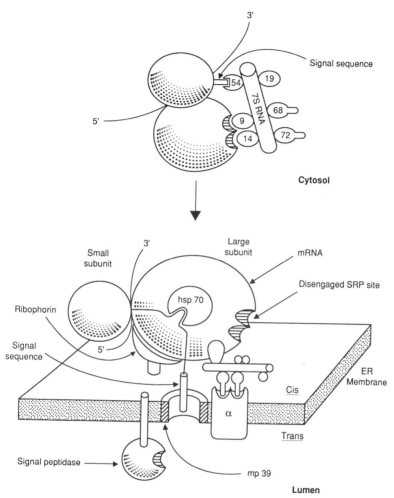

Figure 8.7. Protein synthesis at the ER surface. The signal sequence is recognized by the SRP (upper illustration). After cytosolic movement, the complex becomes bound to ER membranes (lower illustration). This shows a hypothetical model of protein synthesis on the ER's cytosolic face. Several of the participating components are labeled.

8.2.2.2. Delivery of Proteins to the ER's Membrane

Although both the large and small ribosome subunits are near the ER's membrane, only the large subunit is physically linked to the membrane (Unwin, 1977). This linkage is mediated at least in part by the ribosome-associated SRP which binds to SRP receptors of ER membranes. SRP's 68- and 72-kDa proteins act as ligands for the SRP receptor (or docking protein) present in ER membranes. SRP's receptor was first identified by cross-linking experiments (Wiedmann *et al.*, 1987). It soon became apparent that the SRP receptor is composed of at least two subunits, α and β.

Lauffer *et al.* (1985) cloned a cDNA sequence encoding the SRP receptor's α-subunit gene. The deduced amino acid sequence revealed several interesting structural features. The α subunit is a 69-kDa integral ER membrane protein. It is attached to membranes via two putative transmembrane α-helices. A charged residue is apparently buried within the membrane's hydrophobic core. In analogy with many integral membrane proteins participating in transport, such as lactose permease, bacteriorhodopsin, and band 3, this residue may play a direct functional role or trigger specific interactions among other transmembrane domains. A large COOH-terminal domain at the ER's *cis* face possesses many charged residues. A portion of this domain is homologous to nucleic acid binding proteins. This suggests that SRP's 7 S RNA binds to this region of the SRP receptor.

In addition to possible nucleic acid–protein interactions, protein–protein interactions between SRP's 68- and 72-kDa proteins and the SRP receptor also contribute binding affinity. Recent studies have shown that the α subunit binds GTP (Connolly and Gilmore, 1989) and that GTP is necessary for signal sequence release from SRP. Bernstein *et al.* (1989) have speculated that homotypic interactions between SRP's 54-kDa G protein and the SRP receptor's G protein may take place. The coordinated interaction of these G proteins is important in delivery of proteins to the ER. Multiple physical interactions between SRP and its receptor apparently take place.

The SRP receptor's β subunit is a 30-kDa integral protein. It has two or three transmembrane strands. A gene that may correspond to a yeast cell's SRP receptor β chain has been cloned (Deshaies and Schekmann, 1989). This protein contains an unusually basic COOH-terminus. In analogy with the receptor's α subunit, this basic region may potentiate the binding of the acidic 7 S RNA molecule of SRP to its receptor.

The SRP receptor is believed to mediate the delivery of ribosomes and nascent polypeptide chains to the ER. However, the receptor *per se* is not thought to form the transmembrane channel that allows polypeptide passage across the ER's membrane (Walter and Lingappa, 1986). The SRP receptor alternatively cycles through SRP binding and delivery of the nascent protein to the transmembrane delivery channel, which is the subject of the following section.

8.2.2.3. Transmembrane Migration of Proteins

Few areas of modern biology are as poorly understood as protein translocation across ER membranes. One of the difficulties in this area of research has been an inability to reconstitute membrane translocation in a completely defined *in vitro* system. There are two basic approaches used to attack the problem: biochemical and genetic. We will now discuss what these approaches have contributed to our understanding of protein translocation.

After arrival at an ER membrane, a signal sequence is inserted into a translocation pore complex. Translocation pores are made of proteins; this has been shown by their sensitivity to proteolysis and alkylation (e.g., Nicchitta and Blobel, 1989). A proteinaceous transmembrane channel also finds support in extraction studies. These experiments indicated that translocation intermediates could be extracted from membranes without perturbing the lipid bilayer. This suggests that during translocation nascent chains reside in a polar transmembrane environment, presumably a channel or pore constructed from an integral membrane protein(s). Electrophysiological studies using ER components reconstituted in black lipid bilayers have shown the presence of a ribosome-dependent protein translocation pore (Simon and Blobel, 1991). Unfortunately, we do not know the identity of this protein(s) or how it functions.

Krieg *et al.* (1989) have identified a potential protein component of the translocation pore. They constructed an *in vitro* translation system that inserts photoaffinity labels along the entire length of a nascent polypeptide chain. Preprolactin mRNA was truncated at many points along its length. These messages were translated *in vitro* using ribosomes, SRPs, microsomes, aminoacyl tRNAs, and other cofactors. Most importantly, a lysine analogue containing a photoreactive moiety attached to its epsilon amino was included in the translation broth. During translation reactions the photoaffinity group was incorporated into the nascent protein's covalent structure. When photoactivated during translocation, this group became attached to a 39-kDa integral membrane protein (mp39). mp39 was covalently tagged by the photoaffinity group at all of its positions along the secretory protein's length. This suggests that mp39 is in close contact with growing polypeptide chains as they cross the ER's membrane.

The identity and molecular features of mp39 have not been elucidated. Its molecular weight suggests that it may be the SRP receptor's β-subunit. Proteolytic maps of the SRP receptor's β-chain and mp39 should help resolve this issue. It is interesting to note that intramembrane β-structure may be present in the SRP receptor's β-chain; we have previously encountered transmembrane β-sheets within many membrane pores. This possibility is not inconsistent with the hypothesis that the SRP "receptor" only performs a delivery function. The receptor's α-chain has a high affinity for both SRP and its β-subunit. It is possible that the α-subunit's primary function is binding and delivery whereas the β-subunit is responsible for translocation. These and related issues should be sorted out within a few years.

Several proteins, in addition to mp39, may participate in membrane translocation (Walter and Lingappa, 1986). For example, the eukaryotic signal peptidase is composed of six proteins. Ribophorins (I and II) have been identified as ER-specific membrane-bound ribosome-binding proteins. Their deduced amino acid sequences have shown that ribophorins I and II each possess a single transmembrane domain and a large luminal domain (Harnik-Ort *et al.*, 1987; Crimaudo *et al.*, 1987). The ribophorins may allow a ribosome to "push" the nascent polypeptide into a translocation pore. Although ribophorins form complexes with other translocation proteins and participate in translocation (Yu *et al.*, 1990), they may not line the translocation channel.

Chirico *et al.* (1988) have shown that a 70-kDa heat shock protein (hsp70) participates in protein translocation across ER membranes. This protein has also been linked to posttranslational import of proteins into mitochondria (Section 8.2.3.3). hsp70 uses ATP to unfold proteins. This protein was first identified in yeast cytosol extracts as a factor that increased the rate of protein translocation across ER membranes. hsp70 is composed of two very similar proteins encoded by *SSA1* and *SSA2* genes. These genes are constitutively expressed by cells. The hsp70 proteins apparently use ATP to straighten proteins prior to entry into a translocation channel. The slow translocation rate found in the absence of hsp70 is likely due to the presence of kinks that cannot easily pass through the channel's opening.

After emerging at the translocation pore's luminal face, nascent polypeptide chains interact with BiP (binding protein), which is homologous to hsp70 and a member of the hsp70 family of proteins. A groove at BiP's surface, similar to that of HLA (see Figure 4.5), binds to peptide sequences of about seven aliphatic residues (Flynn *et al.*, 1991). Multiple BiPs bind to each protein (Kim *et al.*, 1992), thus preventing protein aggregation and premature folding. BiP is required for membrane translocation (Nguyen *et al.*, 1991). As we will discuss below, BiP also regulates protein retention in the ER.

8.2.2.4. Membrane Proteins: Assumptions of Their Topographies

As Table 2.3 indicates, the NH_2- and COOH-termini of various transmembrane proteins can be found at either the *cis* or *trans* membrane faces. This may seem a bit odd because protein biosynthesis is always in the $NH_2 \rightarrow COOH$ direction. However, the distribution of positive charges in the nascent membrane protein's covalent sequence, not its direction of biosynthesis, directs the NH_2-terminus's membrane orientation. The orientation of any additional transmembrane domains within the same protein are likely determined by the orientation of the first domain. Analyses of prokaryotic and eukaryotic transmembrane proteins have shown that amino acid sequences at their *cis* faces are highly enriched in positively charged Arg and Lys residues (von Heijne, 1986, 1988; Hartmann *et al.*, 1989). In contrast, these residues are largely absent from translocated protein domains. Therefore, as we previously encountered (e.g., Section 2.3.5.1), clusters of positive charges are generally found at a membrane protein's *cis* face. The location of a protein's NH_2-terminus at the *cis* or *trans* membrane faces depends on the presence and positions of positively charged clusters of amino acids.

If positive charges regulate NH_2-terminal position, it should be possible to alter transmembrane orientation by changing a protein's sequence. Data consistent with this hypothesis have been obtained during export targeting studies of alkaline phosphatase (Li *et al.*, 1988) and a fusion protein constructed from the β-lactamase signal sequence and triose phosphate isomerase (Pluckthun and Knowles, 1987). These studies support the idea that positive charges near the signal sequence regulate secretion and NH_2-terminal position.

Recently, von Heijne (1989) has prepared mutants of the *E. coli* membrane protein leader peptidase to directly test the relationship between positive charges and protein topography. Under normal conditions the NH_2-terminus is found at the cytoplasmic membrane's *trans* face. However, when four Lys residues are added to the NH_2-terminal region, it remains at the *cis* face. The entire topography of the protein is flipped, thereby putting the peptidase's active site at the *cis* face. This experiment provides a clear indication that positive charges can affect the location of a membrane protein's NH_2-terminus. The orientation of any additional transmembrane domains are therefore set in alternating fashion (*trans–cis–trans–cis* . . .) until all transmembrane domains have been woven into the membrane.

A more fundamental issue is: how do positive charges trigger retention at the *cis* face? The positive charges do not seem to be recognized by a specific retention receptor. Furthermore, the heterogeneity in the position, composition, and neighboring and intervening sequences make retention receptors an unlikely option. The retention of positive charges at the *cis* face may be explained on simple physicochemical grounds.

To understand how elementary physical forces may influence membrane protein topography, we must identify the physical limitations of translocation. Positively charged clusters are hydrophilic. However, the hydrophilic nature of the positive charges *per se* must be unimportant because negatively charged residues are equivalent to neutral residues in membrane polypeptide translocation. A transmembrane potential must also be an unimportant factor since: (1) the ER does not have a significant transmembrane potential and (2) potential-destroying ionophores have no effect on NH_2-terminal disposition. The dipoles of integral membrane *proteins* may not be important in determining NH_2-terminal disposition. This is likely since: (1) the positive-*cis* rule would suggest that other ER membrane proteins should also have any positive clusters at the *cis* face (just the reverse of how an ER protein dipole should catalyze insertion) and (2) translocation-associated membrane proteins are thought to have positive charges at their *cis* face to facilitate the binding of SRP. Acidic RNA molecules and phospholipids may contribute to the retention of positively charged membrane protein domains in the cytosol.

Internal membrane potentials, particularly dipole potentials, may contribute to the retention of positively charged domains at a membrane's *cis* face. To cross an entire bilayer, each translocated amino acid must pass by the bilayer's center. However, due to the spatial orientation of lipid head groups, carbonyl carbons, and neighboring water molecules, a dipole potential of roughly $+250$ mV is present at the center of *all* phospholipid bilayers (Honig *et al.*, 1986)! Hydrophobic anions diffuse across bilayers whereas hydrophobic cations cannot. Furthermore, the ability of a nascent polypeptide to pass the center dramatically rises as two or more positive charges simultaneously experience a dipole potential. As ER membrane-bound ribosomes extend a growing polypeptide chain, the nascent chain's positive charges will feel a large repulsive force due to the dipole potential. The nascent polypeptide will follow the path of least resistance back to the membrane's *cis* face. In the absence of positive charges, the NH_2-terminus is predicted to follow a "default" pathway to the *trans* membrane face. Although experimental tests of this hypothesis are necessary, the dipole potential may explain the selective barrier for positive charges and the heterogeneity in type and spatial arrangement of positive charges.

8.2.3. Assembly and Addition of Carbohydrates to Membrane Proteins

In Section 3.1 we discussed membrane carbohydrates. Membrane oligosaccharides are assembled in a stepwise fashion on ER and Golgi membranes. *N*-linked glycosylation begins in the ER during membrane protein synthesis (Rothman and Lodish, 1977). *O*-linked oligosaccharide chains are constructed in the Golgi. This section will explore the biosynthesis of membrane carbohydrates.

8.2.3.1. Construction and Transfer of the *N*-linked Oligosaccharide Precursor on
          ER Membranes

All *N*-linked oligosaccharide chains of membrane proteins are initially assembled on the carrier lipid dolichol (Dol) in the ER (Hirschberg and Snider, 1987; Chojnacki and Dallner, 1988). Dolichol is synthesized by ER membrane-bound enzymes. It is an unsaturated long-chain isoprenoid alcohol built from 15–21 isoprene units. Yeast cells typically have fewer isoprene units than higher eukaryotes. Dolichol's long axis extends for up to 92 carbon atoms, sufficient to traverse a bilayer's hydrophobic core three times. However, its membrane-bound conformation is unknown. A dolichol molecule attached to the precursor oligosaccharide chain ($Glc_3Man_9GlcNAc_2$-PP-Dol) is shown in Figure 8.8.

Dolichol's unusual covalent structure endows it with some special physicochemical properties (Lennarz, 1987). For example, it abolishes the phase transition of DMPC bilayers. Dolichol also increases the fluidity of PE bilayers and promotes formation of the hexagonal II phase. Membrane destabilization brought about by dolichol may promote the passage of certain carbohydrate chain lengths across a membrane via local discontinuities or defects in bilayer structure, particularly near proteins.

The biosynthetic pathway of $Glc_3Man_9GlcNAc_2$-PP-Dol is shown in Figure 8.9. The enzyme UDP-GlcNAc:Dol-P transferase catalyzes the transfer of GlcNAc-P from UDP-GlcNAc to Dol-P. This reaction is inhibited by the antibiotic tunicamycin. Although this reaction is known to be carried out by ER membranes, it is not known which side of the ER's membrane is involved. Yeast's UDP-GlcNAc:Dol-P transferase has been

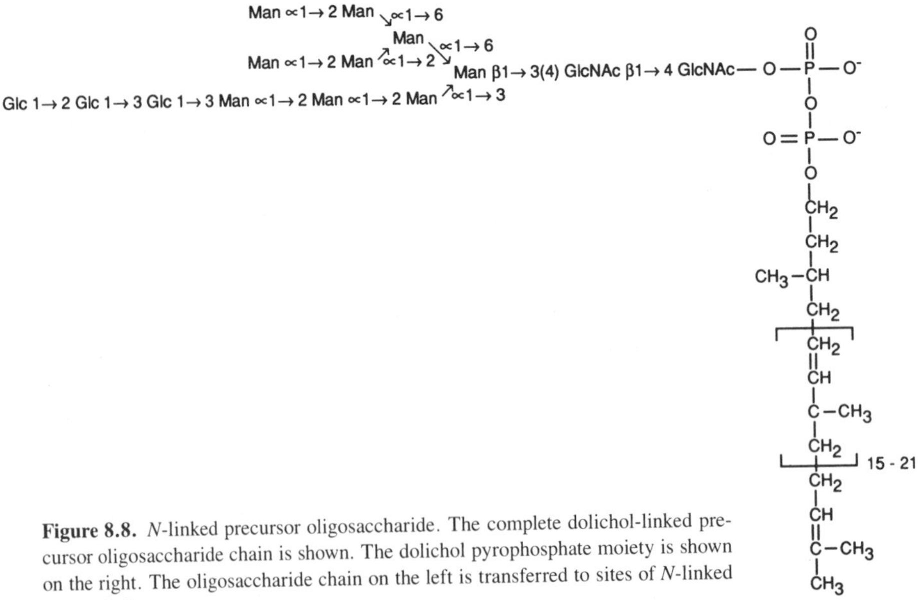

Figure 8.8. *N*-linked precursor oligosaccharide. The complete dolichol-linked precursor oligosaccharide chain is shown. The dolichol pyrophosphate moiety is shown on the right. The oligosaccharide chain on the left is transferred to sites of *N*-linked glycosylation.

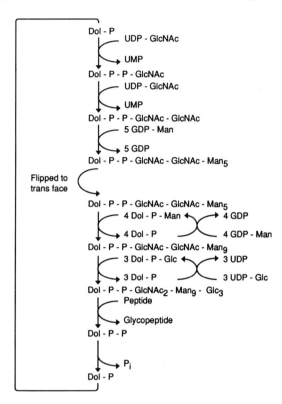

Figure 8.9. Biosynthesis of *N*-linked carbohydrate precursor chain. The steps involved in the synthesis of the *N*-linked oligosaccharide chain precursor, its transfer to a peptide, and the recycling of Dol-P are illustrated.

cloned (Kukuruzinska *et al.*, 1987). It has two putative membrane-spanning domains. When this enzyme's active site and topology are known with certainty, it should be possible to unambiguously determine the location of $GlcNAc_2$-PP-Dol formation. This transferase's function is absolutely essential since prolonged exposure to tunicamycin or disruption of its gene is lethal.

The enzyme β-1,4 mannosyltransferase catalyzes the reaction:

$$GlcNAc_2\text{-PP-Dol} + GDP\text{-Man} \rightarrow Man\beta1 \rightarrow 4GlcNAc_2\text{-PP-Dol} \quad (8.14)$$

This enzyme also possesses two apparent transmembrane domains. Four additional Man residues are then transferred to the precursor oligosaccharide chain. These reactions are thought to occur at the ER membrane's *cis* face since: (1) biochemical studies have localized $Man_{3-5}GlcNAc_2$-PP-Dol to the *cis* face and (2) GDP-Man cannot be transported across the ER's membrane.

The $Man_5GlcNAc_2$-PP-Dol molecule flip-flops from the *cis* to the *trans* face of the ER's membrane. Although $Man_5GlcNAc_2$-PP-Dol is found at the *cis* face, all subsequent forms are exclusively found at the *trans* face. Since saccharide-Dol molecules are unable to spontaneously flip-flop across phospholipid vesicles as measured by ESR spectroscopy, their movement is likely catalyzed by a flippase. When $Man_5GlcNAc_2$-PP-Dol reaches the ER's *trans* face, additional Man and Glc residues are donated by Man-P-Dol and Glc-P-Dol, respectively.

Man-P-Dol is synthesized by Dol-P-mannosyltransferase, an essential integral membrane protein (30 kDa) of the ER. The deduced amino acid sequence of this enzyme revealed a single transmembrane domain and a potential site for PKA-mediated phosphorylation (Orlean *et al.*, 1988). The COOH-terminal side of its transmembrane domain is exclusively composed of bulky hydrophobic residues. This region has been suggested to interact with the dolichol chain. One of the most interesting aspects of Dol-P-mannosyltransferase is that Man-P-Dol is synthesized at the *cis* face but utilized at the *trans* face. Although the topology of the transferase is uncertain, its active site is presumably exposed at the ER's *cis* face. This suggestion is based on the observations that: (1) transferase activity is sensitive to proteases at the *cis* face, (2) the substrate GDP-Man is only available at the *cis* face, and (3) the consensus PKA phosphorylation site, which is on the same side of the transmembrane domain as the catalytic site, must be on the *cis* face to be functional. After Man-P-Dol is synthesized, the mannosyltransferase is believed to flip-flop it to the *trans* ER face (Haselbeck and Tanner, 1982). These investigators showed that the transferase catalyzes Man-P-Dol flip-flop when reconstituted into lipid vesicles composed of PC and dolichol. When Man-P-Dol is released at the *trans* face, it participates in the addition of Man to precursor oligosaccharide chains.

The formation of Man-P-Dol and Glc-P-Dol are regulated by cAMP (Banerjee *et al.*, 1987). The β-adrenergic receptor stimulates the Dol-P-mannosyltransferase activity of several cell types by 40% to 80%. Its K_m is unchanged while the V_{max} doubles. The effect is due to phosphorylation since enzyme activity can be increased or decreased *in vitro* by phosphorylation or dephosphorylation reactions, respectively. Extracellular signaling events regulate PKA, phosphorylation patterns, and consequently Man-P-Dol and *N*-linked oligosaccharide synthesis.

As described in the preceding paragraphs, the first seven carbohydrate residues are thought to be added to dolichol at the ER's *cis* face. The last seven residues of the precursor oligosaccharide are added at the *trans* face (Figure 8.9). Four Man-P-Dol molecules donate their Man residues to $Man_5GlcNAc_2$-PP-Dol to form $Man_9GlcNAc_2$-PP-Dol. A variety of studies have indicated that these Man additions only occur at the ER's *trans* face (Hirschberg and Snider, 1987). Three Glc molecules are then linked to $Man_9GlcNAc_2$-PP-Dol to form $Glc_3Man_9GlcNAc_2$-PP-Dol (Figure 8.9). Glc residues donated by Glc-P-Dol are added by two enzymes to the growing oligosaccharide chain. Although the mechanism of Glc-P-Dol synthesis is uncertain, it is presumed to be similar to that of Man-P-Dol. These Glc molecules are added when the Man additions are finished or nearly finished. The series of membrane reactions leading to $Glc_3Man_9GlcNAc_2$-PP-Dol are common to all eukaryotes. The great evolutionary pressure that has maintained this rather inelegant biosynthetic route has not been adequately explained.

The oligosaccharide precursor is now ready to be transferred to a protein. The enzyme oligosaccharide transferase catalyzes the *en bloc* transfer of $Glc_3Man_9GlcNAc_2$ from Dol-pyrophosphate to a peptide chain at the ER's *trans* surface. A phosphatase cleaves Dol-PP to Dol-P which then begins another round of oligosaccharide precursor assembly.

This oligosaccharide chain is added at the amino acid acceptor sequence Asn-X-Ser(Thr) (Section 3.1). The X represents any amino acid residue with the possible exceptions of Asp and Pro (Hubbard and Ivatt, 1981). The carbohydrate chain is usually cotranslationally added to a nascent polypeptide, although posttranslational transfer has been observed in a few exceptional cases. As we have previously observed, not all consensus N-linked acceptor sites become glycosylated. Steric factors are believed to account for the consistent glycosylation of certain acceptor sites but not others. The small hydrophobic tripeptide acetyl-Asn-Leu-Thr-NHCH$_3$ diffuses across the ER membrane where it becomes glycosylated and trapped in the lumen (Welply et $al.$, 1983). This result indicates that: (1) oligosaccharide transferase is functionally active at the luminal face and (2) the tripeptide sequence is the $signal$ for oligosaccharide addition, not just the position of linkage. Since there are no steric constraints on the tripeptide, they are good substrates for glycosylation. However, some N-linked acceptor sites may be sterically excluded from the oligosaccharide transferase's active site. This is supported by the fact that nonglycosylated acceptor sites can become glycosylated after protein denaturation. Furthermore, when the glycosylation sites and x-ray crystal structures of several proteins were compared, the glycosylation sites were generally found at β-turns or loops. Therefore, N-linked acceptor sites are glycosylated when they are accessible to the active site of oligosaccharide transferase.

The oligosaccharide transferase's active site interacts with both the Asn and Ser(Thr) residues. When translation occurs in the presence of β-hydroxynorvaline, an analogue of Thr without a methylene, glycosylation is reduced. This suggests that the distance of the Ser or Thr residue's OH group from the peptide backbone is an important factor in forming N-linkages (Hubbard and Ivatt, 1981). Similarly, minor modifications of Asn abolish N-linked glycosylation. For example, in a methylamine derivative of Asn a hydrogen atom is replaced by a methyl, thereby resulting in an absence of activity (see Welply et $al.$, 1983). Severe structural restrictions are placed on the oligosaccharide acceptor's site, which contributes to a substrate's steric constraints discussed above.

8.2.3.2. Processing of N-linked Oligosaccharides

After the oligosaccharide precursor is linked to a protein, it is trimmed. Yeast cells remove three Glc and one Man residue whereas higher eukaryotes cleave three Glc and six Man residues away. Integral membrane enzymes exposed at the $trans$ face carry out these reactions. The Glc molecules are removed first within the ER by glucosidase I which cleaves the precursor's terminal Glc and glucosidase II-III which cleaves the two remaining Glc residues. In yeast, an ER membrane-bound mannosidase removes the last trimmed residue. Higher eukaryotes remove six Man residues from the precursor with mannosidase I and mannosidase II, two enzymes associated with the Golgi. Golgi enzymes sculpt the remaining common core oligosaccharide into its many mature forms.

Glycosyltransferases at the Golgi's $trans$ face synthesize glycolipids, O-linked oligosaccharides of glycoproteins, and the terminal portion of N-linked glycoproteins. The Golgi's many glycosyltransferase activities are all exposed at its luminal surface.

Many independent lines of investigation support this conclusion (Hirschberg and Snider, 1987). Moreover, this enzyme topography is almost expected because the oligosaccharide chains undergoing modification face the lumen. However, the membrane topography of these enzymes requires that their nucleotide-sugar substrates be transported into the lumen.

With the exception of CMP-NANA, which is synthesized in the nucleus, all nucleotide-sugars are made in the cytosol. To participate in oligosaccharide synthesis, these nucleotide-sugars must be transported into the Golgi. Double radiolabeling experiments have shown that both the nucleotide and sugar moieties are translocated into the Golgi's lumen. Several integral Golgi membrane proteins catalyze the facilitated diffusion of nucleotide-sugars. These translocators act as antiports; for example, the GDP-fucose translocator exchanges GDP-fucose for GMP, a molecule derived from glycosylation reactions. These translocators are highly specific and saturable carriers with K_m's of about 1 to 10 μM. Their activities are decreased or abolished by reduced temperatures or protease treatments (Hirschberg and Snider, 1987). *In vitro* transport is inhibited by adding free nucleotide phosphates but not sugars. When these nucleotide-sugars reach the lumen, their carbohydrate moiety is added to a growing chain by a glycosyltransferase.

One cycle of a transport–glycosylation process of the *trans* Golgi is illustrated in Figure 8.10. In a living cell many different kinds of chemical reactions, including glycosylation reactions, occur simultaneously in the Golgi. Glycosyltransferases are heterogeneously distributed in Golgi membranes. For example, fucosyltransferase and sialyltransferase are predominantly found in the *trans* Golgi stack whereas N-acetylglucosamine-1-phosphotransferase is found in the *cis* Golgi. It has been speculated that

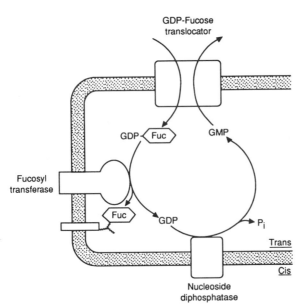

Figure 8.10. Fucosylation in the Golgi. The coupling of transport and glycosylation for fucose, a representative glycosylation substrate, is shown. A Golgi membrane translocator catalyzes the exchange of GMP for GDP-Fuc. A fucosyltransferase catalyzes the addition of fucose to an oligosaccharide chain. GDP is cleaved to GMP and P_i. The GMP promotes the entry of additional substrate.

some glycosyltransferases may form multienzyme complexes, much like some proteins of the mitochondrial inner membrane, to facilitate substrate glycosylation.

8.2.3.3. Biosynthesis of O-linked Oligosaccharides of Glycoproteins and Glycolipids

The two preceding sections described the assembly of N-linked carbohydrates. In addition, O-linked glycosidic bonds between a saccharide's anomeric (number 1) carbon atom and a hydroxyl group of a membrane-associated lipid or protein are also observed. As previously described (Sections 2.1.3 and 2.1.4), this type of linkage is found in glycosyl diacylglycerols of plants and glycosphingolipids of animals. O-glycosidic linkages are also found between oligosaccharide chains and membrane proteins. We shall begin by considering the synthesis of the carbohydrate chains of glycolipids.

All glycosphingolipids are derived from the cerebrosides GalCer and GlcCer. These molecules are synthesized in the ER by the transfer of Gal or Glc from its nucleotide sugar to Cer (Sweeley, 1986). For example, the enzyme UDP-Glc:ceramide glycosyltransferase catalyzes the synthesis of GlcCer [Eq. (8.15)] with inversion of the asymmetric carbon's configuration ($\alpha \rightarrow \beta$).

$$\text{UDP-Glc} + \text{Cer} \rightarrow \text{Glc}(\beta1\rightarrow1')\text{Cer} + \text{UDP} \qquad (8.15)$$

This reaction has not been definitively localized in cells. The enzyme has been reported in both the ER and Golgi compartments; its topography (*cis* or *trans*) is unknown.

Most common glycosphingolipids are derived from LacCer (Figure 2.6). UDP-Gal:glucosylceramide galactosyltransferase, a Golgi enzyme, catalyzes the synthesis of LacCer from UDP-Gal and GlcCer:

$$\text{UDP-Gal} + \text{GlcCer} \rightarrow \text{Gal}(\beta1\rightarrow4)\text{GlcCer} + \text{UDP} \qquad (8.16)$$

The product, LacCer, is a substrate for other glycosyltransferases within the Golgi. These several glycosyltransferases account for the variety of glycolipids encountered in cell membranes. Some glycosyltransferases append sugars to both lipid and protein oligosaccharide chains. This explains why some membrane glycolipids and glyco-proteins possess ABO(H) blood group antigens (Section 8.2.3.4).

O-type glycosidic linkages are almost always found between GalNAc and either Ser or Thr (Figure 3.3). A small fraction ($\sim$5%) of O-linkages found in higher eukaryotes utilize the sugar GlcNAc. The O-linkage in yeast is always formed with mannose. The rare modified amino acids hydroxylysine and hydroxyproline can be found glycosylated in animals and plants, respectively. We will now consider the biosynthesis of O-linked oligosaccharide chains.

In yeast O-linked oligosaccharide synthesis begins in the ER. This conclusion is supported by the fact that a secretory temperature-sensitive mutant (*sec18*) deficient in ER-to-Golgi transport supports chain initiation at a nonpermissive temperature (Kuku-ruzinska *et al.*, 1987; Sections 8.3.6 and 8.3.8). In addition to its role in N-linked chain assembly, Man-P-Dol also donates its mannose moiety to a Ser or Thr residue with inversion of the anomeric carbon's configuration. Mannose is apparently cotranslationally added to polypeptide chains. Additional mannose residues are then linked to this saccharide moiety in the Golgi's lumen.

Oligosaccharides of O-linked glycoproteins in higher eukaryotes are more complex than the simple mannose chains of yeast. They are often composed of GalNAc, Gal, NANA, and Fuc. O-linked chains tend to be shorter than N-linked oligosaccharides with less branching. O-linked oligosaccharide chains are assembled at the Golgi's luminal face by glycosyltransferases. Some of these glycosyltransferase activities may be shared among the various pathways of glycolipid and glycoprotein synthesis. In contrast to N-linked oligosaccharides, O-linked chains of higher eukaryotes are assembled post-translationally in the Golgi without the participation of lipid intermediates. The same pool of nucleotide sugars are utilized for both N-linked chain extension and O-linked chain synthesis.

The primary O-linked chains of glycophorin were previously discussed (Section 3.1.3). Its simplest chain, Gal($\beta1\rightarrow3$)GalNAc-Ser/Thr is synthesized by the sequential actions of two glycosyltransferases. This disaccharide can be further modified by the addition of one or two NANA residues. In the following section we will discuss the ABO(H) blood group system, a family of O-linked membrane oligosaccharide structures.

Steric factors apparently contribute to the selection of O-linked glycosylation sites. Young *et al.* (1979) have studied the O-linked glycosylation of myelin basic protein, a peripheral protein of myelin sheaths, and synthetic oligopeptides corresponding to portions of its covalent structure. These investigators found that O-linked glycosylation reactions were sensitive to an acceptor's covalent structure. Just as in the intact protein, certain residues could or could not be glycosylated. The residues at a peptide's NH_2-terminus were not glycosylated. Moreover, residues with surrounding prolines were more readily modified. Presumably, acceptor accessibility is a key determinant of O-linked glycosylation.

8.2.3.4. Golgi Glycosyltransferases Determine Blood Group Specificity

Individuals within and among species vary in the glycosyltransferases they inherit and express. Consequently, secretory and membrane-associated oligosaccharides of different individuals can vary in structure. This is best illustrated by human blood group antigens.

At the turn of the century Ehrlich and Landsteiner discovered the presence of blood group antigens in animals and people, respectively. Blood group antigens are important in many medical circumstances such as blood transfusion and tissue transplantation. The principal blood group type is the ABO(H) antigen system which is present on almost all human cells. These antigens represent different oligosaccharide structures expressed at cell surfaces. One should bear in mind that these antigens are *not* restricted to: (1) any particular type of cell or tissue or (2) any particular glycoprotein or glycolipid.

Figure 8.11 shows several glycosphingolipids of the LacCer family that possess A blood group activity (Sweeley, 1986; Hughes, 1983). This illustrates that different kinds of glycosphingolipids possess blood group antigens. The physiologically impor-tant structural difference between blood group substances is their terminal residue:

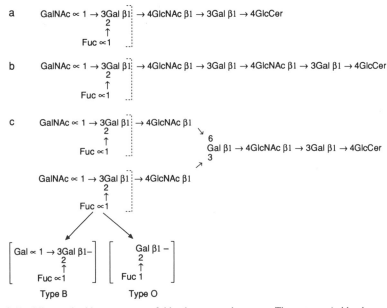

Figure 8.11. Oligosaccharide structures of blood group substances. Three type A blood-group-active glycosphingolipids are shown (a–c). The blood-group-active regions are at the left-hand sides of the dotted lines. If the structures within the brackets are replaced by those given at the bottom of the figure, the type B or O phenotypes are observed.

GalNAc for blood group A or Gal for blood group B (Figure 8.11, insert). When this terminal position is vacant, the blood type O is found. Type O individuals synthesize only the immediate oligosaccharide precursor of the A and B antigens.

Nearly everyone has *H* genes which encode the physical precursor H structure; phenotypically it is the absence of A and B. Some individuals inherit the *A* gene which encodes an *N*-acetylglucosaminyltransferase. When present, this Golgi enzyme catalyzes the addition of GalNAc to the H structure. The *B* gene is an allele of the *A* gene that encodes a galactosyltransferase. This enzyme's action leads to an alternate structure (Figure 8.11, insert) that is the B blood group antigen. An individual's blood group is a reflection of his or her glycosyltransferase activity. In addition to their presence on glycolipids, ABO blood group antigens are found on secretory proteins such as those found in serum or milk and, occasionally, on membrane proteins.

8.2.4. Posttranslational Protein Transport across Mitochondrial Membranes

Most mitochondrial proteins are synthesized in a cell's cytosol and then escorted to mitochondria by molecular chaperones. Chaperones, such as hsp70, prevent mitochondrial precursor proteins from prematurely folding in the cytosol. Mitochondrial proteins of the matrix, intermembrane space, and inner membrane must first be translocated from the cytosol to the outer membrane and then to these final destinations. We will now discuss how proteins cross the mitochondrial permeability barrier.

8.2.4.1. Mitochondrial Binding

The first step in mitochondrial membrane translocation is the association of a precursor protein with a mitochondrion. An NH_2-terminal signal sequence of 20–80 amino acids directs a protein to a mitochondrion. Deletion studies have shown that the initial 12 residues are absolutely required for transport. Genetically engineered fusion proteins containing this signal peptide and a nonmitochondrial "passenger" protein have shown that the signal peptide's presence is a sufficient condition for mitochondrial protein import.

The mitochondrial signal peptide is rich in hydrophobic and basic amino acid residues. When this sequence forms an α-helix, the basic residues extend from one side of the helix while the hydrophobic residues extend from the opposite face. The positive side of an amphipathic signal sequence may bind to anionic lipids on the outer mitochondrial membrane. Initial binding is purely electrostatic in nature and has been reconstituted in lipid vesicles and lipid monolayers containing anionic lipids (Myers et al., 1987; Tamm, 1986). Lipid vesicles composed of 5:1 DPPC:DPPG demonstrate a K_d of 0.43 μM for the mitochondrial signal sequence, although binding to mitochondrial membranes may be of lower affinity. The spatial arrangement of the presequence's positive charges allows for high-affinity receptor-like binding to simple model membranes. The binding of the presequence to lipid membranes affects the packing of nearby hydrocarbon chains. This interaction enhances the rate of mitochondrial protein translocation by: (1) increasing the local concentration of presequence-containing proteins about the outer membrane and (2) restricting the presequence to a two-dimensional surface and thus increasing greatly its rate of engaging the receptor/translocation machinery. This binding reaction may also potentiate protein unfolding.

After binding to a mitochondrion's outer membrane, a protein rapidly migrates along this surface. Mitochondrion-targeted proteins accumulate at contact sites between the inner and outer membranes. These contact sites participate in protein translocation into the mitochondrial matrix. Translocation intermediates spanning the contact site have been observed by both electron microscopic and biochemical methods (e.g., Schleyer and Neupert, 1985). Membrane proteins at contact sites are believed to be responsible for translocation.

8.2.4.2. General Insertion Sites

General insertion sites are composed of a set of membrane proteins found at contact sites. To date, four of these proteins have been identified: ISP42, p32, MAS70, and MOM17 (Kiebler et al., 1990; Baker and Schatz, 1991). These membrane proteins are assembled to form a hetero-oligomeric complex capable of mediating mitochondrial protein recognition and translocation at contact sites. Figure 8.12 shows a hypothetical model of a mitochondrial contact site.

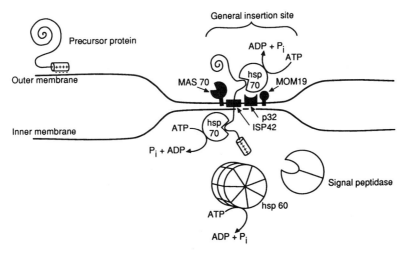

Figure 8.12. Mitochondrial membrane passage. A protein possessing an amphipathic sorting signal binds to the outer mitochondrial membrane (left). It accumulates at general insertion sites where it is unwound and translocated into the mitochondrial matrix (center). The protein's signal sequence is removed by a signal peptidase in the matrix, where it refolds.

Two integral membrane proteins of the outer mitochondrial membrane (p32 and MAS70) are receptors for mitochondrial precursor proteins. p32 recognizes immature mitochondrial proteins possessing an NH_2-terminal amphipathic signal sequence (Pain et al., 1988; Baker and Schatz, 1991). The rat's mitochondrial signal sequence receptor binds with a K_d of 2 μM (Gillespie, 1987); it retains binding ability when reconstituted in liposomes. The *MIR1* gene of yeast encodes the p32 receptor. Its deduced amino acid sequence is similar to the mitochondrial phosphate transporter of mammals. MAS70 is also an apparent translocation receptor in outer mitochondrial membranes. However, only one protein, the ADP/ATP translocator, is known to be targeted via this receptor (Sollner et al., 1990). It recognizes an internal signal sequence of its precursor protein, not an amphipathic sequence. Its deduced amino acid sequence suggests that it is attached to outer membranes by a single transmembrane domain.

Another protein, MOM19, (*m*itochondrial *o*uter *m*embrane; *N. crassa* nomenclature), is also a component of general insertion sites. It is apparently anchored to the outer membrane by one transmembrane domain (Schneider et al., 1991). It is linked tightly to MOM38 (or ISP42 of yeast). MOM19 may participate in precursor recognition, translocation, and/or the assembly of general insertion sites.

ISP42 (*i*mport *s*ite *p*rotein) is a 42-kDa integral membrane protein of outer mitochondrial membranes. It is another component of general insertion sites. ISP42 plays a central role in import. *ISP42* gene disruption experiments have shown that its product is absolutely required for cell viability. Biochemical studies using genetically engineered precursor proteins containing a photoaffinity label have demonstrated that these proteins pass ISP42 on their journey across the outer membrane. The deduced amino acid sequence of ISP42 indicated the presence of a region containing many negative charges; this domain might interact with the positively charged signal sequence of mitochondrial precursor proteins. However, the deduced sequence did not reveal an unambiguous transmembrane domain. As mentioned above, one might expect β structure, not transmembrane α-helices, in proteins performing a pore-like function. It will be interesting to compare the sequences of ISP42 (or its homologue MOM38) with mp39 and SRP's β subunit when they become available.

8.2.4.3. Membrane Passage

Membrane translocation requires a large inside-negative electrochemical potential gradient. Although a transmembrane potential is not a sufficient condition to trigger translocation, the inner membrane's potential will attract the signal sequence's high positive charge. Although membrane translocation is not simple electrophoretic migration, this attraction will at least substantially lower the energy barrier for translocation across a membrane.

Proteins cannot be translocated across mitochondrial membranes in their native three-dimensional globular state; they must first be unwound and then threaded across a general insertion site. There is now compelling evidence indicating that only unfolded proteins are translocated. As mentioned above, mitochondrion-targeted proteins have translocation intermediates with domains in both the cytosol and mitochondrial matrix. This suggests that proteins are translocated in a piecemeal fashion. To demonstrate the importance of unfolding, Eilers and Schatz (1986) used genetic engineering to prepare a fusion protein containing sequences derived from a mitochondrial signal sequence and dihydrofolate reductase. When the fusion protein was mixed with energized mitochondria *in vitro*, it was translocated into the mitochondrial matrix. Methotrexate is a drug that binds to dihydrofolate reductase with high affinity. When this ligand is bound to the fusion protein, it cannot undergo translocation. Since methotrexate stabilizes the conformation of dihydrofolate reductase, it is unable to unfold, thus inhibiting translocation. On the other hand, if a protein's conformation is destabilized by the introduction of point mutations, its translocation is accelerated (Vestweber and Schatz, 1988). Therefore, protein unfolding is an integral part of protein translocation across mitochondrial membranes.

Cytosolic ATP participates in moving proteins across mitochondrial contact sites. Its role, however, seems to be limited to protein unfolding. Unfolded proteins are able to cross mitochondrial contact sites in the absence of ATP. For example, nascent unfolded proteins still attached to tRNA do not require ATP for mitochondrial entry. Cytosolic ATP is used by hsp70 and to unfold mitochondrial precursor proteins (Deshaies *et al.*, 1988; Kang *et al.*, 1990). Cells genetically deficient in hsp70 are unable to translocate proteins into their mitochondria.

We have identified the energy sources required for precursor protein unfolding in the cytosol and presequence entry into the matrix. However, this does not explain why proteins continue to move across a contact site. Why don't proteins just stop when the signal arrives in the matrix? Matrix proteins, including a mitochondrial version of

hsp70, and matrix ATP molecules may provide an additional driving force for translocation and ensure the proper folding of delivered proteins.

8.2.4.4. Matrix Proteins Participating in Translocation and Folding

In addition to the outer membrane proteins discussed above, matrix proteins are
also involved in precursor protein translocation and folding. Figure 8.12 shows matrix
proteins known to participate in translocation. These include hsp70, hsp60, and a signal
peptidase.

Mitochondrial hsp70 is nearly identical to cytosolic hsp70, except that its precursor
contains an amphipathic mitochondrial targeting sequence. It is encoded by the nuclear
SSC1 gene. hsp70 is the first matrix protein to bind precursors as they emerge from a
contact site; it is required for both translocation and folding (Kang *et al.*, 1990). hsp70
releases bound precursors in the presence of ATP. It may participate in pulling a
precursor protein across a membrane and in maintaining a precursor's loose conformation prior to delivery of hsp60.

Another heat shock protein, hsp60, participates in precursor folding. hsp60 is an
oligomer of 14 identical subunits. It maintains precursors in loose conformations within
the matrix and releases them in an ATP-dependent manner that may help catalyze
appropriate precursor folding. During or after hsp60 binding, the amphipathic signal
sequence is removed by a signal peptidase. The mitochondrial signal peptidase is a
heterodimer of two ~50 kDa hydrophilic proteins encoded by the unlinked nuclear
genes *MAS1* and *MAS2* (Yang *et al.*, 1991). Both genes are required for cell viability.
In the presence of Mn^{2+}, these two proteins form a functional signal peptidase.

8.2.5. Cotranslational and Posttranslational Protein Insertion in Bacterial Membranes

We have deferred our discussion of bacterial membrane protein insertion to this
point since bacteria exhibit a wide range of insertion mechanisms, including all of those
previously discussed for eukaryotes. These insertion mechanisms include cotranslational, posttranslational, and, apparently, some insertion mechanisms unique to bacteria. In addition to their cytoplasm, gram-negative bacteria must route proteins to the
inner membrane, the periplasmic space, the outer membrane, and the extracellular
environment (Figure 8.13, panel a) (Saier *et al.*, 1989; Hirst and Welch, 1988). This
section examines some bacterial membrane protein insertion mechanisms.

Periplasmic, outer membrane, and most secretory protein precursors are synthesized with a signal sequence. These signal sequences are a string of 20 hydrophobic
amino acids with a positively charged NH_2-terminus. In addition to a signal sequence,
protein translocation requires an electrochemical membrane potential, ATP, and several
proteins which participate in translocation.

Several components of the membrane translocation machinery have been identified

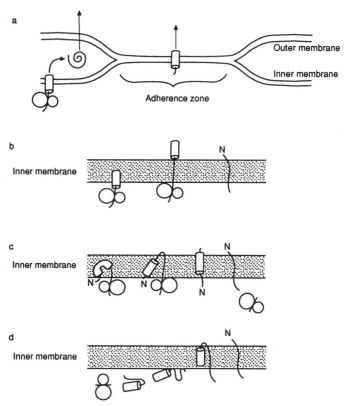

Figure 8.13. Translocation into and across bacterial membranes. A survey of potential protein insertion mechanisms is presented. Panel a illustrates two mechanisms of protein secretion. Panels b, c, and d show conventional cotranslational insertion, loop cotranslational insertion, and posttranslational trigger insertion.

by studying temperature-sensitive mutants of *E. coli*. The protein secAp has been identified in *secA* mutants. (sec mutants of prokaryotes are indicated with a letter whereas those of eukaryotes are identified with a number.) The deduced amino acid sequence of this protein (M_r = 101,902) indicates that it does not have an apparent transmembrane domain (Schmidt *et al.*, 1988). This is in agreement with the identification of secAp as a peripheral membrane protein (Saier *et al.*, 1989). secAp binds to membranes and translocates pro-porin molecules across inner membranes in the presence of ATP and trigger factor, a protein that unfolds nascent polypeptides.

secYp is an integral cytoplasmic membrane protein of bacteria. Its deduced amino acid sequence suggests that it has ten transmembrane domains (Cerretti *et al.*, 1983). This protein apparently interacts with signal sequences. It has been suggested that secYp forms a translocation channel. secYp is homologous to sec61p of eukaryotes.

In Sections 8.2.2.3 and 8.2.4.2 we discussed how protein unfolding is an integral part of protein translocation across ER and mitochondrial membranes. Three unfolding proteins have been identified in *E. coli*. Trigger factor, a 63-kDa protein, was previously mentioned. The protein secBp is a 90-kDa oligomer built from 17-kDa subunits. A third protein, GroEL, is a tetradecamer composed of 65-kDa subunits. GroEL is an ATPase that resembles hsp60. These three proteins participate in nascent polypeptide unfolding in bacteria. These functionally similar proteins appear to act on distinct nascent chains.

A few secretory proteins of *E. coli* do not possess an apparent signal sequence. On the basis of indirect evidence, it has been suggested that secretory proteins without signal sequences are translocated across bacterial membranes at zones of adherence between the inner and outer membranes (Hirst and Welch, 1988). This has some similarity with translocation across mitochondrial membranes.

Most of the *E. coli* inner membrane proteins lack recognizable signal sequences. These include, for example, lactose permease, ATP synthase's F_0 subunits, photosynthetic pigment proteins, bacteriorhodopsin, secYp, signal peptidase, and others. Interestingly, mannitol permease has an amphipathic signal sequence resembling those found for mitochondrial proteins. Bacteriorhodopsin has a short unique NH_2-terminal extension whose function is unknown. secAp and sometimes secYp are required for insertion of cytoplasmic membrane proteins. The detailed mechanism(s) of protein insertion is uncertain.

Several models of inner membrane protein insertion have been proposed (Reithmeier, 1985). Since these models may apply to different proteins or circumstances, more than one model may be operational *in vivo*. Periplasmic, outer membrane, and most secretory proteins must first insert into the cytoplasmic membrane. Inouye and colleagues proposed a loop model to account for the insertion of the *E. coli* lipoprotein into cytoplasmic membranes (Figure 8.13, panel c). This model suggests that the positive charge of the NH_2-terminal amino acids remains at the inner surface of the cytoplasmic membrane. As the protein is synthesized, a loop is extended across the inner membrane. The signal sequence is cleaved at the periplasmic surface of the cytoplasmic membrane. It has been suggested that outer membrane proteins such as lipoprotein migrate from the inner to the outer membranes via zones of adherence between the two membranes. Wickner has proposed the trigger hypothesis of protein insertion into the cytoplasmic membrane (Figure 8.13, panel d). This hypothesis suggests that the signal sequence's principal role is to alter a protein's folding pathway. The protein spontaneously inserts into the cytoplasmic membrane co- or posttranslationally. Several lines of evidence suggest that this may be an important means of insertion for the bacteriophage M-13 coat protein into the *E. coli* inner membrane (Wickner, 1979; Kuhn *et al.*, 1986; Wickner and Lodish, 1985).

Although the insertion of secretory, periplasmic, and outer membrane proteins has been explored in some detail, the insertion of resident cytoplasmic membrane proteins is not well understood. Gene fusion studies (e.g., Silhavy *et al.*, 1983) indicate that information contained within a membrane protein, other than an NH_2-terminal signal sequence, is required for its routing. Presumably, internal "signals" are important in the insertion of many proteins into the cytoplasmic membrane.

8.3. BIOSYNTHETIC MEMBRANE FLOW

Most of a typical cell's membranes are on the move. One of the best examples of membrane flow is found in the biosynthetic secretory apparatus of eukaryotes. In contrast to a prokaryote's cytoplasmic membrane, the components of eukaryotic membranes are generally constructed at distant sites and then translocated to their final destinations. After their synthesis in the ER, membrane proteins and some lipids must navigate through numerous membrane-bound compartments and domains to reach their prescribed locations. This section will focus on the properties of secretory endomembranes and membrane flow during biosynthetic and secretory processes. We shall begin by considering the origins of this line of investigation back in the 1960s.

8.3.1. Palade's Paradigm: Biosynthetic Flow among Cell Compartments

The introduction of electron microscopy to cell biology brought many new insights in cell structure. One outstanding contribution to our understanding of cell structure and function was Palade's subcellular analysis of biosynthetic flow. He studied the pancreatic exocrine cell because it specializes in protein biosynthesis and secretion. Its very structure suggests the biosynthetic secretory pathway. The organelles' spatial locations suggest a flow of secretory molecules from the ER→Golgi→secretory granules→ plasma membrane and the extracellular environment. Palade's work proved this pathway of biosynthetic flow (Palade, 1975).

Figure 8.14 shows a map of the biosynthetic flow among cell compartments. Although Palade's work focused on the secretion of luminal proteins, membrane-associated components are also translocated along this pathway. Palade (1975) has recognized the six major steps of secretion as synthesis, segregation, transport, concentration, storage, and release. The subcellular pathway and kinetics of secretory flow were established using radioisotopic pulse–chase experiments. Cells, tissues, or animals were briefly exposed to a radiolabeled precursor, such as an amino acid. The samples were then "chased" with unlabeled precursors. After various lengths of time individual organelles were isolated by ultracentrifugation. The temporal order of appearance of radioactive secretory proteins in each cell fraction revealed the route they followed. After fixation, electron microscopic autoradiography was used to morphologically identify the subcellular compartments containing radiolabel. In this way biochemical and ultrastructural approaches were combined to characterize the mechanism of secretion.

8.3.2. Defining Organelles and Their Membranes: Where to Draw the Line between the ER and Golgi?

Organelles are distinct membrane-bound compartments. They are defined according to their structural and biochemical characteristics. However, the morphological and

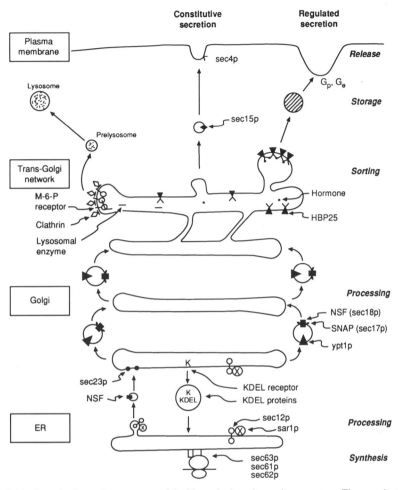

Figure 8.14. Organization and components of the biosynthetic endomembrane system. The organization of the biosynthetic organelles is shown. The organelles are identified on the left and their principal functions are listed on the right. The several proteins known to participate in endomembrane trafficking within the secretory apparatus are listed (see text for details). The components are not drawn to scale.

biochemical boundaries of the ER and Golgi overlap. Careful experiments showed that "purified" Golgi contained high levels of ER-"specific" marker enzymes. To simultaneously probe the biochemical and structural properties of Golgi, Palade and co-workers (for a review see Farquhar and Palade, 1981) combined traditional biochemical and electron microscopic approaches. Antibodies directed against ER enzymes, such as glucose-6-phosphatase and NADPH-cytochrome P450 reductase, were covalently attached to polyacrylamide beads. Purified Golgi cisternae and vesicles bound to these beads. Bead-associated materials were indeed elements of the Golgi because: (1) they were obtained from the Golgi fraction after ultracentrifugation, (2) they contained Golgi marker enzymes, and (3) when bead-associated material was examined by electron microscopy, it looked like Golgi. Yet this same material contained enzymes and antigens specific for the ER. Therefore, the Golgi must contain regions that are very similar in composition to the ER. The Golgi cisternae generally bound to the beads via their dilated rims, thus suggesting that these areas contain ER components. These ER-like regions of the Golgi likely communicate with the ER through vesicles that shuttle membrane and luminal components between these organelles.

8.3.3. The Rate of Flow among Biosynthetic Compartments

Secretory materials and membrane components flow along the biosynthetic pathway as outlined above. Pulse–chase experiments indicated that "typical" secretory proteins require about 2 hr to complete their biosynthesis and extracellular release. However, these experiments measured the overall rate of release of all biosynthesized materials. More recent studies have measured these rates for individual proteins (Pfeffer and Rothman, 1987). These rates vary significantly between different cell types and secretory materials. For example, under certain conditions albumin leaves the ER with a half-time of only 30 min whereas complement component C3 lingers, with a $t_{1/2} = 2\frac{1}{2}$ hr in this organelle. Before we address the heterogeneity of individual transport rates, we must understand the bulk flow rate among biosynthetic organelles.

Wieland et al. (1987) devised an approach to measure the bulk flow rate through the biosynthetic apparatus based on earlier studies of N-linked glycosylation (Section 8.2.3.1). The N-acyltripeptide N-acyl-Asn-[^{125}I]-Tyr-Thr was synthesized with various short fatty acyl chains. Due to their hydrophobicity, these compounds enter cells by simple diffusion. They contain the N-linked glycosylation site sequence of Asn-X-Thr/Ser. As N-acyl tripeptides diffuse into the ER they become glycosylated and trapped within this organelle. By measuring the efflux of ^{125}I-labeled N-acylglycotripeptides, the rate of bulk flow through the secretory apparatus can be inferred. The measurement of N-acylglycotripeptide flow through these organelles is very much like using a free buoy to measure the speed of a river's current. Wieland et al. (1987) found that the half-time for peptide transport from the ER to the extracellular environment was 5–10 min. This suggests that the fluxes of many luminal proteins are retarded by other factors. The surface-to-volume ratios of the ER, Golgi, and transport vesicles are similar (roughly 80 to 100 μM^{-1}; Wieland et al., 1987). This suggests that the kinetics of membrane flux should be similar to that of volume flux. Kaplan and Simoni (1985) have measured the transport rate of cholesterol from its site of synthesis in the ER to the plasma membrane. They found that cholesterol's half-time for transport was 10 min. Cholesterol flux required an energy supply. When metabolic poisons such as potassium cyanide are added, cholesterol transport rapidly stops. Cholesterol could leave the ER by several potential pathways (e.g., Figure 8.5). Since cholesterol transport is abolished by energy poisons, it is likely moving through the secretory endomembrane system. The bulk flow rate of both luminal components and membrane cholesterol through the secretory apparatus is about 10 min.

Proteins are slowed in their secretory migration by a variety of factors. Several ER luminal proteins retain secretory proteins while they encourage secretory proteins to adopt their physiologically relevant conformations. Furthermore, glycosylation, proteolytic processing, receptor binding events, and aggregation may all contribute to the reduced transport rates of secretory proteins. These several factors will be considered in the following paragraphs.

8.3.4. KDEL Spells Protein Retention in the ER's Lumen

If matter is continually flowing among secretory organelles, how do these compartments retain their unique identities? Receptor-like interactions provide one means of maintaining the unique protein compositions of organelles.

Several membrane-bound and luminal proteins are known to be retained within the ER. Two well-studied luminal ER proteins are BiP and protein disulfide isomerase. BiP binds to newly synthesized proteins and is believed to promote their proper folding. When the sequences of several luminal ER proteins were compared, a common sequence of four amino acids, KDEL (Lys-Asp-Glu-Leu), was found at the COOH-termini (Pelham, 1989). The KDEL sequence (or HDEL in yeast) has been shown to promote protein retention in the ER. When the KDEL sequence is deleted from BiP, it behaves like a secretory protein. Conversely, when KDEL is added to the COOH-terminal domain of the secretory protein lysozyme, it remains in the ER. This indicates that the KDEL sequence is a sufficient condition to keep soluble proteins in the ER. Furthermore, it indirectly suggests that specific signaling is not required for protein secretion since KDEL-deletion mutants of BiP, which should not possess such a signal, are secreted proteins.

KDEL-containing proteins are not anchored to the ER's *trans* face. Ceriotti and Colman (1988) have shown that BiP and BiP's KDEL-deficient mutant both diffuse in the ER's lumen. Furthermore, electron microscopic cytochemistry has shown that BiP is not associated with the ER's membrane. These results suggest that BiP is not physically held in the ER. Alternatively, KDEL-containing proteins are immediately returned to the ER when they enter the *cis* Golgi. Pelham (1988) has demonstrated this by expressing a genetically engineered protein in cells. The KDEL sequence was attached to the lysosomal enzyme cathepsin D. Cathepsin D and many other lysosomal enzymes are immediately modified by the addition of mannose-6-phosphate upon entry into the Golgi (Section 8.3.9). When KDEL-labeled cathepsin D is expressed, it becomes modified by mannose-6-phosphate. This result indicates that KDEL-containing cathepsin D is retrieved from the Golgi's *cis* side. Therefore, as shown in Figure 8.14, KDEL-containing proteins flow from the ER to the Golgi (anterograde transport) but are then recycled back to the ER (retrograde transport). The recycling of certain ER components from the Golgi accounts for the overlap in the biochemical and morphological boundaries of these organelles.

A receptor for the KDEL sequence is believed to be responsible for recognizing KDEL-containing proteins near the *cis* Golgi and returning them to the ER. A putative

KDEL receptor has recently been identified by Vaux *et al*. (1990). These investigators prepared antibodies against peptides possessing a COOH-terminal KDEL sequence. A second set of antibodies recognizing the KDEL binding site of the anti-KDEL antibodies was prepared. These anti-anti-KDEL antibodies, or anti-idiotype antibodies, bind to KDEL receptors just like the KDEL sequence but with a higher affinity. These anti-idiotype antibodies recognize a 72-kDa transmembrane protein that is found at the interface between the ER and Golgi. The purified KDEL receptor binds physiological KDEL-containing substrates with a K_d of 23 μM. Therefore, this 72-kDa integral membrane protein displays all of the properties expected for a KDEL receptor. The identification of the KDEL receptor raises the next important questions: how does the KDEL receptor with its bound KDEL-containing ligand know where to go and how does it avoid the massive flow of material exiting the ER?

8.3.5. Maintaining Organelle Membrane Composition: The Retention of Resident ER and Golgi Membrane Proteins

Investigators have only recently begun to study the mechanism of membrane protein retention in the ER and Golgi. Using an adenovirus model system, Nilsson *et al*. (1989) have reported that the COOH-terminal sequence DEKKMP (Asp-Glu-Lys-Lys-Met-Pro) at a membrane protein's *cis* face potentiates ER retention. Adenoviruses escape immune defense mechanisms because they produce an ER membrane protein, E3/19, that traps HLA class I molecules in the ER. E3/19's ability to remain in the ER depends critically on its COOH-terminus. Deletion experiments have shown that the DEKKMP hexapeptide participates in ER membrane retention. Recent site-directed mutagenesis studies (Gabathuler and Kvist, 1990) have shown that the Met-Pro sequence at the COOH-terminus and a Ser-Phe-Ile (SFI) sequence preceding the DEKKMP sequence also contribute to ER retention. Furthermore, Shin *et al*. (1991) have shown that two positive charges near the COOH-terminus (e.g., -KKXX and -RKXX) are essential in ER membrane protein retention. The twin positively charged COOH-terminal motif has only been found on transmembrane proteins known to be retained in the ER. Therefore, these positively charged sequences form another type of ER retention signal. Although the manner in which this sequence confers retention is unknown, it is believed to be physically restrained in ER membranes in contrast to the rapid recycling of KDEL proteins. Perhaps this signal pattern binds to cytosolic proteins that anchor ER membrane proteins in place. The close association between the ER and microtubules raises the possibility that some ER membrane proteins may be tethered to these tubules. However, the KDEL sequence is not used by ER membrane proteins. This raises the question of how other ER membrane proteins not possessing two positive charges near their COOH-terminus (e.g., ribophorins) are held in the ER.

Specific sequences of Golgi membrane proteins contribute to their retention in this organelle. Golgi retention signals have been associated with the cytoplasmic face of Golgi membranes (e.g., Kex2p and DPAP A). A putative Golgi retention sequence has been identified in the E1 protein of avian infectious bronchitis virus (IBV) (Machamer and Rose, 1987). IBV, like other studied coronaviruses, buds from the Golgi. This provides coronaviruses with the ability to evade immune destruction by accumulating in the very hospitable environment of a cell's cytosol. Therefore, the E1 protein which collects in the Golgi after infection or transfection is a useful model of Golgi membrane retention. The E1 protein contains three transmembrane domains. When the first two transmembrane domains of the E1 protein are removed by mutagenesis, the altered protein flows into the plasma membrane. In contrast, if the second and third transmembrane domains are removed, the protein is retained by the Golgi. This suggests that something special about the first transmembrane domain of E1 allows its retention in the Golgi. Since E1 is not detected at plasma membranes, it seems that its most likely mechanism of retention is physical restraint. Whether E1 interacts with proteins such as clathrin is unknown.

Recently, Aoki *et al.* (1992) have identified a specific Golgi retention sequence within the Golgi enzyme galactosyltransferase, which has a single transmembrane domain. The retention sequence was localized to the cytoplasmic side of its transmembrane domain. Within this region, Cys and His residues are necessary but insufficient for Golgi retention. For example, after site-directed mutagenesis of Cys-29→ Ser-29 or His-32→Arg-32, galactosyltransferase was not retained in the Golgi. Furthermore, introduction of the Cys-X-X-His sequence into the transferrin receptor did not cause Golgi retention. However, some Golgi retention could be observed when the cytoplasmic half of galactosyltransferase's transmembrane domain was inserted into the transferrin receptor. In the next few years Golgi retention sequences and their transmembrane links to clathrin or nonclathrin coats should be clarified.

A clear link between membrane skeletal proteins and endomembrane protein retention has been obtained for the Golgi. Payne and Schekman (1989) have studied a mutant yeast cell deficient in clathrin's heavy chain (*chc1Δ*). These cells secrete an immature form of α-factor, a pheromone. In wild-type cells the α-factor is proteolytically processed by the endoprotease Kex2p, a 120-kDa transmembrane enzyme found in the *trans* Golgi. However, in clathrin-deficient cells this enzyme is found in plasma membranes, thus accounting for the secretion of α-factor's precursor. Clathrin could mediate Kex2p retention by physically anchoring it to Golgi membranes or it could be rapidly retrieved from plasma membranes by clathrin-coated vesicles. To help distinguish between these possibilities, Seeger and Payne (1992) examined *chc1-ts* cells, which carry a temperature-sensitive allele of *CHC1*. When shifted to a nonpermissive temperature, the Golgi membrane proteins Kex2p and dipeptidylaminopeptidase A (DPAP A) rapidly appeared at the plasma membrane of *chc1-ts*, but not *wt*, cells. Two lines of evidence support the idea that the temperature-sensitive clathrin molecule releases the proteins from the Golgi. Although the trafficking of Golgi proteins is affected rapidly after a temperature shift, vacuolar protein trafficking—which would be expected to participate in the endocytotic recycling of plasma membrane proteins—was unaffected for 2 hr. Second, in cells expressing the double mutant *sec1chc1-ts*, which interfere simultaneously with both exocytotic secretion and clathrin function, Kex2p and DPAP A do not appear at the plasma membrane. Therefore, clathrin molecules hold these Golgi-specific proteins in the Golgi.

The role of clathrin and nonclathrin membrane-associated proteins in maintaining

Golgi identity and endomembrane trafficking in higher eukaryotes finds support in a second independent line of research using the drug brefeldin A (BFA), a fungal metabolite. Clathrin is linked to membranes of the *trans* Golgi network via HA-1 adaptors, which include the γ and β'-adaptins. The protein β-COP, which is homologous to β-adaptin (Duden *et al.*, 1991b), likely plays a similar role in linking nonclathrin coats to the *cis* medial, and *trans* regions of the Golgi. BFA interacts with an as yet undefined site thus blocking γ-adaptin and β-COP, but not α-adaptin of plasma membranes, from binding to their membrane sites (Wong and Brodsky, 1992). Within a few seconds: protein transport through the secretory apparatus stops, the Golgi disappears, and the Golgi's components redistribute and mix with those of the ER (Doms *et al.*, 1989). The redistribution process requires energy and microtubules, further supporting the idea of retrograde transport from the Golgi to the ER (Lippincott-Schwartz *et al.*, 1990). Coat proteins are therefore thought of as a framework which maintains the structure of the Golgi (Duden *et al.*, 1991a).

8.3.6. Genetic Dissection of the Secretory Pathway: ER-to-Golgi Transport

The molecular mechanisms responsible for secretory flow are largely unknown. To identify proteins participating in the secretory pathway, secretory mutants of yeast (*Saccharomyces cerevisiae*) were selected (e.g., Novick *et al.*, 1980). Eukaryotic yeast cells were chosen because they are easy to grow and can be subjected to genetic manipulation. They were mutagenized with nitrous acid or ethyl methane sulfonate (EMS). Temperature-sensitive secretory mutants were isolated by density gradient centrifugation. At the restrictive temperature (37°C) the mutants failed to secrete proteins such as invertase and acid phosphatase; these proteins accumulated in cytoplasmic vesicles and thereby increased a mutant cell's density. Genetic analyses of hundreds of separate mutant clones have shown that 27 or more different genes are involved in secretion. The gene products participating in the secretory pathway and their physiological roles are beginning to be described.

In Section 8.2.2 we described how proteins were delivered to the ER's lumen. After their biosynthesis and folding, secretory and membrane proteins must be shuttled from the ER to the Golgi. Genetic selection experiments have shown that about a dozen genes participate in ER-to-Golgi communication.

Some of the vesicle membrane proteins participating in ER-to-Golgi transfer have been identified. Genetic studies have identified two proteins, sec12p and sar1p, that play essential roles in secretory flow from the ER to the Golgi (Nakano *et al.*, 1988; Nakano and Muramatsu, 1989). At their nonpermissive temperatures, these mutant yeast cells accumulate material in cytoplasmic vesicles. These two membrane proteins directly interact to bring about secretory flow between the ER and Golgi.

sec12p* is an integral membrane protein of the ER and Golgi. Its deduced amino acid sequence indicates that it is a 70-kDa protein with a single transmembrane domain.

---

*We will follow the uniform conventions in this field when referring to mutations, genes, and gene products. For example, *sec12* is a mutation, *SEC12* is a gene, and sec12p is a gene product.

It has a large domain exposed at the ER membrane's cytosolic face. This domain could, for example, play a role in recognizing *cis* stacks of Golgi membranes. sec12p closely collaborates with at least one other protein, sar1p, to affect transport. The *SAR1* gene has also been cloned (Nakano and Muramatsu, 1989). It encodes an $M_r = 21,450$ protein with two putative transmembrane domains. Interestingly, sar1p is another G protein. This is consistent with the finding that nonhydrolyzable GTP analogues (e.g., GMPPCP, compound **5** of Chapter 7) block transport between the ER and Golgi. Although these experiments have been of a descriptive nature, it is tempting to make mechanistic inferences. The broad features of ER-to-Golgi transport may be analogous to the behavior of RGC receptors (Section 7.3). sec12p could carry out the recognition/binding functions while its G-protein partner performs a regulatory role. This speculation may or may not pan out on close scrutiny.

Another protein, sec18p, is required for ER-to-Golgi transport. Wilson *et al.* (1989) have identified this yeast protein as a homologue of the mammalian NEM-sensitive fusion (NSF) protein. NSF functions in several types of endomembrane fusion events. We will return to this protein in Section 8.3.8.

Our next stop along the secretory membrane pathway is the Golgi's *cis* face. Transitional vesicles containing membrane-bound sec12p/sar1p are charged with secretory materials and membrane components destined for other organelles. Specific recognition and triggering events are thought to bring out their fusion with the Golgi. One Golgi membrane component important in this process has been identified by the *sec23* mutant of yeast. Since transitional vesicles cannot interact with *sec23* mutants at their nonpermissive temperature, it is hypothesized that the *SEC23* gene product is important in this process. After dumping their secretory materials in the Golgi, sec12p and sar1p are thought to be recycled to the ER, although the recycling of ER membrane proteins, in contrast to ER luminal proteins, is not well understood.

8.3.7. Genetic Analysis of Golgi-to-Plasma Membrane Transport

As mentioned above (Section 8.2.3), the Golgi is actively engaged in the processing of secretory and membrane proteins. The Golgi is comprised of several compartments with distinct enzyme activities and functions. Golgi-associated vesicles are believed to mediate the molecular communication within a Golgi's stacks. Genetic studies have identified the *YPT1* gene as a key participant in vesicle transport through the Golgi (Segev *et al.*, 1988). At a nonpermissive temperature these yeast cells glycosylate and secrete invertase very slowly. Morphological studies have shown that this secretory inhibition is accompanied by the cytoplasmic accumulation of Golgi-derived vesicles. There is complete confidence that this phenotype is linked to *YPT1* since *YPT1*-containing plasmids restore the wild-type phenotype. Yeast's *YPT1* gene is located on chromosome VI between the β-tubulin and actin genes; it encodes a 23-kDa G protein. This is not surprising since nonhydrolyzable GTP analogues also block intra-Golgi transport. Two COOH-terminal Cys residues, an unusual feature of proteins, are found on ypt1p. The loss of these residues is a lethal mutation. The Cys residues may be linked to a membrane by fatty acylation (Section 2.3.1). In addition, anti-ypt1p antibodies have identified a similar protein in the Golgi of murine cells. ypt1p is apparently exposed on a Golgi vesicle's *cis* face where it participates in membrane trafficking.

The genetic analysis of secretory mutants has been extended to temperature-sensitive mutants of exocytosis. At a nonpermissive temperature, *sec4* mutants accumulate large numbers of secretory vesicles and are deficient in invertase secretion (Salminen and Novick, 1987). The *SEC4* gene has been unambiguously linked to these phenotypic properties. *SEC4* encodes a 23.5-kDa G protein, highly homologous to ypt1p. The protein sec4p is found at the *cis* face of secretory vesicles. Its deduced amino acid sequence revealed no apparent transmembrane domain. It shares double Cys residues at its COOH-terminus with ypt1p. Sec4p is likely associated with membranes via fatty acyl groups. Both ypt1p and sec4p must be covalently modified by the *BET2* gene product, a protein prenyltransferase, for functional activity (Rossi *et al.*, 1991).

Salminen and Novick (1989) have identified another structural gene, *SEC15*, whose product is a late participant in secretion. Furthermore, genetic studies indicate that *SEC4* duplication can compensate for mutant sec15p proteins with diminished activities. Based on this and other evidence, Salminen and Novick suggest that sec4p, a putative G protein, regulates the functional activity of sec15p. *SEC15* encodes a 100-kDa protein with no apparent transmembrane domain. However, sec15p is a peripheral protein that associates with membranes in a pH-dependent fashion. Although its putative regulatory partner sec4p cycles between secretory and plasma membranes, sec15p is not associated with plasma membranes. The biochemical role of sec15p is unknown.

Although genetic studies have identified many proteins participating in secretion, they have not yet passed the descriptive stage. Since other G proteins are regulatory in nature, putative biosynthetic G proteins (Table 8.1) are presumed to be regulatory elements. With the exceptions of sec18p and sec17p, the physiological roles of these proteins (e.g., sec12p and sec15p) are unknown. Are these proteins similar to ligands and receptors or perhaps homophilic receptors (like NCAM) for cell compartments? Alternatively, might these proteins function like viral fusion proteins (e.g., Section 8.1.2.3) or control phospholipases to potentiate fusion? The number of interesting new

Table 8.1.
G proteins in Biosynthesis and Secretion

Gene product(s)	Function(s)
Ribosome	Initiation
	Elongation
SRP, 54-kDa subunit	Signal sequence binding/delivery to ER
SRP receptor, α subunit	SRP binding
sar1p	ER-to-Golgi flow
ypt1p	Intra-Golgi flow
sec4p	Constitutive secretion
G_p	Regulated secretion
G_e	Regulated secretion

mutants is beginning to wane; attention should now be focused on what these proteins do. Additional studies combining information from these genetic approaches with the cellular reconstitutions discussed in the following section will help identify the functional roles of these gene products.

8.3.8. Subcellular and Cellular Reconstitutions of Secretory Flow

Although a complete *in vitro* reconstitution of a single fusion step in secretory transport is a few years away, subcellular and cellular reconstitutions of individual steps have been successful. The objective of these reconstitution experiments is to identify the functional components of the secretory pathway. During subcellular reconstitutions, individual organelle fractions or their subfractions are tested for their ability to support protein maturation *in vitro*. The aims of cellular reconstitution are the same; however, these experiments use permeabilized or semi-intact cells to study the requirements of secretory transport.

Semi-intact cells were prepared by osmotic shock or mechanical disruption (Beckers *et al.*, 1987). Large holes were torn in the cell's plasma membranes. CHO cells with temperature-sensitive ER export were infected by VSV. At a nonpermissive temperature these cells incorporated radioactive amino acids into VSV G protein but did not transport them to the Golgi. Semi-intact cells were then prepared to test the conditions required for ER-to-Golgi transport. VSV G protein arrival at the *cis* Golgi was detected by the formation of $Man_5GlcNAc_2$ from its high-mannose precursor on the VSV G protein. At a nonpermissive temperature no transport in these semi-intact cells was detected, in agreement with the cells' phenotype. At a permissive temperature, transport required ATP and cytosolic factors, primarily proteins. Therefore, ER-to-Golgi transport is an energy-dependent process that requires cytosolic factors.

Rothman and colleagues have developed a subcellular *in vitro* system to study the intercisternal transport of proteins within the Golgi (e.g., Wattenberg *et al.*, 1986). Golgi were isolated from mutant and wild-type CHO cells. The donor Golgi fraction was isolated from VSV-infected CHO cells deficient in GlcNAc transferase activity. Since these cells were unable to incorporate GlcNAc into oligosaccharides, glycoproteins were unable to mature. Acceptor Golgi cisternae were isolated from uninfected wild-type cells. When incubated with [³H]-GlcNAc, the transfer of VSV G protein from donor to acceptor could be measured by its uptake of radioactivity. During the transport process vesicles coated with an unidentified protein other than clathrin bud from the donor Golgi membrane. The nonclathrin coat disassembles followed by smooth vesicle binding and fusion with the acceptor Golgi membrane. Early experiments with this assay showed that ATP and cytosol were required for intercisternal transport. The cytosol was found to contain several activities that contribute to transport.

One protein participating in intercisternal transport was identified by exploiting its sensitivity to *N*-ethylmaleimide (NEM). NEM did not block the budding of vesicles from donor Golgi. In the presence of NEM, uncoated vesicles accumulated about the periphery of acceptor Golgi cisternae. To identify the NEM-sensitive factor (NSF), fractions of unmodified cytosolic extracts were added to NEM-inactivated Golgi membranes. In this way a 76-kDa NSF required for vesicle-to-Golgi membrane fusion was found. This protein exists as a tetrameric complex in cells.

A cDNA containing the NSF gene has been isolated (Wilson *et al.*, 1989). The cDNA sequence revealed a protein of 83 kDa with no apparent transmembrane domains. Its lack of transmembrane domains is consistent with the relative ease with which it can be dislodged from membranes. NSF contains a consensus ATP-binding sequence. Several structural and functional criteria have established that NSF of mammals is sec18p of yeast (Wilson *et al.*, 1989). In addition to its role in intercisternal transport in the Golgi, NSF is also required for ER-to-Golgi transport in semi-intact cells (Beckers *et al.*, 1989). Therefore, NSF appears to be of broad importance in promoting fusion events between endomembranes.

In addition to NSF, several additional participants in endomembrane fusion have been identified. NSF binds to Golgi membranes in a specific and saturable manner. Weidman *et al.* (1989) have defined some of the requirements for the specific reassociation of purified NSF to Golgi membranes. At least three additional proteins participate in the NSF–membrane binding interaction. These are: (1) the NSF receptor, (2) SNAP, a series of three soluble NSF attachment proteins, and (3) factor B, another protein promoting NSF–membrane binding (Weidman *et al.*, 1989). SNAP proteins, which have been linked with sec17p in genetic studies, are peripheral membrane proteins that link NSF to membranes (Clary *et al.*, 1990). NSF and SNAP are present at about 10^5 copies/cell. About 200,000 NSF receptors are present in each cell. These three components form a high-affinity ternary prefusion complex on Golgi membranes. Under these conditions the affinity of NSF for Golgi membranes is very high ($K_d \simeq 6 \times 10^{-11}$ or lower). This high-affinity complex triggers endomembrane fusion, although the mechanism of fusion remains unknown.

Recently, vesicle release from the *trans* Golgi network has been reconstituted in an *in vitro* system (de Curtis and Simons, 1989). Cells were virally infected for $3\frac{1}{2}$ hr, pulse-labeled for 5 min with [^{35}S]-Met, and then incubated at 19°C for $1\frac{1}{2}$ hr. Incubations at 19°C dramatically decrease the rate of protein export from the *trans* Golgi network, thus inhibiting secretory transport at this point. Since infection diverts protein synthesis to the viral proteins, the incubations tend to clear nonviral proteins from the secretory pathway. The cells were then fractionated by ultracentrifugation. In the presence of ATP and cytosol at 37°C, vesicles bud from the Golgi membranes. These vesicles contain a post-Golgi (mature or processed) form of the viral glycoproteins. These vesicles can be further purified by density gradient centrifugation and immunoadsorption on substrates containing antiviral antibodies. These vesicles do not contain the *trans* Golgi markers galactosyltransferase and sialyltransferase. In addition to viral proteins, eight prominent proteins were observed. This vesicle protein pattern is much simpler than that of other endomembranes. The Golgi proteins enriched in the vesicles may participate in translocation. These experiments have begun to define the conditions and participants in constitutive secretion. Further studies should define the function of vesicle components and reconstitute membrane fusion between vesicles and plasma membranes.

8.3.9. Sorting Things Out at the Golgi's *Trans* Face

The heterogeneous collection of materials reaching the Golgi's *trans* face must be sorted out to ensure that they arrive at their correct destinations. In a typical cell, materials are sorted into at least three types of packages. This sorting function is performed by Golgi membrane-associated transport receptors. The three principal types of export packages are: (1) materials such as hormones destined for regulated secretion, (2) enzymes targeted to the lysosomes, and (3) constitutively secreted materials.

Chung *et al.* (1989) have identified some of the components that sequester proteins participating in regulated secretion. These Golgi membrane transport receptors specifically recognize proteins of the regulated secretory pathway. Polypeptide hormones such as prolactin were covalently coupled to Sepharose beads. Instead of isolating plasma membrane-bound receptors, as we described in the preceding chapter, these investigators fractionated cells and then discarded the plasma membrane fraction. The purified Golgi membranes were solubilized in nonionic detergent and then passed over a prolactin-conjugated affinity chromatography column. When a loaded column was eluted with an acidic buffer, several proteins of about 25 kDa were obtained. These proteins are called HBP25s (hormone binding proteins). HBP25s do not bind to affinity columns prepared from constitutively secreted proteins (albumin and immunoglobulin) or nonsecreted proteins (hemoglobin and myoglobin). However, they do bind to insulin–Sepharose affinity columns. This result was expected since insulin, like prolactin, is secreted via the regulated pathway. These results suggest that the HBPs of Golgi membranes recognize various types of regulated secretory proteins.

During sorting in the *trans* Golgi, regulated secretory proteins bind to the membrane's luminal face and then aggregate into clusters. Since clustering is a way to sort out molecules, it is important to understand how it happens. We have previously seen that receptor cross-linking induces patching of ligands and receptors (see Con A, Section 3.1). To test the role of membrane cross-linking in Golgi sorting, Chung *et al.* (1989) bound HBP25 to insulin–Sepharose affinity columns. When prolactin was then added to the column, it became specifically bound to the HBP25. Since the HBP25s were simultaneously bound to the insulin-containing matrix *and* prolactin, they must be multivalent. Therefore, the segregation of regulated secretory proteins in the Golgi's *trans* cisternae may be due to the cross-linking of multivalent receptors. In many cases the lower pH of the *trans* compartment and/or the high local concentrations of secretory proteins such as insulin will cause their spontaneous oligomerization. Therefore, protein sorting, a complex physiological process, can be understood by the biochemical and biophysical principles we have already encountered.

HBP25s apparently recognize a particular region or sequence of regulated secretory proteins. Moore and Kelly (1986) have used gene fusion experiments to show that a secretory protein's destination can be altered by changing its primary structure. A fusion protein was constructed from genes encoding the human growth hormone (hGH) and VSV G protein. When a truncated VSV G protein is expressed in transfected cells, it is secreted in a constitutive fashion. When separately transfected into cells, hGH collects in regulated secretory granules. The fusion protein was composed of the truncated VSV G protein and hGH's COOH-terminal domain. When the fusion protein is expressed in cells, it accumulates in regulated secretory granules. After the addition of cAMP analogues the fusion protein is exocytosed. This shows that the COOH-terminal domain of a regulated secretory protein can reroute a constitutively secreted protein to the regulated pathway. The molecular interactions between this domain and its putative binding protein are unknown.

As briefly mentioned above, proteins destined for the lysosomes are covalently modified in the *cis* Golgi by the addition of mannose-6-phosphate (M-6-P). Lysosomal proteins possess a tertiary structural feature or signal patch that triggers the addition of M-6-P. This has been suggested by experiments showing that M-6-P addition is highly sensitive to a target protein's conformation, although the precise nature of this signal patch is unknown. Lysosomal enzymes enter the *cis* Golgi with attached high-mannose residues. The enzyme *N*-acetylglucosamine-1-phosphotransferase transfers *N*-acetyl-glucosamine-1-phosphate from UDP-GlcNAc to certain mannose residues of lysosomal enzymes. The enzyme *N*-acetylglucosamine phosphodiesterase then cleaves the *N*-acetylglucosamine away, leaving a M-6-P moiety on the lysosomal enzyme. The M-6-P-tagged lysosomal enzyme is then sequentially transferred among the Golgi's compartments until it reaches the *trans* Golgi network.

The *trans* Golgi network contains two types of M-6-P receptors. Deduced amino acid sequences have been obtained for both receptor types (Dahms *et al.*, 1987; Lobel *et al.*, 1988). These receptors can be distinguished by their divalent cation requirements. The cation-independent (CI) M-6-P receptor has a mass of 270 kDa whereas the cation-dependent (CD) receptor is a 30-kDa protein. Both receptors have a single trans-membrane domain and homologous luminal-facing ligand binding domains. These receptors are cycled through the *trans* Golgi network, coated vesicles, endosomes, and plasma membranes. When M-6-P-tagged enzymes reach the *trans* Golgi network, they bind to the M-6-P receptors.

Several factors may contribute to the sorting of lysosomal enzymes within the *trans* Golgi network. Receptor cross-linking is likely one important contributor to sorting. Receptor cross-linking and aggregation are thought to occur because: (1) the CD M-6-P receptor spontaneously dimerizes or oligomerizes, (2) each lysosomal enzyme generally contains multiple M-6-P moieties, and (3) the slightly acidic environment of the *trans* Golgi network stimulates enzyme aggregation. Golgi membrane-associated clathrin also contributes to lysosomal enzyme sorting. The cytosolic tail of the CI M-6-P receptor specifically binds HA-1 adaptors, which are in turn linked to clathrin (see Section 6.2.1.3 for a discussion of HA-2 adaptor linkages to plasma membranes). HA-1 adaptors are only found on Golgi membranes (Glickman *et al.*, 1989); they promote the delivery of certain membrane proteins into their domain within Golgi membranes. M-6-P receptors can be clustered at several cellular membrane sites by either HA-1 or HA-2 adaptors. Therefore, at least four different factors contribute to the sorting of lysosomal enzymes.

Clathrin-coated vesicles loaded with lysosomal enzymes bud from the Golgi's membrane. These vesicles discharge their contents into an acidic prelysosomal compartment. M-6-P receptors are not found in mature lysosomes. When M-6-P receptors originating from the Golgi or recycled from the plasma membrane encounter the acidic environment of a prelysosomal or endosomal compartment, their conformation changes and thereby releases the lysosomal enzymes. The enzymes' M-6-P residues are then removed within lysosomes.

The sorting of lysosomal enzymes and regulated secretory proteins is affected by drugs and in certain genetic diseases. The antimalarial drug chloroquine causes both lysosomal enzymes and regulated secretory proteins to be released in a constitutive fashion (e.g., Moore *et al.*, 1983). This may be due to chloroquine's ability to neutralize acidic cell compartments. Lysosomal enzymes are also constitutively secreted in a genetic disorder called I cell disease. These patients are deficient in *N*-acetylglucosamine phosphotransferase. Consequently, lysosomal enzymes are neither tagged by M-6-P nor sorted into their appropriate compartment.

Under certain conditions lysosomal enzymes behave as regulated secretory proteins. In some diseases such as arthritis, vasculitis, and nephritis, antibody molecules are deposited at local tissue sites or surfaces. When immune cells such as neutrophils recognize these bound antibodies, they attempt to phagocytose the tissue (see Section 6.2.1). Unable to internalize the surface, neutrophils discharge their hydrolytic enzymes in an attempt to kill the false target. This response causes the tissue damage and inflammation seen in these diseases.

Intracellular sorting and delivery are affected by hormone binding. Glucocorticoids, a family of steroid hormones, regulate the ability of murine mammary tumor virus (MMTV) membrane protein to leave the *trans* Golgi (Haffar *et al.*, 1988). In a normal state, rat cells infected with MMTV do not allow the transport of MMTV's gp70 membrane protein to the plasma membrane. The presence of glucocorticoids stimulates the release of gp70 to cell surfaces. The molecular explanation of this regulatory process is not certain. However, it seems likely that other signal/regulation pathways will be uncovered.

When secretory proteins are not sorted out, they flow to the cell surface by the constitutive pathway. Constitutive secretion is therefore considered as a "default" pathway. If a luminal protein does not possess a retention or sorting signal, it must be a constitutively secreted molecule. We have encountered this rule in many different contexts above. For example, experiments concerning glycotripeptide flow, KDEL sequence deletion, E1 transmembrane domains, E3/19 sequences, HBP25s, and M-6-P receptors all support a default constitutive flow pathway for luminal and membrane proteins (Sections 8.3.3, 8.3.4, 8.3.5, and this section).

REFERENCES AND FURTHER READING

Fusion

Conner, J., and Huang, L. 1985. Efficient delivery of a fluorescent dye by pH-sensitive immunoliposomes. *J. Cell Biol.* **101**:582–589.

Conner, J., *et al.* 1984. pH-sensitive liposomes: Acid-induced liposome fusion. *Proc. Natl. Acad. Sci. USA* **81**:1715–1718.

Correa-Freire, M. C., *et al*. 1984. Introduction of HLA-A/B antigens into lymphoid cell membranes by cell liposome fusion. *J. Immunol*. **132:**69–75.

Duzgunes, N., *et al*. 1981. Studies on the mechanism of membrane fusion: Role of headgroup composition in calcium- and magnesium-induced fusion of mixed phospholipid vesicles. *Biochim. Biophys. Acta* **642:**182–195.

Gething, M. J., *et al*. 1986. Studies on the mechanism of membrane fusion: Site-specific mutagenesis of the hemagglutinin of influenza virus. *J. Cell Biol*. **102:**11–23.

Kao, K. N., and Michayluk, M. R. 1974. A method for high-frequency intergeneric fusion of plant protoplasts. *Planta* **115:**355–367.

Kirk, G. L., *et al*. 1984. A transmembrane model of the lamellar to inverse hexagonal phase transition of lipid membrane–water systems. *Biochemistry* **23:**1093–1102.

Maggio, B., *et al*. 1976. Poly(ethylene glycol), surface potential and cell fusion. *Biochem. J*. **158:** 647–650.

Meer, G. van, and Simons, K. 1983. An efficient method for introducing defined lipids into the plasma membrane of mammalian cells. *J. Cell Biol*. **97:**1365–1374.

Meer, G. van, *et al*. 1985. Parameters affecting low-pH-mediated fusion of liposomes with the plasma membrane of cells infected with influenza virus. *Biochemistry* **24:**3593–3602.

Pagano, R. E., and Weinstein, J. N. 1978. Interactions of liposomes with mammalian cells. *Annu. Rev. Biophys. Bioeng*. **7:**435–468.

Papahadjopoulos, D., *et al*. 1974. Membrane fusion and molecular segregation in phospholipid vesicles. *Biochim. Biophys. Acta* **352:**10–28.

Prujansky-Jakobovits, A., *et al*. 1980. Alteration of lymphocyte surface properties by insertion of foreign functional components of plasma membrane. *Proc. Natl. Acad. Sci. USA* **77:**7247–7251.

Rand, R. P., and Parsegian, V. A. 1986. Mimicry and mechanism in phospholipid models of membrane fusion. *Annu. Rev. Physiol*. **48:**201–212.

Roos, D. S., and Choppin, P. W. 1985. Biochemical studies on cell fusion II. Control of fusion response by lipid alteration. *J. Cell Biol*. **101:**1591–1598.

Scheid, A., and Choppin, P. W. 1974. Identification of biological activities of paramyxovirus glycoproteins. Activation of cell fusion, hemolysis, and infectivity by proteolytic cleavage of an inactive precursor protein of Sendai virus. *Virology* **57:**475–490.

Schramm, M. 1979. Transfer of glucagon receptor from liver membranes to a foreign adenyl-cyclase by membrane fusion procedure. *Proc. Natl. Acad. Sci. USA* **76:**1174–1178.

Shimizu, K., and Ishida, N. 1975. The smallest protein of Sendai virus: Its candidate function of binding nucleocapsid to envelope. *Virology* **67:**427–437.

Siegel, D. P. 1984. Inverted micellar structures in bilayer membranes: Formation rates and half-lives. *Biophys. J*. **45:**399–420.

Siegel, D. P., *et al*. 1989. Intermediates in membrane fusion and bilayer/nonbilayer phase transitions imaged by time-resolved cryo-transmission electron microscopy. *Biophys. J*. **56:** 161–169.

Stegmann, T., *et al*. 1990. Intermediates in influenza induced membrane fusion. *EMBO J*. **9:**4231–4241.

Szoka, F., *et al*. 1980. Fluorescence studies on the mechanism of liposome–cell interactions in vitro. *Biochim. Biophys. Acta* **600:**1–8.

Szoka, F., *et al*. 1981. Use of lectins and polyethylene glycol for fusion of glycolipid-containing liposomes with eukaryotic cells. *Proc. Natl. Acad. Sci. USA* **78:**1685–1689.

Uchida, T., *et al*. 1979. Reconstitution of lipid vesicles associated with HVJ (Sendai virus) spikes. *J. Cell Biol*. **80:**10–20.

Volsky, D. J., and Loyter, A. 1978. An efficient method for reassembly of fusogenic Sendai virus envelopes after solubilization of intact virions with Triton X-100. *FEBS Lett*. **92:**190–194.

Volsky, D. J., *et al.* 1979. Implantation of the isolated human erythrocyte anion channel into plasma membranes of Friend erythroleukemic cells by use of Sendai virus envelopes. *Proc. Natl. Acad. Sci. USA* **76:**5440–5444.

Wang, C.-Y., and Huang, L. 1984. Polyhistidine mediates an acid-dependent fusion of negatively charged liposomes. *Biochemistry* **23:**4409–4416.

Biosynthesis and Assembly

Backer, J. M., and Dawidowicz, E. A. 1987. Reconstitution of a phospholipid flippase from rat liver microsomes. *Nature* **327:**341–343.

Baker, K. P., and Schatz, G. 1991. Mitochondrial proteins essential for viability mediate protein import into yeast mitochondria. *Nature* **349:**205–208.

Banerjee, D. K., *et al.* 1987. cAMP-mediated protein phosphorylation of microsomal membranes increases mannoylphosphodolichol synthase activity. *Proc. Natl. Acad. Sci. USA* **84:**6389–6393.

Berstein, H. D., *et al.* 1989. Model for signal sequence recognition from amino-acid sequence of 54K subunit of signal recognition particle. *Nature* **340:**482–486.

Bishop, W. R., and Bell, R. M. 1985. Assembly of the endoplasmic reticulum phospholipid bilayer: The phosphatidylcholine transporter. *Cell* **42:**51–60.

Bishop, W. R., and Bell, R. M. 1988. Assembly of phospholipids into cellular membranes: Biosynthesis, transmembrane movement and intracellular translocation. *Annu. Rev. Cell Biol.* **4:**579–610.

Cabelli, R. J., *et al.* 1988. SecA protein is required for secretory protein translocation into *E. coli* membrane vesicles. *Cell* **55:**683–692.

Carman, G. M., and Henry, S. A. 1989. Phospholipid biosynthesis in yeast. *Annu. Rev. Biochem.* **58:**635–669.

Cerretti, D. P., *et al.* 1983. The *spc* ribosomal protein operon of *Escherichia coli*: Sequence and cotranscription of the ribosomal protein genes and a protein export gene. *Nucleic Acids Res.* **11:**2599–2616.

Chirico, W. J., *et al.* 1988. 70K heat shock related proteins stimulate protein translocation into microsomes. *Nature* **332:**805–810.

Chojnacki, T. Z., and Dallner, G. 1988. The biological role of dolichol. *Biochem. J.* **251:**1–9.

Chung, K.-N., *et al.* 1989. Molecular sorting in the secretory pathway. *Science* **243:**192–197.

Collier, D. N., *et al.* 1988. The antifolding activity of SecB promotes the export of the *E. coli* maltose-binding protein. *Cell* **53:**273–283.

Colman, A., and Robinson, C. 1986. Protein import into organelles: Hierarchical targeting signals. *Cell* **46:**321–322.

Connolly, T., and Gilmore, R. 1989. The signal recognition particle receptor mediates the GTP-dependent displacement of SRP from the signal sequence of the nascent polypeptide. *Cell* **57:**599–610.

Connolly, T. *et al.* 1991. Requirement of GTP hydrolysis for dissociation of the signal recognition particles from its receptor. *Science* **252:**1171–1173.

Crimaudo, C., *et al.* 1987. Human ribophorins I and II: The primary structure and membrane topology of two highly conserved rough endoplasmic reticulum-specific glycoproteins. *EMBO J.* **6:**75–82.

Dalbey, R. E., and Wickner, W. 1987. Leader peptidase of *Escherichia coli*: Critical role of a small domain in membrane assembly. *Science* **235:**783–787.

Dawidowicz, E. A. 1987. Dynamics of membrane lipid metabolism and turnover. *Annu. Rev. Biochem.* **56**:43–61.

Deshaies, R. J., and Schekman, R. 1989. SEC62 encodes a putative membrane protein required for protein translocation into yeast endoplasmic reticulum. *J. Cell Biol.* **109**:2653–2664.

Deshaies, R. J., *et al.* 1988. A subfamily of stress proteins facilitates translocation of secretory and mitochondrial precursor polypeptides. *Nature* **332**:800–804.

Devaux, P. F. 1988. Phospholipid flippases. *FEBS Lett.* **234**:8–12.

Di Rienzo, J. M., *et al.* 1978. The outer membrane proteins of gram negative bacteria: Biosynthesis, assembly, and functions. *Annu. Rev. Biochem.* **47**:481–532.

Eilers, M., and Schatz, G. 1986. Binding of a specific ligand inhibits import of a purified precursor protein into mitochondria. *Nature* **322**:228–232.

Eilers, M., and Schatz, G. 1988. Protein unfolding and the energetics of protein translocation across biological membranes. *Cell* **52**:481–483.

Flugge, U.-I., *et al.* 1991. The major chloroplast envelope polypeptide is the phosphate translocator and not the protein import receptor. *Nature* **353**:364–367.

Flynn, G. C., *et al.* 1991. Peptide-binding specificity of the molecular chaperone BiP. *Nature* **353**:726–730.

Gillespie, L. L. 1987. Identification of an outer mitochondrial membrane protein that interacts with a synthetic signal peptide. *J. Biol. Chem.* **262**:7939–7942.

Griffiths, G., and Simons, K. 1986. The trans Golgi network: Sorting at the exit site of the Golgi complex. *Science* **234**:438–443.

Guan, J.-L., *et al.* 1985. Glycosylation allows cell-surface transport of an anchored secretory protein. *Cell* **42**:489–496.

Harnik-Ort, V., *et al.* 1987. Isolation and characterization of cDNA clones for rat ribophorin I: Complete coding sequence and in vitro synthesis and insertion of the encoded product into endoplasmic reticulum membranes. *J. Cell Biol.* **104**:855–863.

Hartmann, E., *et al.* 1989. Predicting the orientation of eukaryotic membrane-spanning proteins. *Proc. Natl. Acad. Sci. USA* **86**:5786–5790.

Haselbeck, A., and Tanner, W. 1982. Dolichyl phosphate-mediated mannosyl transfer through liposomal membranes. *Proc. Natl. Acad. Sci. USA* **79**:1520–1524.

Hirata, F., and Axelrod, J. 1978. Enzymatic synthesis and rapid translocation of phosphatidylcholine by two methyltransferases in erythrocyte membranes. *Proc. Natl. Acad. Sci. USA* **75**:2348–2352.

Hirschberg, C. B., and Snider, M. D. 1987. Topography of glycosylation in the rough endoplasmic reticulum and Golgi apparatus. *Annu. Rev. Biochem.* **56**:63–87.

Hirst, T. R., and Welch, R. A. 1988. Mechanisms for secretion of extracellular proteins by gram-negative bacteria. *Trends Biochem Sci.* **13**:265–269.

Honig, B. H., *et al.* 1986. Electrostatic interactions in membranes and proteins. *Annu. Rev. Biophys. Biophys. Chem.* **15**:163–193.

Hubbard, S. C., and Ivatt, R. J. 1981. Synthesis and processing of asparagine-linked oligosaccharides. *Annu. Rev. Biochem.* **50**:555–583.

Hughes, R. C. 1983. *Glycoproteins.* Chapman & Hall, London.

Hurt, E. C., and Schatz, G. 1987. A cytosolic protein contains a cryptic mitochondrial targeting signal. *Nature* **325**:499–503.

Kabat, E. V. 1976. *Structural Concepts in Immunology and Immunochemistry.* Holt, Rinehart & Winston, New York.

Kang, P.-J., *et al.* 1990. Requirement for hsp70 in the mitochondrial matrix for translocation and folding of precursor proteins. *Nature* **248**:137–143.

Kaplan, M. R., and Simoni, R. D. 1985. Intracellular transport of phosphatidylcholine to the plasma membrane. *J. Cell Biol.* **101:**441–445.

Kiebler, M., *et al.* 1990. Identification of a mitochondrial receptor complex required for recognition and membrane insertion of precursor proteins. *Nature* **348:**610–616.

Kim, P. S., *et al.* 1992. Transient aggregation of nascent thyroglobulin in the endoplasmic reticulum: Relationship to the molecular chaperone, BiP. *J. Cell Biol.* **118:**541–549.

Kodaki, T., and Yamashita, S. 1987. Yeast phosphatidylethanolamine methylation pathway. Cloning and characterization of two distinct methyltransferase genes. *J. Biol. Chem.* **262:** 15428–15435.

Kok, J. W., *et al.* 1989. Salvage of glycosylceramide by recycling after internalization along the pathway of receptor-mediated endocytosis. *Proc. Natl. Acad. Sci. USA* **86:**9896–9900.

Koval, M., and Pagano, R. E. 1989. Lipid recycling between the plasma membrane and intracellular compartments: Transport and metabolism of fluorescent sphingomyelin analogues in cultured fibroblasts. *J. Cell Biol.* **108:**2169–2181.

Krieg, U. C., *et al.* 1989. Protein translocation across the endoplasmic reticulum membrane: Identification by photocrosslinking of a 39-kD integral membrane glycoprotein as part of a putative translocation tunnel. *J. Cell Biol.* **109:**2033–2043.

Kuhn, A., *et al.* 1986. The cytoplasmic carboxy terminus of M13 procoat is required for the membrane insertion of its central domain. *Nature* **322:**335–339.

Kukuruzinska, M. A., *et al.* 1987. Protein glycosylation in yeast. *Annu. Rev. Biochem.* **56:** 915–944.

Lauffer, L., *et al.* 1985. Topology of signal recognition particle receptor in endoplasmic reticulum membrane. *Nature* **318:**334–338.

Lehnhardt, S., *et al.* 1987. The differential effect on two hybrid proteins of deletion mutations within the hydrophobic region of the *Escherichia coli* OmpA signal peptide. *J. Biol. Chem.* **262:**1716–1719.

Lennarz, W. J. 1987. Protein glycosylation in the endoplasmic reticulum: Current topological issues. *Biochemistry* **26:**7205–7210.

Li, P., *et al.* 1988. Alteration of the amino terminus of the mature sequence of a periplasmic protein can severely affect export in *Escherichia coli*. *Proc. Natl. Acad. Sci. USA* **85:**7685–7689.

Lingappa, V. R. 1991. More than just a channel: Provocative new features of protein traffic across the ER membrane. *Cell* **65:**527–530.

Martin, J., *et al.* 1991. Chaperonin-mediated protein folding at the surface of groEL through a "molten globule"-like intermediate. *Nature* **352:**36–42.

Myers, M., *et al.* 1987. Thermodynamic characterization of interactions between ornithine transcarbamylase leader peptide and phospholipid bilayer membranes. *Biochemistry* **26:** 4309–4315.

Nguyen, T. H., *et al.* 1991. Binding protein BiP is required for translocation of secretory proteins into the endoplasmic reticulum in *Saccharomyces cerevisiae*. *Proc. Natl. Acad. Sci. USA* **88:** 1565–1569.

Nicchitta, C. V., and Blobel, G. 1989. Nascent secretory chain binding and translocation are distinct processes: Differentiation by chemical alkylation. *J. Cell Biol.* **108:**789–795.

Nikawa, J., *et al.* 1987a. Nucleotide sequence and characterization of the yeast PSS gene encoding phosphatidylserine synthase. *Eur. J. Biochem.* **167:**7–12.

Nikawa, J., *et al.* 1987b. Primary structure and disruption of the phosphatidylinositol synthase gene of *Saccharomyces cerevisiae*. *J. Biol. Chem.* **262:**4876–4881.

Nunnari, J. M., *et al.* 1991. Characterization of the rough endoplasmic reticulum ribosome-binding activity. *Nature* **352:**638–640.

Orlean, P., *et al.* 1988. Cloning and sequencing of the yeast gene for dolichol phosphate mannose synthase, an essential protein. *J. Biol. Chem.* **263:**17499–17507.

Pagano, R. E. 1988. What is the fate of diacylglycerol produced at the Golgi apparatus? *Trends Biochem. Sci.* **13:**202–205.

Pagano, R. E., and Sleight, R. G. 1985. Defining lipid transport pathways in animal cells. *Science* **229:**1051–1057.

Pain, D., *et al.* 1988. Identification of a receptor for protein import into chloroplasts and its localization to envelope contact zones. *Nature* **331:**232–237.

Pfeffer, S. R., and Rothman, J. E. 1987. Biosynthetic protein transport and sorting by the endoplasmic reticulum and Golgi. *Annu. Rev. Biochem.* **56:**829–852.

Pluckthun, A., and Knowles, J. R. 1987. The consequences of stepwise deletion from the signal processing site of β-lactamase. *J. Biol. Chem.* **262:**3951–3957.

Randall, L. L. 1992. Peptide binding by chaperone SecB: Implications for recognition of nonnative structure. *Science* **257:**241–245.

Reithmeier, R. A. F. 1985. Assembly of proteins into membranes. In: *Biochemistry of Lipids and Membranes* (D. E. Vance and J. E. Vance, eds.), Benjamin–Cummings, Menlo Park, Calif, pp. 503–558.

Romisch, K., *et al.* 1989. Homology of 54K protein of signal-recognition particle, docking protein and two *E. coli* proteins with putative GTP-binding domains. *Nature* **340:**478–482.

Rothman, J. E., and Lodish, H. F. 1977. Synchronised transmembrane insertion and glycosylation of a nascent membrane protein. *Nature* **269:**775–780.

Rothman, R. E., *et al.* 1988. Construction of defined polytopic integral transmembrane proteins. *J. Biol. Chem.* **263:**10470–10480.

Ruohola, H., *et al.* 1988. Reconstitution of protein transport from the endoplasmic reticulum to the Golgi complex in yeast: The acceptor Golgi compartment is defective in the *sec23* mutant. *J. Cell Biol.* **107:**1465–1476.

Saier, M. H., *et al.* 1989. Insertion of proteins into bacterial membranes: Mechanism, characteristics, and comparisons with the eucaryotic process. *Microbiol. Rev.* **53:**333–366.

Schleyer, M., and Neupert, W. 1985. Transport of proteins into mitochondria: Translocation intermediates spanning contact sites between outer and inner membranes. *Cell* **43:**339–350.

Schmidt, M. G., *et al.* 1988. Nucleotide sequence of the *sec A* gene and *secA(TS)* mutations preventing protein export in *Escherichia coli. J. Bacteriol.* **170:**3404–3414.

Schneider, H., *et al.* 1991. Targeting of the master receptor MOM19 to mitochondria. *Science* **254:**1659–1662.

Silhavy, T. J., *et al.* 1983. Mechanisms of protein localization. *Microbiol. Rev.* **47:**313–344.

Simon, S. M., and Blobel, G. 1991. A protein-conducting channel in the endoplasmic reticulum. *Cell* **65:**371–380.

Sjostrom, M., *et al.* 1987. Signal peptide amino acid sequences in *Escherichia coli* contain information related to final localization. A multivariate data analysis. *EMBO J.* **6:**823–831.

Skerjanc, I. S., *et al.* 1988. Identification of hydrophobic residues in the signal sequence of mitochondrial preornithine carbamyltransferase that enhance the rate of precursor import. *J. Biol. Chem.* **263:**17233–17236.

Snider, M. D., and Rogers, O. C. 1984. Transmembrane movement of oligosaccharide-lipids during glycoprotein synthesis. *Cell* **36:**753–761.

Snider, M. D., and Rogers, O. C. 1987. Membrane traffic in animal cells: Cellular glycoproteins return to the site of Golgi mannosidase. *J. Cell Biol.* **103:**265–276.

Sollner, T., *et al.* 1989. MOM19, an import receptor for mitochondrial precursor proteins. *Cell* **59:**1061–1070.

Sollner, T., *et al.* 1990. A mitochondrial import receptor for the ADP/ATP carrier. *Cell* **62:**107–115.

Sweeley, C. C. 1986. Sphingolipids. In: *Biochemistry of Lipids and Membranes* (D. E. Vance and J. E. Vance, eds.), Benjamin–Cummings, Menlo Park, Calif., pp. 361–403.

Tamm, L. K. 1986. Incorporation of a synthetic mitochondrial signal peptide into charged and uncharged phospholipid monolayers. *Biochemistry* **25**:7470–7476.

Unwin, P. T. N. 1977. Three-dimensional model of membrane-bound ribosomes obtained by electron microscopy. *Nature* **269**:118–122.

Vance, D. E. 1986. Phospholipid metabolism in eucaryotes. In: *Biochemistry of Lipids and Membranes* (D. E. Vance and J. E. Vance, eds.), Benjamin–Cummings, Menlo Park, Calif., pp. 242–270.

van Meer, G. 1989. Lipid traffic in animal cells. *Annu. Rev. Cell Biol.* **5**:247–275.

Verner, K., and Schatz, G. 1988. Protein translocation across membranes. *Science* **241**:1307–1313.

Vestweber, D., and Schatz, G. 1988. Point mutations destabilizing a precursor protein enhance its post-translational import into mitochondria. *EMBO J.* **7**:1147–1151.

von Heijne, G. 1986. The distribution of positively charged residues in bacterial inner membrane proteins correlates with the transmembrane topology. *EMBO J.* **5**:3021–3027.

von Heijne, G. 1988. Transcending the impenetrable: How proteins come to terms with membranes. *Biochim. Biophys. Acta* **947**:307–333.

von Heijne, G. 1989. Control of topology and mode of assembly of a polytopic membrane protein by positively charged residues. *Nature* **341**:456–458.

Vrije, T. de, *et al.* 1988. Phosphatidylglycerol is involved in protein translocation across *Escherichia coli* inner membranes. *Nature* **334**:173–175.

Walter, P., and Lingappa, V. 1986. Mechanism of protein translocation across the endoplasmic reticulum membrane. *Annu. Rev. Cell Biol.* **2**:499–516.

Watanabe, M., *et al.* 1986. *In vitro* synthesized bacterial outer membrane protein is integrated into bacterial inner membranes but translocated across microsomal membranes. *Nature* **323**:71–73.

Welply, J. K., *et al.* 1983. Substrate recognition by oligosaccharyltransferase. Studies on glycosylation of modified asn-x-thr/ser tripeptides. *J. Biol. Chem.* **258**:11856–11863.

Wessels, H. P., and Spiess, M. 1988. Insertion of a multispanning membrane protein occurs sequentially and requires only one signal sequence. *Cell* **55**:61–70.

Wickner, W. 1979. The assembly of proteins into biological membranes: The membrane trigger hypothesis. *Annu. Rev. Biochem.* **48**:23–45.

Wickner, W., 1988. Mechanisms of membrane assembly: General lessons from the study of M13 coat protein and *Escherichia coli* leader peptidase. *Biochemistry* **27**:1081–1086.

Wickner, W. T., and Lodish, H. F. 1985. Multiple mechanisms of protein insertion into and across membranes. *Science* **230**:400–407.

Wiedmann, M., *et al.* 1987. A signal sequence receptor in the endoplasmic reticulum membrane. *Nature* **328**:830–833.

Wolin, S. L., and Walter, P. 1989. Signal recognition particle mediates a transient elongation arrest of preprolactin in reticulocyte lysate. *J. Cell Biol.* **109**:2617–2622.

Yamane, K., *et al.* 1987. *In vitro* translocation of protein across *Escherichia coli* membrane vesicles requires both proton motive force and ATP. *J. Biol. Chem.* **262**:2358–2362.

Yang, M., *et al.* 1991. The *MAS*-encoded processing protease of yeast mitochondria. *J. Biol. Chem.* **266**:6416–6423.

Yonath, A., *et al.* 1987. A tunnel in the ribosomal subunit revealed by three-dimensional image reconstruction. *Science* **236**:813–816.

Young, J. D., *et al.* 1979. Enzymic O-glycosylation of synthetic peptides from sequences in basic myelin protein. *Biochemistry* **18**:4444–4448.

Yu, Y., *et al.* 1990. Antiribophorin antibodies inhibit targeting to the ER membrane of ribosomes containing nascent secretory polypeptides. *J. Cell Biol.* **111**:1335–1342.

Membrane Flow and Communication between Organelle Membranes

Aoki, D., *et al.* 1992. Golgi retention of a trans-Golgi membrane protein, galactosyltransferase, requires cysteine and histidine residues within the membrane-anchoring domains. *Proc. Natl. Acad. Sci. USA* **89:**4319–4323.

Beckers, C. J. M., *et al.* 1987. Semi-intact cells permeable to macromolecules: Use in reconstitution of protein transport from the endoplasmic reticulum to the Golgi complex. *Cell* **50:** 523–534.

Beckers, C. J. M., *et al.* 1989. Vesicular transport between the endoplasmic reticulum and the Golgi stack requires the NEM-sensitive fusion protein. *Nature* **339:**397–400.

Ceriotti, A., and Colman, A. 1988. Binding to membrane proteins within the endoplasmic reticulum cannot explain the retention of the glucose-regulated protein GRP78 in *Xenopus* oocytes. *EMBO J.* **7:**633–638.

Chung, K.-N., *et al.* 1989. Molecular sorting in the secretory pathway. *Science* **243:**192–197.

Clary, D. O., *et al.* 1990. SNAPs, a family of NSF attachment proteins involved in intracellular membrane fusion in animals and yeast. *Cell* **61:**709–721.

Dahms, N., *et al.* 1987. 46kd mannose 6-phosphate receptor: Cloning, expression, and homology to the 215 kd mannose 6-phosphate receptor. *Cell* **50:**181–192.

de Curtis, I., and Simons, K. 1989. Isolation of exocytic carrier vesicles from BHK cells. *Cell* **58:** 719–727.

Doms, R. W., *et al.* 1989. Brefeldin A redistributes resident and itinerant Golgi proteins to the endoplasmic reticulum. *J. Cell Biol.* **109:**61–72.

Donaldson, J. G., *et al.* 1991. Guanine nucleotides modulate the effects of brefeldin A in semipermeable cells: Regulation of the association of a 110-kD peripheral membrane protein with the Golgi apparatus. *J. Cell Biol.* **112:**579–588.

Duden, R., *et al.* 1991a. Involvement of β-COP in membrane traffic through the Golgi complex. *Trends Cell Biol.* **1:**14–19.

Duden, R., *et al.* 1991b. β-COP, a 110kd protein associated with non-clathrin-coated vesicles and Golgi complex shows homology to β-adaptin. *Cell* **64:**649–665.

Farquhar, M. G., and Palade, G. E. 1981. The Golgi apparatus (complex)—From artifact to center stage. *J. Cell Biol.* **91:**77s–103s.

Gabathuler, R., and Kvist, S. 1990. The endoplasmic reticulum retention signal of the E3/19K protein of adenovirus type 2 consists of three separate amino acid segments at the carboxy terminus. *J. Cell Biol.* **111:**1803–1810.

Glickman, J. N., *et al.* 1989. Specificity of binding of clathrin adaptors to signals on the mannose-6-phosphate/insulin-like growth factor II receptors. *EMBO J.* **8:**1041–1047.

Haffar, O. K., *et al.* 1988. Glucocorticoid-regulated localization of cell surface glycoproteins in rat hepatoma cells is mediated within the Golgi complex. *J. Cell Biol.* **106:**1463–1474.

Hoschstenbach, F., *et al.* 1992. Endoplasmic reticulum resident protein of 90 kilodaltons associates with the T- and B-cell antigen receptors and major histocompatibility antigens during their assembly. *Proc. Natl. Acad. Sci. USA* **89:**4734–4738.

Hsu, V. W., *et al.* 1991. A recycling pathway between the endoplasmic reticulum and the Golgi apparatus for retention of unassembled MHC class I molecules. *Nature* **352:**441–494.

Ishihara, A., *et al.* 1988. Analysis of lateral redistribution of a monoclonal antibody complex plasma membrane glycoprotein which occurs during cell locomotion. *J. Cell Biol.* **106:**329–343.

Kao, C.-Y., and Draper, R. K. 1992. Retention of secretory proteins in an intermediate compartment and disappearance of the Golgi complex in an END4 mutant of Chinese hamster ovary cells. *J. Cell Biol.* **117:**701–715.

Kobayashi, T., and Pagano, R. E. 1988. ATP-dependent fusion of liposomes with the Golgi apparatus of perforated cells. *Cell* **55**:797–805.

Kornfeld, S. 1987. Trafficking of lysosomal enzymes. *FASEB J.* **1**:462–468.

Ktistakis, N. T., *et al.* 1991. PtK1 cells contain a nondiffusible, dominant factor that makes the Golgi apparatus resistant to brefeldin A. *J. Cell Biol.* **113**:1009–1023.

Lippincott-Schwartz, J., *et al.* 1989. Rapid redistribution of Golgi proteins into the ER in cells treated with brefeldin A: Evidence for membrane cycling from Golgi to ER *Cell* **56**:801–813.

Lippincott-Schwartz, J., *et al.* 1990. Microtubule-dependent retrograde transport of proteins into the ER in the presence of brefeldin A suggests an ER recycling pathway. *Cell* **60**:821–836.

Lobel, P., *et al.* 1988. Cloning and sequence analysis of the cation-independent mannose 6-phosphate receptor. *J. Biol. Chem.* **263**:2563–2570.

Machamer, C. E. 1991. Golgi retention signals: Do membranes hold the key? *Trends Cell Biol.* **1**: 141–144.

Machamer, C. E., and Rose, J. K. 1987. A specific transmembrane domain of a coronavirus E1 glycoprotein is required for its retention in the Golgi region. *J. Cell Biol.* **105**:1205–1214.

Messner, D. J., *et al.* 1989. Isolation and characterization of membranes from bovine liver which are highly enriched in mannose-6-phosphate receptors. *J. Cell Biol.* **108**:2149–2162.

Moore, H. P., and Kelly, R. B. 1986. Re-routing of a secretory protein by fusion with human growth hormone sequences. *Nature* **321**:443–446.

Moore, H. P., *et al.* 1983. Chloroquine diverts ACTH from a regulated to a constitutive secretory pathway in AtT-20 cells. *Nature* **302**:434–436.

Nair, J., *et al.* 1990. Sec2 protein contains a coiled-coil domain essential for vesicular transport and a dispensable carboxy terminal domain. *J. Cell Biol.* **110**:1897–1909.

Nakano, A., and Muramatsu, M. 1989. A novel GTP-binding protein, Sar1p, is involved in transport from the endoplasmic reticulum to the Golgi apparatus. *J. Cell Biol.* **109**:2677–2691.

Nakano, A., *et al.* 1988. A membrane glycoprotein, Sec12p, required for protein transport from the endoplasmic reticulum to the Golgi apparatus in yeast. *J. Cell Biol.* **107**:851–863.

Nilsson, T., *et al.* 1989. Short cytoplasmic sequences serve as retention signals for transmembrane proteins in the endoplasmic reticulum. *Cell* **58**:707–718.

Novick, P., *et al.* 1980. Identification of 23 complementation groups required for post-translational events in the yeast secretory pathway. *Cell* **21**:205–215.

Palade, G. 1975. Intracellular aspects of the process of protein synthesis. *Science* **189**:347–358.

Payne, G. S., and Schekman, R. 1989. Clathrin: A role in the intracellular retention of a Golgi membrane protein. *Science* **245**:1358–1365.

Pelham, H. R. B. 1988. Evidence that luminal ER proteins are sorted from secreted proteins in a post-ER compartment. *EMBO J.* **7**:913–918.

Palham, H. R. B. 1989. Control of protein exit from the endoplasmic reticulum. *Annu. Rev. Cell Biol.* **5**:1–23.

Pfeffer, S. R., and Rothman, J. E. 1987. Biosynthetic protein transport and sorting by the endoplasmic reticulum and Golgi. *Annu. Rev. Biochem.* **56**:829–852.

Rossi, G., *et al.* 1991. Dependence of Ypt1 and Sec4 membrane attachment on Bet2. *Nature* **351**: 158–161.

Rothman, J. E. 1981. The Golgi apparatus: Two organelles in tandem. *Science* **213**:1212–1219.

Ruohola, H., *et al.* 1988. Reconstitution of protein transport from the endoplasmic reticulum to the Golgi complex in yeast: The acceptor Golgi compartment is defective in *sec23* mutant. *J. Cell Biol.* **107**:1465–1476.

Salminen, A., and Novick, P. J. 1987. A ras-like protein is required for a post-Golgi event in yeast secretion. *Cell* **49**:527–538.

Salminen, A., and Novick, P. J. 1989. The Sec15 protein responds to the function of the GTP binding protein, Sec4, to control vesicular traffic in yeast. *J. Cell Biol.* **109:**1023–1036.

Seeger, M., and Payne, G. S. 1992. Selective and immediate effects of clathrin heavy chain mutations on Golgi membrane protein retention in *Saccharomyces cerevisiae*. *J. Cell Biol.* **118:**531–540.

Segev, N., *et al.* 1988. The yeast GTP-binding YPT1 protein and a mammalian counterpart are associated with the secretion machinery. *Cell* **52:**915–924.

Shin, J., *et al.* 1991. Signals for retention of transmembrane proteins in the endoplasmic reticulum studied with CD4 truncation mutants. *Proc. Natl. Acad. Sci. USA* **88:**1918–1922.

Vaux, D., *et al.* 1990. Identification by anti-idiotype antibodies of an intracellular membrane protein that recognizes a mammalian endoplasmic reticulum retention signal. *Nature* **345:** 495–502.

Wattenberg, B. W., and Rothman, J. E. 1986. Multiple cytosolic components promote intra-Golgi protein transport. *J. Biol. Chem.* **261:**2208–2213.

Wattenberg, B. W., *et al.* 1986. A novel prefusion complex formed during protein transport between Golgi cisternae in a cell-free system. *J. Biol Chem.* **261:**2202–2207.

Weidman, P. J., *et al.* 1989. Binding of an N-ethylmaleimide sensitive fusion protein to Golgi membranes requires both a soluble protein(s) and an integral membrane receptor. *J. Cell Biol.* **108:**1589–1596.

Wieland, F. T., *et al.* 1987. The rate of bulk flow from the endoplasmic reticulum to the cell surface. *Cell* **50:**289–300.

Wilson, D. W., *et al.* 1989. A fusion protein required for vesicle mediated transport in both mammalian cells and yeast. *Nature* **339:**355–359.

Wong, D. H., and Brodsky, F. M. 1992. 100-kD proteins of the Golgi and trans-Golgi network-associated coated vesicles have related but distinct membrane binding properties. *J. Cell Biol.* **117:**1171–1179.

Chapter 9

Membranes in Cancer

The term *cancer* is used to describe many diseases that display aberrant regulation of cell growth. Historically, cancer research has been largely descriptive in nature. That is, investigators cataloged the differences between normal cells and their oncogenic counterparts. For many years the effects were much clearer than the causes. Recent molecular studies on a few types of cancer have begun to explain the origins of tumor cells, how they evade chemotherapy, how they metastasize, and how they can be destroyed by immune cells.

Since tumor cells are defective in intercellular communication, one might anticipate that plasma membranes participate in their abnormal behavior. Indeed, the topics presented below recapitulate the principles of membrane structure and function learned in the preceding chapters. Oncogenes are cancer-causing genes whose products represent modified forms of normal growth-promoting transmembrane signal transduction elements—including ligands, receptors, and cytosolic signaling machinery. These gene products form one of the best, yet still incomplete, models of signal transduction. Membrane transporters in certain tumor cells protect them from chemotherapeutic drugs. Inappropriate expression of adherence-promoting membrane proteins leads to metastasis. The intracellular trafficking of tumor antigens or their fragments followed by their expression at plasma membranes can trigger immune destruction. In this chapter we will present these and other molecular features of cancer cell membranes to illustrate and extend your knowledge of membranes.

9.1. PHENOTYPIC PROPERTIES OF CANCER CELLS AND THEIR MEMBRANES

In this section several studies describing the *in vitro* properties of tumor or transformed cells will be discussed (Table 9.1). Many altered properties of tumor or transformed cells can be attributed to changes in gene expression. Several of these genetic changes will be presented in the next section.

Cancer-causing agents can act *in vivo* or *in vitro*. Cells originating from an *in vivo* tumor are referred to as tumor cells. On the other hand, cells treated with carcinogens

Table 9.1.
Altered Properties of Cells after Transformation

Cell behavior and communication	Membrane properties
Cells become smaller and spherical	Increased sugar transport
Decreased cell adhesion	Display new antigens
Increased growth and movement	Increased surface charge density
Decreased electrical communication (gap junctions)	Decreased membrane potential
	Altered surface carbohydrates
Increased agglutinability with lectins	Altered cytochemical staining for carbohydrates
Enhanced endocytotic activity	Changes in cytoskeletal structures
	Changes in transmembrane signaling

in vitro can adopt the phenotype of tumor cells; these are called transformed cells. When placed *in vivo*, transformed cell lines become metastatic or nonmetastatic tumors. This allows one to sort out the various properties of normal, nonmetastatic, and metastatic cells. Indeed, some inconsistencies in the scientific literature may be due to failure to recognize differences in tumor origin (e.g., leukemia versus sarcoma) and metastatic ability. From an experimental standpoint, transformed cells (metastatic or nonmetastatic) yield unambiguous data since their pretransformation parent cell lines serve as good controls. From a clinical perspective, transformed cells are far removed from an *in vivo* environment and therefore may provide irrelevant data. A judicious approach is to study both tumor and transformed cells with their limitations clearly in mind.

One of the most apparent differences between normal and transformed cells is their morphology. Whereas normal surface-adherent cells display a flattened morphology, transformed cells possess a more rounded shape (Ambros *et al.*, 1975). Figure 9.1 shows scanning electron micrographs of "normal" and transformed cells. In this study, cells infected by a temperature-sensitive mutant of Rous sarcoma virus were employed to

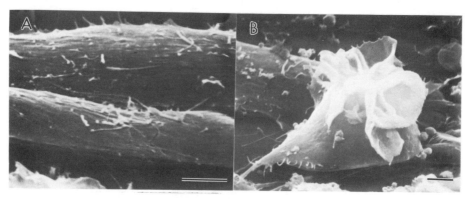

Figure 9.1. Scanning electron micrographs of Rous sarcoma virus-infected fibroblasts. Panel A shows the morphology of normal cells. These elongated cells have few surface ruffles or folds. Panel B shows transformed counterparts of those in panel A. These cells exhibit a more spherical morphology with many surface ruffles. Bars = 3 μm. (From Ambrose *et al.*, 1975, *Proc. Natl. Acad. Sci. USA* **72**:3144.)

illustrate the normal and transformed phenotypes. Transformed cells display many more surface ruffles than control cells.

The pronounced changes in cell morphology during transformation are accompanied by changes in cytoskeletal structures. Figure 9.2 shows fluorescence photomicrographs of a normal and transformed cell pair stained with fluorescent antiactin antibodies. Both normal and transformed cells display submembranous actin staining. Normal cells have a latticework of microfilaments. They are straight with many branch points and frequently span a cell's entire length. In contrast, transformed cells have fewer and thinner microfilaments. These diminished microfilaments do not form a lattice. Most importantly, the stress fibers found in normal cells are absent from transformed cells. The reorganization of actin may account for differences in cell morphology. For example, the disappearance of stress fibers and the accompanying centralization of actin may explain the loss of a flattened morphology and the appearance of membrane ruffling, respectively.

Recent studies have suggested a molecular origin for these microfilament changes. Cell transformation is often accompanied by increased protein kinase activity (Section 9.2, Table 9.2). Indeed, phosphotyrosine levels increase by eight- to tenfold in transformed cells. The phosphotyrosine level of vinculin, an actin-binding protein, jumps dramatically after transformation. Since vinculin links actin cables to plasma membranes (Figures 3.12 and 9.6), its phosphorylation may release microfilaments from a membrane, thus leading to stress fiber disassembly. Therefore, changes in cell morphology and anchorage-dependent growth might be traced in part to enzyme reactions at a membrane's *cis* face.

There is no major change in microtubular structure after cell transformation. Microtubules appear less organized because tumor cells are more spherical (Tucker *et al.*, 1978). Intermediate filaments are another major class of cytoskeletal structures. Figure 9.3 shows intermediate filaments of normal and transformed cells. Normal cells possess a lattice of intermediate filaments which become less fibrillar after transforma-

Figure 9.2. Distribution of microfilaments in normal and transformed cells. Normal (panel A) and transformed (panel B) cells were stained with a fluorescent antiactin antibody. The disappearance of stress fibers in transformed cells should be noted. Bar = 10 μm. (From Tucker *et al.*, *Cell* **13**:629, copyright © 1978 by Cell Press.)

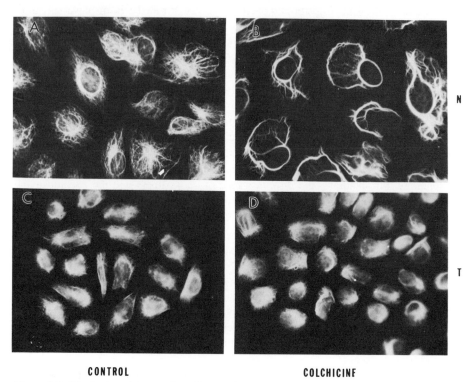

CONTROL   COLCHICINE

Figure 9.3. Distribution of intermediate filaments in normal and transformed cells in the absence and presence of colchicine. Cells were stained with a fluorescent antibody directed against intermediate filaments. Normal cells (N) were observed in the absence and presence of colchicine in panels A and B, respectively. Panels C and D show transformed cells (T) in the absence and presence of colchicine. Changes in both the lattice of intermediate filaments and their response to colchicine are observed. Cells are roughly 10 μm in size. (From Hynes and Destree, *Cell* **13**:151, copyright © 1978 by Cell Press.)

tion (panels A and C). The distinction between intermediate filaments of normal and transformed cells is more apparent after colchicine treatment (panels B and D). Colchicine treatment induces a coiling of intermediate filaments about the nucleus and cell periphery. In sharp contrast, the intermediate filaments of transformed cells clump near the nucleus (Hynes and Destree, 1978).

Early cytochemical studies suggested that transformation alters a cell's outer carbohydrate coat. Subsequent studies showed that many tumor and transformed cells are more easily agglutinated by lectins. The probable origins of these changes are found in the chemistry of cell surface carbohydrates. The sialic acid content of human tumor cells, but not leukemic cells, is increased. However, variable sialic acid levels have been reported for transformed cells.

Another feature of membrane carbohydrates correlated with transformation is the production of larger *N*-linked oligosaccharide chains. These chains are more highly branched at the trimannosyl core due to the addition of $\beta1\rightarrow6$-linked lactosamine. This is due to an increase in activity of GlcNAc transferase V (Yamashita et al., 1984). There are several biochemical consequences of this alteration. These larger chains provide additional sites for the addition of sialic acid. Furthermore, they provide additional lectin binding sites, some specific for transformation such as $\beta1\rightarrow6$ core linkages (Dennis et al., 1987). Although these altered membrane carbohydrates are likely of minimal importance in aberrant growth control, they are of great importance in cell–cell interactions participating in metastasis (Section 9.4).

The glycosphingolipid and glycoprotein composition of transformed and tumor cell membranes have also been evaluated (Srinivas and Colburn, 1982; Nicolau, 1978). Although there are no remarkable changes in phospholipid composition, glycosphingo-lipids are significantly altered by transformation. These differences are due to changes in the glycosyltransferase activities participating in membrane biosynthesis. As a rule, tumor or transformed cells possess fewer complex gangliosides or complex neutral glycosphingolipids. The expression of simpler glycosphingolipids is increased. For example, the amount of II^3-NANA-Gg_3Cer is decreased while the quantity of II^3-NANA-lactosyl-Cer is increased. Similarly, the content of simple neutral glycosphingolipids such as lactosyl-Cer is increased. Some blood group antigens are lost because they are composed of these outer saccharide residues. Interestingly, antibodies directed against these blood group precursors have been found in some lung cancer patients (Schrump et al., 1988). The structural modifications of these glycosphingolipids likely account for their increased accessibility to enzymes and lectins and some aspects of their antigenicity.

There have been numerous studies of cell surface proteins on normal and transformed cells. There is general agreement that upon transformation, fibronectin, a large ($\sim$250 kDa) external peripheral protein, disappears. Tumor cells apparently lose fibronectin because proteolytic activities are released that digest it (Chen and Chen, 1987; Section 9.4). The destruction of fibronectin at the trans membrane face and the release of microfilaments at the cis face, as discussed earlier, severs completely transmembrane adhesive sites (Figures 3.12 and 9.6). When fibronectin is added to transformed cell cultures, it can temporarily restore partial adherence and a flattened morphology (Yamada et al., 1976).

Additional quantitative changes in membrane protein composition have been reported. For example, the appearance of new antigens is sometimes linked to the expression of viral genes. Other quantitative changes may be related to differences in cellular gene expression, although their significance may not be clear.

The electrical properties of cell surfaces have sometimes been correlated with transformation. For example, many investigators have reported that tumor and transformed cells possess a higher net surface charge density. However, numerous counter-examples can be found and no firm conclusions can be reached. Unfortunately, the various cell types, conditions, and cellular metastatic propensities contribute to the difficulty in interpreting these results. Cationized ferritin binding studies have provided evidence that anionic sites tend to cluster on highly metastatic cells (Raz et al., 1980). Changes in the number and distribution of anionic sites might influence cell–cell interactions.

Decreased resting transmembrane potentials have apparently correlated with transformation (Lai *et al.*, 1984). Similarly, proliferating normal cells have diminished membrane potentials. The change in transmembrane potential could be due to metabolic alterations brought about by transforming proteins. In addition, increases in surface charge density would also be expected to decrease transmembrane potentials. Decreases in transmembrane potentials are accompanied by diminished cell–cell communications since gap junctions are voltage-sensitive.

9.2. ABERRANT MEMBRANE SIGNALS OF ONCOGENE PRODUCTS POTENTIATE CELL GROWTH

Cancer has many origins. Unfortunately, these origins are poorly understood. Although we know that certain conditions (e.g., ultraviolet light, certain chemicals, radiation, and some viruses) can cause cancer, we do not know *how* they cause cancer. Recently, the exciting discovery of over 40 human oncogenes (cancer-causing genes) has raised the hope that some forms of cancer are understandable.

Oncogenes arose from proto-oncogenes, their earliest form. Proto-oncogenes are believed to be important in normal cell proliferation and differentiation. Overexpression or certain mutations of proto-oncogenes result in transformation. The designation of these genes is usually preceded by *c* to indicate their cellular origin. Oncogenes were first discovered in the genomes of oncogenic viruses. These viruses are capable of transforming cells *in vitro* and causing tumors *in vivo*. Viral oncogenes are preceded by a *v* to indicate their association with viruses. For example, the *v-erbB* oncogene of avian erythroblastosis viruses is derived from the cellular gene *c-erbB* that encodes the epidermal growth factor receptor.

Proto-oncogene and oncogene products operate at several sites within cells. In some cases oncogenes encode defective membrane proteins; in all cases they apparently involve defective intracellular growth signals. We shall now consider a few selected oncogene products that directly involve plasma membranes.

Figure 9.4 shows three molecular membrane-related pathways that result in excessive cell growth and transformation. All of these mechanisms involve altered transmembrane signals. Cell transformation can result from the amplification or overexpression of normal growth regulatory genes. Three examples of this shown in Figure 9.4 are the overexpression of EGF receptors, normal *c-ras* G proteins, and *c-sis* gene products. In these cases "normal" growth signals yield abnormal growth responses. Point mutations of the *c-ras* and *neu* genes can create *c-ras* and *neu* oncogenes that code for transforming proteins which produce excessive growth signals. Many oncogenes such as *v-erbB* and *v-fms* code for membrane proteins that are similar to growth factor receptors. The v-erbB protein has lost its regulatory domains. When the *fms* gene is expressed in nonhemopoietic tissues, cell transformation is a result. Other oncogene products such as *v-sis* are proteins that act as ligands for growth factor receptors. The final class of membrane-related oncogenes interfere with normal signal transduction. For example, *ras* and *v-src* genes encode a G protein and a tyrosyl kinase, respectively. We will now consider these molecular membrane origins of cancer in detail.

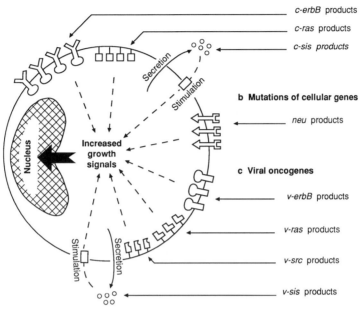

Figure 9.4. Mechanisms of cell transformation by membrane-associated proteins. Three general mechanisms of cell transformation by proto-oncogenes and oncogenes are shown. Representative examples of each class of transformation are given.

9.2.1. Some Oncogene Products Resemble Receptors

Many oncogenes encode membrane-associated proteins that contain a tyrosine-specific protein kinase domain (Table 9.2). These tyrosyl kinase-containing oncogene products can be divided into two groups. The first group is made up of integral membrane proteins that resemble growth factor receptors (e.g., *v-erbB*, *v-fms*, *v-kit*, and *neu*). The second group contains oncogenes such as *v-src* and *v-abl* that are associated with a membrane via fatty acyl groups. We shall now discuss the role of three growth factor receptors in tumorigenesis.

The *c-erbB* gene encodes the epidermal growth factor (EGF) receptor. EGF is a mitogenic protein that stimulates cell division. When EGF binds to an EGF receptor, it triggers the enzyme activity of an intrinsic tyrosine-specific protein kinase. This is followed by a cascade of signaling events that culminate in the clustering and internalization of ligand–receptor complexes and the expression of growth-promoting genes.

The improper expression of EGF receptors leads to cancer by two general routes. The first pathway involves the overexpression of the *c-erbB* proto-oncogene (Figure 9.4) which has been observed in several human malignancies (Libermann *et al.*, 1985). The deranged expression of *c-erbB* is due to gene amplification and/or rearrangement. Viral integration near *c-erbB* and insertion of a viral promoter near it can lead to gene amplification. Rearrangements or multiple copies of human chromosome 7 can also

Table 9.2.
Survey of Cell Surface Transforming Proteins

Viral oncogene	Oncogenic viruses					Cognate cellular gene				
	Virus	Species	Tumorigenicity	Type	Gene product	Proto-oncogene	Human chromosome	Function	Type	Gene product
Tyrosyl protein kinases										
v-erbB	Avian erythroblastosis	Chicken	Erythroleukemia	Integral	gp65	c-erbB	7	Epidermal growth factor receptor	Integral	gp170
v-fms	SM feline sarcoma	Cat	Sarcoma	Integral	gp140	c-fms		CSF-1 receptor	Integral	gp170
v-mpl	Myeloproliferative mouse leukemia virus	Mouse	Leukemia	Integral	MPL	c-mpl		Cytokine-like receptor	Integral	MPLK/MPLP
Not applicable						neu	17	Heregulin-α receptor	Integral	p185
Not applicable						trkA		Nerve growth factor-like receptor	Integral	p140
Not applicable						met	7	Hepatocyte growth factor	Integral	Met
v-kit	HZ4 feline sarcoma	Cat	Fibrosarcoma	Fatty acid	p80	c-kit	4	Unknown receptor	Integral	p145
v-src	Rous sarcoma	Chicken	Sarcoma	Fatty acid	pp60	c-src	20	Signaling	Fatty acid	
v-abl	Abelson leukemia	Cat and mouse	B-cell lymphoma			c-abl	9	Signaling		
v-fps/v-fes		Cat and chicken	Sarcoma	Fatty acid	p140/p85	c-fps/c-fes				p98/p92
G proteins										
v-Ha-ras	Harvey	Rat and mouse	Sarcoma and erythroleukemia	Fatty acid	p21	Harvey c-ras	11	Signaling	Fatty acid	p21
v-Ki-ras	Kirsten	Rat	Sarcoma and erythroleukemia	Fatty acid	p21	Kirsten c-ras	12	Signaling	Fatty acid	p21
Ligand										
v-sis	Simian sarcoma	Monkey	Sarcoma	Binds soluble ligand		c-sis	22	PDGF ligand		

lead to increased expression of *c-erbB* and transformation. The second route involves a viral derivative of this proto-oncogene called *v-erbB*. The oncogenic activity of *v-erbB* arises from its altered amino acid sequence.

The *v-erbB* gene is found in avian erythroblastosis viruses. It encodes a transforming membrane protein known as gp65[v-erbB] (M_r = 65,000). Many NH$_2$- and COOH-terminal sequences were deleted during the transduction of *c-erbB* into *v-erbB*. The truncated gene product has lost both the extracellular ligand-binding domain that controls receptor function and an intracellular tyrosine-containing sequence that down-regulates kinase activity. The mechanism of action of gp65[v-erbB] is not certain. Although it has a tyrosyl kinase-like domain, its enzyme activity has been difficult to detect. The unregulated gp65[v-erbB] presumably leads to avian cell transformation by generating constant growth signals.

The *c-fms* gene encodes the macrophage colony-stimulating factor-1 (CSF-1) receptor (Sherr *et al.*, 1985). This 170-kDa transmembrane glycoprotein possesses a tyrosyl kinase domain at a plasma membrane's *cis* face (Figure 9.5). When CSF-1 binds to its receptor on immature bone marrow cells, it stimulates them to form colonies of monocytes. The McDonough strain of feline sarcoma virus (SM-FeSV) has acquired most of the *c-fms* gene. The viral oncogene encodes a transmembrane glycoprotein (M_r = 140,000; gp140[v-fms]). The expression of gp140[v-fms] at cell surfaces is required for transformation. In the oncogene product 40 COOH-terminal amino acids are replaced with 11 dissimilar residues. This alteration removes a tyrosine phosphorylation site, which allows downregulation of the normal *c-fms* product.

The *neu* gene, also known as *erbB2*, has been found in rat and human genomic DNA; an oncogenic viral counterpart has not been discovered. It encodes a transmembrane glycoprotein (M_r = 185,000; p185) that appears to be a growth factor receptor (Bargmann *et al.*, 1986). Recent studies have suggested that heregulin-α is its biological ligand (Holmes *et al.*, 1992), although other potential ligands have also been identified (Yarden and Peles, 1991). The *neu* gene product is 50% identical with the EGF receptor, including an intracellular tyrosyl kinase domain and an extracellular Cys-rich domain.

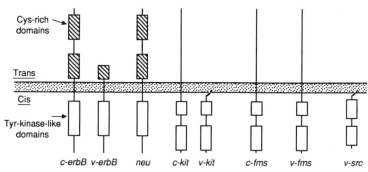

Figure 9.5. Structural models of transforming membrane proteins. The structural relationships among eight proto-oncogenes and oncogenes are illustrated. Straight lines within the models indicate protein domains whereas the curved wavy line represents fatty acyl groups that attach the *v-kit* and *v-src* gene products to membranes. Crosshatched areas are Cys-rich domains. Tyrosine kinase-like domains are found at a membrane's *cis* face.

Cancer-causing chemicals can mutate the *neu* gene leading to neuroblastomas and glioblastomas. A single point mutation (Val-664 to Glu-664) confers malignancy (Bargmann *et al.*, 1986). This mutation is within the protein's transmembrane domain. Sixteen contiguous hydrophobic amino acids lie between the mutation site and the *cis* membrane face. A stretch of 16 hydrophobic amino acids is sufficient to traverse a bilayer (Section 2.3.3). The mutation's location suggests that the transmembrane domain is more than a passive membrane anchor. The proto-oncogene and oncogene products are found on normal and tumor cell surfaces, respectively, at the same density. Therefore, overexpression cannot explain tumorigenicity in humans, although *neu* overexpression transforms cultured cells. The mutation likely stabilizes p185, the *neu* gene product, in an active form leading to continuous growth stimulation. The mutant form of p185 has been found to aggregate in membranes whereas its normal counterpart is not aggregated under identical conditions (Weiner *et al.*, 1989). This suggests several possibilities. For example, p185's altered transmembrane domain could trigger its clustering and growth stimulation. The tyrosyl kinase domain likely participates in oncogenesis. However, the mechanism of interaction between p185's transmembrane and tyrosyl kinase domains is unknown. The insertion of a charged amino acid could change the conformation of the transmembrane domain and/or alter local lipid packing. Alternatively, aggregation may directly trigger tyrosyl kinase activity.

9.2.2. The *sis* Oncogene Product Resembles a Ligand

As described in the preceding section, the overexpression of growth factor receptors or mutations within them lead to cell transformation. The aberrant expression of ligands can also lead to the malignant phenotype. The *sis* family of gene products is an example of ligand-promoted cell transformation.

Platelet-derived growth factor (PDGF) is a potent mitogen for cells of mesenchymal origin. It is composed of two homologous chains, A and B, linked by a disulfide bridge. Both chains possess mitogenic activity. The B subunit is encoded by the *c-sis* gene. When PDGF binds to cell surface PDGF receptors, it activates the receptor's tyrosyl kinase activity. Under normal conditions this ligand–receptor interaction plays an important role in inflammation and wound healing.

The simian sarcoma virus carries an analogue of the *c-sis* gene called *v-sis*. The *v-sis* gene encodes a protein that is nearly identical to PDGF's B chain. The initial translation product of *v-sis*, p29[sis], dimerizes and then becomes proteolytically processed to yield the mature transforming protein p24[sis] (Waterfield *et al.*, 1983; Heldin and Westermark, 1984). The p24[sis] mechanism of action is uncertain, but it probably involves secretion and binding to PDGF receptors (Figure 9.4). This idea is consistent with several experimental observations. First, when added to cell cultures *in vitro*, PDGF can reversibly induce a transformed cell-like phenotype. Moreover, human tumors release PDGF-like growth factors. This autocrine stimulation pathway may contribute to human disease since *c-sis* genes are overexpressed in several human cancers.

9.2.3. Transforming Membrane Proteins Resembling Signal Transduction Elements: Tyrosyl Kinase Oncogenes

Several oncogenes such as *v-src* and *v-abl* encode membrane-associated tyrosine-specific protein kinases. The *v-src* oncogene is carried by the Rous sarcoma virus, the first cancer-causing virus to be discovered. The *v-src* oncogene product, pp60[v-src], has been localized to the *cis* face of plasma membranes using electron microscopic cytochemistry. Interestingly, it is clustered in the vicinity of membrane ruffles or adhesion sites such as gap junctions.

In a clever study, Jakobovits *et al.* (1984) used a hormonal promoter to express the *v-src* oncogene at different levels in transfected cells. At low constitutive levels of pp60[v-src] expression, cells retained a normal phenotype. As the expression level of *v-src* was raised, cells rapidly became transformed. This indicates that a critical threshold of this membrane protein must be reached to induce transformation. At a dose just sufficient to induce transformation, no enhanced phosphorylation of p36, a major target of pp60[v-src], was found. p36 is an abundant protein at a plasma membrane's *cis* face that may participate in cytoskeletal control. This experiment suggests that phosphorylation of p36 accompanies but is not required for transformation. Apparently, the oncogenic pathway is not related to p36 phosphorylation.

The normal cellular counterpart of *v-src* is *c-src*. The *c-src* proto-oncogene is found in many animals, including humans. It has a very high degree of homology to *v-src*. Similar to many other proto-oncogenes, it is expressed in normal cells. The gene product is a 60-kDa tyrosyl kinase. In contrast to some of the proto-oncogene products discussed above, the overexpression of pp60[c-src] does not transform cells. This suggests that the differences in amino acid sequences of pp60[c-src] and pp60[v-src], not the number of copies, is the factor determining transformation. The p36 protein is also a substrate for pp60[c-src], although its relevance to growth is uncertain. The addition of PDGF to cells causes phosphorylation of pp60[c-src]; this suggests that the proto-oncogene product participates in normal growth control.

9.2.4. The Transforming Membrane Protein Middle T Antigen Subverts the Function of the *c-src* Gene Product

Viruses can cause cancer directly by coding for a viral oncogene or indirectly by causing the overexpression of a proto-oncogene. A third mechanism relies on subversion of the normal host signal transduction system(s). This strategy is used by polyoma viruses to transform cells *in vitro* and cause tumors *in vivo*. The polyoma virus encodes a membrane protein known as the middle T antigen that is required for transformation. This 57-kDa phosphoprotein is expressed at a plasma membrane's *cis* face in infected cells. One hydrophobic sequence of amino acids is buried in the membrane. The middle T antigen binds to pp60[c-src] (Courtneidge and Smith, 1983). Antibodies to middle T antigen co-immunoprecipitate pp60[c-src]. Similarly, antibodies to pp60[c-src] co-immuno-precipitate middle T antigen. The tyrosyl kinase activity of both polyoma virus-infected cells and middle T antigen–pp60[c-src] complexes is enhanced. Since there is no increase in the amount of pp60[c-src], its kinase activity must be increased. Transformation-defective mutants of polyoma virus are unable to bind to pp60[c-src] or stimulate kinase activity. Therefore, middle T antigen is able to bind to pp60[c-src], thereby stimulating its activity and causing transformation.

9.2.5. Transforming Membrane Proteins Resembling Signal Transduction Elements: The *ras* Guanine-Binding Proteins

Ras genes were first discovered in rat retroviruses that cause *rat* sarcomas. They were subsequently found in genomes of many eukaryotes—from yeasts to humans (Barbacid, 1987). During normal conditions, *c-ras* gene products participate in membrane signal transduction during cell proliferation or differentiation. Certain mutations of *c-ras* genes create *c-ras* oncogenes which trigger the transformed phenotype *in vitro* (Bar-Sagi and Feramisco, 1986) and tumor formation *in vivo*. These oncogenes are of keen interest since they have been found in at least 10 to 15% of human cancers.

Humans possess at least three very similar *c-ras* genes. They encode proteins of 21 kDa (p21). These proteins are associated with a plasma membrane's *cis* face (Willingham *et al.*, 1980). They are synthesized in the cytosol and then acylated with palmitic acid at Cys-186 to provide membrane attachment. The palmitate moiety is rapidly turned over; it has a half-time of only 20 min. Small changes in acylation and deacylation reactions could significantly alter membrane signal transduction. This membrane linkage is absolutely required for the oncogene's transforming activity. All ras proteins possess a high degree of homology to G proteins, especially the α subunit (Section 7.3 discusses G proteins). They bind guanine nucleotides and exhibit GTPase activity, as do G proteins. The location, amino acid sequence, and activity of ras proteins identify them as a class of G proteins.

Under most normal conditions, *ras* gene products are in an inactive state. In this state GDP is bound to p21 and no signals are generated. When a stimulus is received from another source such as a receptor, the GDP molecule at the nucleotide binding site is exchanged for GTP. A conformational change in p21 then transmits its proliferation or differentiation signal to another protein(s) such as GAP (see below). The active p21 conformation is then rapidly lost because its intrinsic GTPase activity cleaves GTP to GDP.

Subtle alterations of normal *ras* genes can lead to disastrous consequences for a cell and organism. A single point mutation within a *ras* gene creates a *ras* oncogene. By comparing nucleotide sequences of *ras* genes and oncogenes, it was found that replacement of Gly-12 with any other amino acid creates the transformed phenotype. The amino acid residue Gln-16 is also highly sensitive to mutation. Both of these residues are in the protein domain that interacts with the phosphoryl group of GTP. These mutations locally perturb an α-helix and thereby decrease the protein's ability to interact with phosphoryl groups during hydrolysis of GTP. Consequently, bound GTP "locks" p21 into an active conformation leading to a continuous stream of growth signals. The inhibition of GTPase activity is the principal means of *in vivo* malignant transformation by *ras* oncogenes. Pharmaceutical companies are now searching for drugs that might stabilize mutant p21 molecules in a resting conformation (e.g., Gibbs, 1991).

Any mutation that favors an active p21 conformation promotes transformation. Genetic engineering experiments have shown that certain modifications of Asp-116 or Asp-119, which reside in the guanine nucleotide binding region, lead to transformation. These mutations dramatically decrease the affinity of p21 for GDP and GTP. Since GDP dissociates much more easily and since GTP is at a much higher cytosolic concentration than GDP, the active conformation becomes favored. Hence, these mutations also lead to malignant cells. An effector molecule of p21, GTPase-activating protein (GAP), has been identified (Trahey *et al.*, 1988). GAP is a 120-kDa protein whose hydrophobic NH_2-terminal sequences may interact with a plasma membrane's *cis* face. GAP binds to normal and oncogenic mutants of p21 at its effector domain (residues 32 to 40). Although GAP greatly stimulates p21's GTPase activity, position 12 mutants of p21 are unaffected by GAP. GAP is clearly an effector molecule of p21. The subsequent steps of signal transmission and cellular transformation beyond GAP are just beginning to be understood.

9.2.6. Variations on an Onco-Theme

As described above, the molecular defects underlying some forms of cancer are beginning to be understood. A common theme among transforming proteins is their participation in normal cell growth. This is true for both viral oncogene-directed tumors and malignancies due to cellular oncogenes activated by chemical mutagens, viral integration near proto-oncogenes, or chromosomal translocations. Plasma membranes play a central role in initiating unregulated cell growth. Oncogenetics has helped to elucidate abnormal and normal membrane signaling events in growth control. For example, ligands (*sis*) binds to receptors (*erbB*, *fms*, *kit*, and *neu*) that undergo conformational changes. At the *cis* face of plasma membranes a signal directly (*erbB*, *fms*, *kit*, and *neu*) or indirectly (*src*, *abl*, *fps*, and *fms*) triggers tyrosine protein kinase activity and G-protein activity (*Ha-ras*, *Ki-ras*, and *N-ras*). The *ras* p21 protein then interacts with GAP whereas *src* proteins interact with p36. However, these two apparent pathways are not mutually exclusive. For example, microinjection of antibodies directed against p21 blocks cell transformation by *src*, *fes*, and *fms*. These various proteins may all be part of a common pathway. Since pp60[v-src] and related proteins are dependent on p21 for transformation, they apparently lie upstream along the same signaling pathway. These studies suggest that membrane-associated transformation pathways may share a common sequence of transmembrane signaling events.

Figure 9.6 illustrates some possible interactions during cancer. On the left is shown an attempt to integrate our incomplete knowledge of transmembrane signaling with our incomplete knowledge of oncogene products. The locations of specific transforming protein lesions in the pathway are indicated by the relevant oncogene name. These tyrosyl kinases and p21 proteins potentiate phosphatidylinositol (PI) turnover (Sugimoto *et al.*, 1984; Fleischman *et al.*, 1986). Cells transformed by *ras* genes have an increased ratio of diacylglycerol (DAG) to phosphatidylinositol 4,5-bisphosphate (PIP_2)

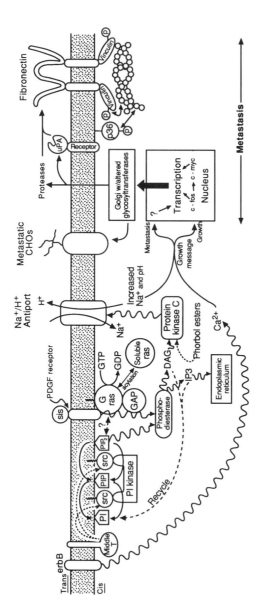

Figure 9.6. Possible interrelationships among oncogene products and signal transduction pathways during aberrant growth, adhesion, and metastasis. This figure summarizes some of the information regarding the expression and role of membrane proteins, lipids, and carbohydrates in triggering abnormal growth, adhesion, and metastasis of cells. Abbreviations are given in the text. Although this figure only represents a crude and speculative summary of oncogenic changes, it does illustrate possible sites which might be exploited in the next generation of anticancer drugs. The following key is used: signals, ~; reactions, ——; interference points, . .; proteins, ○; lipids, ○; and organelles, □.

suggesting that p21 can affect the breakdown of PIP_2. Furthermore, microinjection of anti-PIP_2 antibodies into *ras*-transformed cells reduces cell growth and causes cell morphology to revert to a normal appearance (Fukami *et al.*, 1988). PI turnover can also be augmented by pp60[v-src]. pp60[v-src] apparently interacts with a PI kinase. This PI kinase phosphorylates PI to form PIP_2. This neatly explains why phosphate is rapidly cycled through PI intermediates in *src*-transformed cells. During both of these circumstances, increased quantities of the products DAG and inositol 1,4,5-trisphosphate (IP_3) are produced.

Phorbol esters are chemicals that promote tumors. They bind to protein kinase C, thereby stimulating ion fluxes, phosphorylation reactions, and up-regulation of *c-fos* and *c-myc* proto-oncogenes. The molecular shape of phorbol esters is similar to that of DAG; these compounds bind to the same site on PKC. DAG stimulates PKC in a similar, but regulated, fashion. The second reaction product of PIP_2 cleavage is IP_3. IP_3 binds to the ER, causing the release of Ca^{2+}. The ionic changes mediated by IP_3 and PKC are believed to trigger cell growth.

The analogy between normal and abnormal growth can be extended. Recent studies have shown that the addition of growth factors or phorbol esters to cells leads to a sequential expression of the nuclear proto-oncogenes *c-fos* and *c-myc*. Their gene products are likely transcriptional regulators. A cell's growth program is then altered by these proteins that up-regulate hundreds of other genes. The steady expression of these many genes is manifest as the transformed phenotype. Oncogenes occur at many levels within the same growth regulatory pathway, including the control of nuclear transcription.

9.3. EXPRESSION OF MEMBRANE P-GLYCOPROTEINS ALLOWS CANCER CELLS TO EVADE CHEMOTHERAPY

Chemotherapy is a principal form of cancer treatment. When chemotherapy fails, it is often due to the expression and amplification of *multidrug resistance (mdr)* genes which code for several membrane-associated proteins called P-glycoproteins (also gp170). Most tumor cells die when exposed to a chemotherapeutic drug *in vitro* or *in vivo*. A few cells may survive treatment and then grow into cell colonies *in vitro* or tumors *in vivo*. These surviving cells show resistance to the drug they were treated with *and* to many other chemotherapeutic agents (Endicott and Ling, 1985). In other words, a specific drug can induce resistance to many additional drugs which may or may not be structurally similar to the initial drug. Hence, the name multidrug resistance genes is used.

Recent studies have shown how tumor cells can become resistant to multiple chemotherapeutic agents. These drugs enter cells by passive diffusion (Section 6.1.1) down their concentration gradients. Mdr cells contain abnormally low drug levels. This is due to active transport of the drug out of a resistant tumor cell against its concentration gradient. The P-glycoproteins are a family of proteins participating in the outward pumping of cytotoxic drugs. Their deduced amino acid sequences show two homologous regions that also possess extensive homology with bacterial transport proteins (Chen *et al.*, 1986; Gros *et al.*, 1986). The *mdr* gene products are integral membrane proteins possessing 12 putative transmembrane domains. The sixth transmembrane domain may play a central role in drug recognition (Devine *et al.*, 1992). P-glycoproteins are phosphorylated in living cells. They have two ATP binding sites that provide the energy required to maintain low drug levels.

A second and possibly third large transmembrane *mdr* gene product have been identified. The second human P-glycoprotein is a product of the *mdr3* gene (nomenclature follows sequence homologies with hamster *mdr* genes). Alternative splicing of *mdr3* pre-mRNAs leads to several slightly different membrane proteins (van der Bliek *et al.*, 1987). The multiple *mdr* primary transcripts and their alternate splicing contribute to the diversity of drugs pumped by these proteins.

The transport mechanism(s) of P-glycoproteins is likely to be similar to those of gram-negative bacterial transport systems which utilize periplasmic proteins to transport nutrients from the periplasmic space to the cytoplasm. The presence of cytotoxic drugs selects for cells expressing higher levels of *mdr* genes. In the presence of ATP, these drugs are pumped out of a tumor cell, thus allowing these cells to escape chemotherapy.

Although we are beginning to understand the reasons for chemotherapeutic failures, the very presence of *mdr* genes has been puzzling. Why should higher organisms possess membrane-bound drug pumps? Pastan (1990) has recently identified one apparent physiological role of P-glycoproteins. Using anti-P-glycoprotein antibodies, Pastan and co-workers tested normal human tissues for P-glycoprotein expression. They found P-glycoproteins at the apical surfaces of tissues from the adrenal, kidney, liver, pancreas, uterus, and small and large intestine. P-glycoprotein expression was also found at the blood–brain and blood–testes barriers. The distribution of P-glycoproteins in healthy tissues suggests that its normal function is to remove natural toxins from the body. For example, natural toxins in foodstuffs are likely removed by P-glycoproteins of the gastrointestinal tract. Toxins reaching the blood are removed by the kidney. The brain and testes may be accorded special protection by their local P-glycoproteins. P-glycoprotein expression levels correlate inversely with the likelihood of successful chemotherapeutic treatment. For example, pancreatic tumors, which are very aggressive and difficult to manage clinically, express high levels of P-glycoproteins under normal circumstances. It is more likely that tumor cells derived from the pancreas will express an *mdr* gene; thence their aggressive phenotype. Since many chemotherapeutic drugs were isolated from natural sources, the functional role of P-glycoproteins makes sense. However, it is possible that P-glycoproteins have additional functions *in vivo* (e.g., Valverde *et al.*, 1992). Scientists are now attempting to find ways of manipulating P-glycoproteins to improve chemotherapy (Sorrentino *et al.*, 1992).

9.4. ROLE OF MEMBRANES IN METASTASIS AND INVASION

Unchecked cell growth leads to benign or malignant tumors. Benign tumors displace neighboring tissues, but do not invade them. In general, benign tumors can be easily removed by surgery. On the other hand, malignant tumors invade tissues locally and metastasize to distant sites. It is our inability to treat thousands of metastases that forms the central problem of cancer research. The metastatic process involves several steps including: cellular detachment from the original tumor, penetration into the circulatory (or lymph) system, dissemination throughout an organism via the circulation and/or lymph, and adherence to and penetration through capillary walls to establish new metastatic foci.

The metastatic phenotype has been correlated with the expression of specific oncogenes (Egan *et al.*, 1987). The expression of certain membrane-associated signal transduction oncogene products (*ras*, *fms*, *fes*, and *src*) leads to metastatic cells *in vivo*. On the other hand, some nuclear oncogenes (e.g., *myc*) do not appear to trigger metastasis. It will be important to extend these observations to other oncogenes. If metastasis is a direct result of altered gene expression and if these proto-oncogenes participate in normal growth control, then the intracellular signals giving rise to altered growth and metastasis likely bifurcate at or before the level of transcriptional activation (Figure 9.6).

The cell surface is responsible for mediating the interactions between tumor cells and their environment. Poste and Nicolson (1980) have shown that the metastatic ability of tumor cells is determined by their plasma membranes. Low (F1) and high (F10) metastatic sublines of a mouse B16 melanoma cell line were prepared. When injected intravenously into mice, the latter—but not the former—forms many lung metastases. Plasma membrane vesicles were prepared from both cell sublines. When F10 membrane vesicles were fused with F1 cells, the cells became much more metastatic. This indicates that all of the components required for the metastatic phenotype are carried on plasma membranes. In contrast, F1 plasma membranes could not convey the nonmetastatic phenotype to F10 cells. This suggests that: (1) the relatively small dilution of F10 membranes by F1 membrane vesicles was insufficient to alter metastasis and (2) membrane components inhibiting metastasis apparently do not exist in F1 membranes.

Since the metastatic phenotype resides within plasma membranes, it might be possible to identify membrane components responsible for the metastatic spread of tumor cells and to inhibit metastasis by agents acting on these components. Several lines of evidence indicate that it might be possible to control metastasis by inhibiting tumor cell adherence. Vollmers and Birchmeier (1983) prepared seven monoclonal antibodies specific for B16 melanoma cells that inhibited their adhesion to tissue culture dishes. Four of these antibodies were found to inhibit metastasis when they were injected into mice with or before F10 cells. Recently, B16 melanoma surface integrins have been linked with adhesion to basement membranes (Kramer *et al.*, 1989). As previously described (Section 7.5.1), RGD-containing peptides and proteins, such as fibronectin, participate in many types of integrin-mediated cell adhesive phenomena. When B16–F10 melanoma cells were mixed with RGD-containing peptides and then injected into mice, the cells' metastatic activity was depressed by up to 97% (Humphries *et al.*, 1986a). Control peptides with altered RGD sequences showed little or no efficacy. Therefore, it seems that direct intervention in RGD-promoted cell adherence may contribute to the control of metastases.

In addition to its RGD-containing domain, fibronectin possesses a heparin-binding domain that has also been linked to cell adhesive events and metastasis (e.g., Drake *et al.*, 1992). Fibronectin's heparin-binding domain triggers cell adhesion by binding to cell surface proteoglycans. The leukocyte cell surface integral membrane glycoprotein CD44 interacts with the heparin-binding domain (Jalkanen and Jalkanen, 1992). CD44, which is primarily but not exclusively found on leukocytes, is expressed in various molecular forms ranging from ~85 to 200 kDa depending on alternative splicing and chondroitin sulfate addition. The *trans* face of CD44 binds to fibronectin and collagen whereas its *cis* domain interacts with the cytoskeletal proteins vimentin and ankyrin. CD44 participates in lymphocyte activation, cell trafficking via high endothelial venules (HEV), and binding to the extracellular matrix. A variant form of CD44, CD44v, has been linked to the metastatic spread of two rat carcinomas (Gunthert *et al.*, 1991). CD44v, which contains a 162-amino-acid insert in its extracellular domain, is found on metastatic, but not nonmetastatic, cells. Antibodies directed against the variant portion of CD44v inhibit the metastatic dissemination of cells. When a CD44v construct was expressed in transfected nonmetastatic tumor cells, a full-blown metastatic phenotype was observed after injection into syngeneic rats. The expression and altered molecular structure of CD44v leads to deranged cell adhesive phenomena and circulation through the lymphatic system via HEV, thus promoting metastasis.

We have already learned that cell surface carbohydrates are modified by transformation. The increase in sialylation and $\beta1\rightarrow6$ branching of N-linked oligosaccharides correlates with metastatic ability. Furthermore, mutant metastatic cells deficient in $\beta1\rightarrow6$ GlcNAc transferase V became nonmetastatic. On the other hand, induction of $\beta1\rightarrow6$ branching in nonmetastatic cells increased metastasis. These carbohydrate structures can be altered using enzyme inhibitors. Swainsonine is an inhibitor of α-mannosidase II, a glycosidase of the Golgi apparatus. In its presence, cells synthesize different oligosaccharide chains containing many mannose residues. B16–F10 melanoma cells were able to grow in the presence of swainsonine. However, intravenously injected swainsonine-treated cells were unable to metastasize (Humphries *et al.*, 1986b). These studies show that specific changes in cell surface carbohydrates are associated with metastasis and that interference with these changes inhibits metastasis.

When tumor cells metastasize, they can colonize many different organs. In general, specific tumors metastasize to specific organs. The reasons for this specificity are just beginning to be understood. Membrane proteins and antigens were first correlated with metastatic cell targeting to specific organs. The target organ of some tumors is determined by membrane carbohydrates. For example, lectinlike receptors of hepatocytes normally participate in the clearance of effete erythrocytes. However, these same receptors can bind tumor cells expressing appropriate carbohydrate moieties. The metastatic spread of leukemias and lymphomas to distant lymphoid organs is in some cases linked to the expression of homing receptors such as MEL-14 (Section 7.5.3.2), which is functionally, but not structurally, similar to CD44. When this lectinlike membrane receptor is expressed on malignant lymphocytes, they spread to all lymph nodes (Bargatze *et al.*, 1989). Both metastasis and organ colonization are influenced by membrane carbohydrates and lectinlike receptors.

After metastatic cells bind to distant circulatory sites, they must cross capillary walls to invade nearby tissues. Unfortunately, the invasiveness of tumor cells has been difficult to quantitate because of capricious *in vitro* assays. However, membrane-associated proteases appear to play an important role (Ossowski, 1988; Aoyana and Chen, 1990). These proteases participate in the disruption of barriers such as basement membranes, thus facilitating tumor cell invasion.

Tumor and transformed cells synthesize and secrete the protease urokinase-type plasminogen activator (uPA). The released uPA then binds to urokinase receptors at tumor cell surfaces via its NH_2-terminal domain. The cell surface-bound uPA remains enzymatically active. It catalyzes the formation of plasmin from plasminogen. Plasmin participates in the degradation of basement membranes and activation of collagenase. Several lines of evidence suggest that these processes are important. uPA synthesis is greatly stimulated by transformation. *In vivo* tumors are rich in uPA, presumably due to its heightened synthetic rate. Furthermore, anti-uPA antibodies inhibit metastasis. The invasive potential of tumor cells has been correlated with the amount of surface-found uPA. Successful metastasis requires membrane components to perform both adherence and invasive functions.

9.5. TUMOR-SPECIFIC ANTIGENS

Tumor cells frequently express new or different cellular antigens. These "new" antigens could be due to the expression of tumor virus gene products, oncogenes, fetal antigens not expressed in mature tissues, alterations in oligosaccharide sequences or processing, and mutant gene products. Although many apparent antigenic changes have been cataloged (Woodruff, 1980), the important changes are those that are specific for a tumor cell. Tumor-specific antigens are molecules that are recognized by the immune system and are not associated with any normal cell.

Good evidence supporting the presence of tumor-specific antigens has been accumulating for 35 years. These antigens have generally been detected by phenomenological procedures. For example, many tumor antigens have been identified by transplanting tumors from one animal to another apparently syngeneic animal. However, it is difficult to be absolutely certain that no genetic differences have crept into the "syngeneic" animal population used for study. Ward *et al.* (1989) have provided compelling evidence that tumor-specific antigens exist on spontaneous and ultraviolet light-induced sarcomas in mice. These investigators isolated tumor cells and nonmalignant cells from each tumor-bearing mouse. Monoclonal antibodies and cloned cytotoxic T lymphocytes were generated against tumor cells but not autologous nonmalignant cells. Tumor-specific antigens have been unambiguously identified on some tumors.

9.5.1. Recognition of Tumor Antigens

Tumor-specific antigens are recognized by the immune system in several ways. Foreign antigens can trigger the proliferation of B cells and the production of antibody molecules which bind to targets. In addition, antigens can also elicit antibody-independent killing of tumor cells by T lymphocytes. The latter pathway seems more important in many experimental systems (e.g., Boon *et al.*, 1990: Hersey *et al.*, 1990). Since many important tumor antigens do not stimulate antibody production by B cells, they have been extraordinarily difficult to isolate and study.

How Tumor Cells May Betray Their Oncogenic Status

Recently, the first molecular descriptions of tumor antigens recognized by cytotoxic T cells have been reported (e.g., van der Hoorn *et al.*, 1985; Boon *et al.*, 1990; Hersey *et al.*, 1990; Jung and Schluesener, 1991; van der Bruggen *et al.*, 1991). These tumor antigens are associated with transmembrane proteins, proteins found at the *cis* face of membranes (e.g., p21*ras*), and intracellular proteins. During protein turnover in cells, fragments of constituent proteins are generated by cytosolic enzymatic machinery (Glynne *et al.*, 1991). These fragments are believed to be transported into the ER via mtp (*M*HC-linked *t*ransporter *p*rotein) proteins embedded in the ER membrane (Figure 9.7) (Deverson *et al.*, 1990; Trowsdale *et al.*, 1990; Spies *et al.*, 1990). mtp's are encoded by at least two genes: *mtp-1* and *mtp-2*. The deduced amino acid sequence of mtp-1 indicates that it is a large protein (79 kDa) with ten transmembrane domains and an ATP-binding domain. Interestingly, it is homologous to ATP-dependent transporters such as the mdr proteins. Within the ER, protein fragments (oncopeptides) derived from tumor antigens bind to intracellular HLA molecules that carry them to the cell surface. Cytotoxic T cells recognize the HLA–oncopeptide complex at cell surfaces. Cytotoxic T cells then release a variety of molecules that stimulate tumor cell killing. In this way lymphocytes survey the extracellular and intracellular antigens of neighboring cells for oncogenic transformation.

It seems likely that this mechanism of oncopeptide membrane trafficking and expression contributes to host resistance to tumors. For example, several types of human cancer are known to be associated with a T-cell-dependent antigen (van der Bruggen

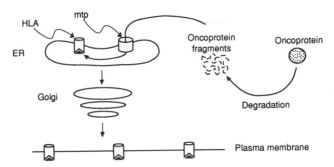

Figure 9.7. A speculative model of antigen trafficking and expression by tumor cells. During their normal turnover in cells, proteins are degraded into peptides. These peptides are transported into the ER by mtp proteins. Peptides containing oncogenic mutations may bind to the HLA molecules inside the ER. The HLA–oncopeptide complex is transported to the cell surface via the secretory pathway. The tumor cells, now carrying oncopeptide on their surfaces, trigger their destruction by cytotoxic T cells of the host.

et al., 1991). Another recent study has shown that healthy people possess T cells reactive with a small peptide derived from p21^{ras} which included its oncogenic point mutation (Jung and Schluesener, 1991). Moreover, the heterogeneous ability of oncopeptides to bind to polymorphic HLA antigens may explain why some families (which would share genetic backgrounds and HLA types) may have a greater or lesser tendency to develop certain forms of cancer. This mechanism suggests the molecular events which could constitute a form of "immune surveillance."

9.5.2. Antigens of Malignant Melanoma: Target or Host Destruction?

Malignant melanoma cells are derived from melanocytes, skin cells that produce the pigment melanin. Consequently, regions of malignant growth appear as dark splotches. Malignant melanoma is a fairly rare form of cancer that occurs most frequently in light-skinned persons. It is often associated with regions of the body such as the face, arms, and legs that are exposed to ultraviolet light. The amount of exposure to sunlight correlates with the likelihood of developing this disease. A patient's prognosis is related inversely to how far the malignant cells have invaded the underlying tissue.

Malignant melanoma is a favorite target for experimental studies because it strongly induces a host's immune response, which affects tumor growth. Microscopic studies of tumors have shown large numbers of T lymphocytes infiltrating regions of tumor regression. But what are the molecular cues recognized by these cells? Hersey *et al.* (1990) have recently reported the preliminary characterization of human melanoma antigens. Melanoma cells were extracted by detergent followed by SDS-PAGE. The separated proteins were transferred to nitrocellulose paper by Western blotting. The nitrocellulose blots were cut into small sections corresponding to different molecular weight regions. Cloned T cells were tested for their ability to proliferate in the presence of each piece of nitrocellulose. Most melanoma cell proteins and all proteins of control cells were unable to trigger proliferation. However, two molecules of 46 to 50 kDa and 36 to 40 kDa were found to trigger lymphocyte proliferation. Using melanoma cell cDNA libraries expressed in *E. coli*, Hersey *et al.* (1990) have found a cDNA clone which specifically stimulates T-cell proliferation. This molecule may be useful in the immunotherapy of melanoma. Indeed, the future of immunotherapy of many forms of cancer seems bright (Marx, 1989).

Melanoma cells that escape immune destruction may metastasize to other organs. The expression of a melanoma antigen called MUC18 correlates with metastasis. As melanoma cells invade underlying tissues, MUC18 expression dramatically rises. Lehmann *et al.* (1989) have recently reported the deduced amino acid sequence of MUC18. This protein has a single transmembrane domain and five extracellular immunoglobulin-like regions. It possesses a significant degree of homology with the neural cell adhesion molecule (NCAM). Since MUC18 has also been found on vascular smooth muscle cells, it is thought to mediate the adherence step in melanoma metastasis via the vascular system.

REFERENCES AND FURTHER READING

Phenotypic Properties of Tumor and Transformed Cells

Ambros, V. R., *et al*. 1975. Surface ruffles as markers for studies of cell transformation by Rous sarcoma virus. *Proc. Natl. Acad. Sci. USA* **72**:3144–3148.

Chen, J.-M., and Chen, W.-T. 1987. Fibronectin-degrading proteases from the membranes of transformed cells. *Cell* **48**:193–203.

Chen, L. B., *et al*. 1976. Correlation between tumor induction and the large external transformation sensitive protein on the cell surface. *Proc. Natl. Acad. Sci. USA* **73**:3570–3574.

Hynes, R.O., and Destree, A.T. 1978. 10 nm filaments in normal and transformed cells. *Cell* **13**: 151–163.

Lai, C. N., *et al*. 1984. Temperature-dependent transmembrane potential changes in cells infected with a temperature-sensitive Moloney sarcoma virus. *J. Cell. Physiol.* **121**:139–142.

Nicolau, C. 1978. *Virus-Transformed Cell Membranes*. Academic Press, New York.

Raz, A., *et al*. 1980. Distribution of membrane anionic sites on B16 melanoma variants with differing lung colonising potential. *Nature* **284**:363–364.

Rieber, M., and Rieber, M. S. 1981. Metastatic potential correlates with cell surface protein alterations in B16 melanoma variants. *Nature* **293**:74–75.

Schrump, D. S., *et al*. 1988. Recognition of galactosylgloboside by monoclonal antibodies derived from patients with primary lung cancer. *Proc. Natl. Acad. Sci. USA* **85**:4441–4445.

Srinivas, G., and Colburn, N. H. 1982. Ganglioside changes induced by tumor promoters in promotable JB6 mouse epidermal cells: Antagonism by an antipromoter. *J. Natl. Cancer Inst.* **68**:469–473.

Tucker, R. W., *et al*. 1978. Tubulin and actin in paired nonneoplastic and spontaneously transformed neoplastic cell lines in vitro: Fluorescent antibody studies. *Cell* **13**:629–642.

Yamada, K. M., *et al*. 1976. Cell surface protein partially restores morphology, adhesiveness, and contact inhibition of movement to transformed fibroblasts. *Proc. Natl. Acad. Sci. USA* **73**: 1217–1221.

Oncogenes

Barbacid, M. 1987. *ras* genes. *Annu. Rev. Biochem.* **56**:779–827.

Bargmann, C. I., *et al*. 1986. Multiple activations of the *neu* oncogene by a point mutation altering the transmembrane domain of p185. *Cell* **45**:649–657.

Bar-Sagi, D., and Feramisco, J. R. 1986. Induction of membrane ruffling and fluid-phase pinocytosis in quiescent fibroblasts by *ras* proteins. *Science* **233**:1061–1068.

Besmer, P., *et al*. 1986. A new acute transforming feline retrovirus and relationship of its oncogene *v-kit* with the protein kinase gene family. *Nature* **320**:415–421.

Bishop, J. M. 1983. Cellular oncogenes and retroviruses. *Annu. Rev. Biochem.* **52**:301–354.

Courtneidge, S. A., and Smith, A. E. 1983. Polyoma virus transforming protein associates with the product of the *c-src* cellular gene. *Nature* **303**:435–439.

Deuel, T. F., *et al*. 1983. Expression of a platelet-derived growth factor-like protein in simian sarcoma virus transformed cells. *Science* **221**:1348–1350.

Egan, S. E., *et al*. 1987. Transformation by oncogenes encoding protein kinases induced the metastatic phenotype. *Science* **238**:202–205.

Fiore, P. P., *et al*. 1987. Overexpression of the human EGF receptor confers an EGF-dependent transformed phenotype. *Cell* **51**:1063–1070.

Fleischman, L. F., *et al*. 1986. *Ras*-transformed cells: Altered levels of phosphatidylinositol-4,5-bisphosphate and catabolites. *Science* **231**:407–410.

Fukami, K., *et al*. 1988. Antibody to phosphatidylinositol 4,5-bisphosphate inhibits oncogene-induced mitogenesis. *Proc. Natl. Acad. Sci. USA* **85**:9057–9061.

Gibbs, J. B. 1991. Ras C-terminal processing enzymes—New drug targets? *Cell* **65**:1–4.

Heldin, C.-H., and Westermark, B. 1984. Growth factors: Mechanism of action and relation to oncogenes. *Cell* **37**:9–20.

Holmes, W. E., *et al*. 1992. Identification of heregulin, a specific activator of p185[erbB2]. *Science* **256**:1205–1210.

Jakobovits, E. B., *et al*. 1984. Hormonal regulation of the Rous sarcoma virus *src* gene via a heterologous promoter defines a threshold dose for cellular transformation. *Cell* **38**:757–765.

Kaech, S., *et al*. 1991. Association of p60c-src with polyoma virus middle-T antigen abrogating mitosis-specific activation. *Nature* **350**:431–433.

Libermann, T. A., *et al*. 1985. Amplification, enhanced expression and possible rearrangement of EGF receptor gene in primary human brain tumours of glial origin. *Nature* **313**:144–147.

Rettenmier, C. W., *et al*. 1985. Transmembrane orientation of glycoproteins encoded by the *v-fms* oncogene. *Cell* **40**:971–981.

Sherr, C. J., *et al*. 1985. The *c-fms* proto-oncogene product is related to the receptor for the mononuclear phagocyte growth factor, CSF-1. *Cell* **41**:665–676.

Sugimoto, Y., *et al*. 1984. Evidence that the Rous sarcoma virus transforming gene product phosphorylates phosphatidylinositol and diacylglycerol. *Proc. Natl. Acad. Sci. USA* **81**:2117–2121.

Trahey, M., *et al*. 1988. Molecular cloning of two types of GAP complementary DNA from human placenta. *Science* **242**:1697–1700.

Varmus, H. E. 1984. The molecular genetics of cellular oncogenes. *Annu. Rev. Genet.* **18**:553–612.

Waterfield, M. D., *et al*. 1983. Platelet-derived growth factor is structurally related to the putative transforming protein p28[sis] of simian sarcoma virus. *Nature* **304**:35–39.

Weiner, D. B., *et al*. 1989. A point mutation in the *neu* oncogene mimics ligand induction of receptor aggregation. *Nature* **339**:230–231.

Willingham, M. C., *et al*. 1980. Localization of the *src* gene product of the Harvey strain of MSV to plasma membrane of transformed cells by electron microscopic immunocytochemistry. *Cell* **19**:1005–1014.

Yarden, Y., and Peles, E. 1991. Biochemical analysis of the ligand for the *neu* oncogenic receptor. *Biochemistry* **30**:3543–3550.

Multidrug Resistance

Chen, C.-J., *et al*. 1986. Internal duplication and homology with bacterial transport proteins in the *mdr*1 (P-glycoprotein) gene from multidrug-resistant human cells. *Cell* **47**:381–389.

Devine, S. E., *et al*. 1992. Amino acid substitutions in the sixth transmembrane domain of P-glycoprotein alter multidrug resistance. *Proc. Natl. Acad. Sci. USA* **89**:4564–4568.

Dhir, R., and Gros, P. 1992. Functional analysis of chimeric proteins constructed by exchanging homologous domains of two P-glycoproteins conferring distinct drug resistance profiles. *Biochemistry* **31**:6103–6110.

Endicott, J. A., and Ling, V. 1989. The biochemistry of P-glycoprotein-mediated multidrug resistance. *Annu. Rev. Biochem.* **58:**137–171.

Gros, P., *et al.* 1986. Mammalian multidrug resistance gene: Complete cDNA sequence indicates strong homology to bacterial transport proteins. *Cell* **47:**371–380.

Kartner, N., *et al.* 1983. Cell surface P-glycoprotein associated with multidrug resistance in mammalian cells. *Science* **221:**1285–1288.

Pastan, I. 1990. Presented at the FASEB meeting June 4–7, New Orleans.

Sorrentino, B. P., *et al.* 1992. Selection of drug-resistant bone marrow cells in vivo after retroviral transfer of human *MDR1*. *Science* **257:**99–103.

Valverde, M. A., *et al.* 1992. Volume-regulated chloride channels associated with the human multidrug-resistance P-glycoprotein. *Nature* **355:**830–833.

Van der Bliek, A. M., *et al.* 1987. The human *mdr3* gene encodes a novel P-glycoprotein homologue and gives rise to alternatively spliced mRNAs in liver. *EMBO J.* **6:**3325–3331.

Metastasis and Invasion

Aoyama, A., and Chen, W.-T. 1990. A 170-kDa membrane-bound protease is associated with the expression of invasiveness by human malignant melanoma cells. *Proc. Natl. Acad. Sci. USA* **87:**8296–8300.

Arch, R., *et al.* 1992. Participation in normal immune responses of a metastasis-inducing splice variant of CD44. *Science* **257:**682–685.

Bargatze, R. F., *et al.* 1987. High endothelial venule binding as a predictor of the dissemination of passaged murine lymphomas. *J. Exp. Med.* **166:**1125–1131.

Dennis, J. W., *et al.* 1987. β1→6 branching of Asn-linked oligosaccharides is directly associated with metastasis. *Science* **236:**582–585.

Drake, S. L., *et al.* 1992. Cell surface phosphatidylinositol-anchored heparin sulfate proteoglycan initiates mouse melanoma cell adhesion to a fibronectin-derived, heparin-binding synthetic peptide. *J. Cell Biol.* **117:**1331–1341.

Gunthert, U., *et al.* 1991. A new variant of glycoprotein CD44 confers metastatic potential to rat carcinoma cells. *Cell* **65:**13–24.

Hardwick, C., *et al.* 1992. Molecular cloning of a novel hyaluronan receptor that mediates tumor cell motility. *J. Cell Biol.* **117:**1343–1350.

Haynes, I. R., *et al.* 1989. CD44—a molecule involved in leukocyte adherence and T cell activation. *Immunol. Today* **10:**423–428.

Humphries, M. J., *et al.* 1986a. A synthetic peptide from fibronectin inhibits experimental metastasis of murine melanoma cells. *Science* **233:**467–470.

Humphries, M. J., *et al.* 1986b. Oligosaccharide modification by swainsonine treatment inhibits pulmonary colonization by B16–F10 murine melanoma cells. *Proc. Natl. Acad. Sci. USA* **83:**1752–1756.

Jalkanen, S., and Jalkanen, M. 1992. Lymphocyte CD44 binds the COOH-terminal heparin-binding domain of fibronectin. *J. Cell Biol.* **116:**817–825.

Kramer, R. H., *et al.* 1989. Melanoma cell adhesion to basement membrane mediated by integrin-related complexes. *Cancer Res.* **49:**393–402.

Ossowski, L. 1988. In vivo invasion of modified chorioallantoic membrane by tumor cells: The role of cell surface-bound urokinase. *J. Cell Biol.* **107:**2437–2445.

Poste, G., and Nicolson, G. L. 1980. Arrest and metastasis of blood-borne tumor cells are modified by fusion of plasma membrane vesicles from highly metastatic cells. *Proc. Natl. Acad. Sci. USA* **77**:399–403.

Saiki, I., *et al*. 1989. Inhibition of metastasis of murine malignant melanoma by synthetic polymeric peptide containing core sequences of cell-adhesive molecules. *Cancer Res.* **49**: 3815–3822.

Vollmers, H. P., and Birchmeier, W. 1983. Monoclonal antibodies inhibit the adhesion of mouse B16 melanoma cells *in vitro* and block lung metastasis *in vivo*. *Proc. Natl. Acad. Sci. USA* **80**: 3729–3733.

Yamashita, K., *et al*. 1984. Comparative study of the oligosaccharides released from baby hamster kidney cells and their polyoma transformant by hydrazinolysis. *J. Biol. Chem.* **259**:10834–10840.

Yogeeswaran, G., and Salk, P. L. 1981. Metastatic potential is positively correlated with cell surface sialylation of cultured murine tumor cell lines. *Science* **212**:1514–1516.

Tumor-Specific Antigens

Boon, T., *et al*. 1990. Genetic analysis of tum⁻ antigens. Implications for T cell mediated immune surveillance. In: *Cellular Immunity and the Immunotherapy of Cancer* (M. T. Lotze and O. J. Finn, eds.), Wiley–Liss, New York, pp. 287–294.

Deverson, E. V., *et al*. 1990. MHC class II region encoding proteins related to the multidrug resistance family of transmembrane transporters. *Nature* **348**:738–741.

Glynne, R., *et al*. 1991. A proteasome-related gene between the two ABC transporter loci in the class II region of the human MHC. *Nature* **353**:357–360.

Hersey, P., *et al*. 1990. Characterization of melanoma antigens recognized by human T cells. In: *Cellular Immunity and the Immunotherapy of Cancer* (M. T. Lotze and O. J. Finn, eds.), Wiley–Liss, New York, pp. 343–349.

Jung, S., and Schluesener, H. J. 1991. Human T lymphocytes recognize a peptide of single point-mutated, oncogenic ras proteins. *J. Exp. Med.* **173**:273–276.

Lehmann, J. M., *et al*. 1989. MUC18, a marker of tumor progression in human melanoma, shows sequence similarity to the neural cell adhesion molecules of the immunoglobulin superfamily. *Proc. Natl. Acad. Sci. USA* **86**:9891–9895.

Marx, J. L. 1989. Cancer vaccines show promise at last. *Science* **245**:813–815.

Spies, T., *et al*. 1990. A gene in the human major histocompatibility complex class II region controlling the class I antigen presentation pathway. *Nature* **348**:744–747.

Trowsdale, J., *et al*. 1990. Sequences encoded in the class II region of the MHC related to 'ABC' superfamily of transporters. *Nature* **348**:741–744.

van der Bruggen, *et al*. 1991. A gene encoding an antigen recognized by cytolytic T lymphocytes on human melanoma. *Science* **254**:1643–1647.

van der Hoorn, F. A., *et al*. 1985. Characterization of gp85^{gag} as an antigen recognized by Moloney leukemia virus-specific cytolytic T cell clones that function in vivo. *J. Exp. Med.* **162**: 128–144.

Ward, P. L., *et al*. 1989. Tumor antigens defined by cloned immunological probes are highly polymorphic and are not detected on autologous normal cells. *J. Exp. Med.* **170**:217–232.

Woodruff, M. F. A. 1980. *The Interaction of Cancer and Host: Its Therapeutic Significance*. Grune & Stratton, New York.

Index

Page numbers given in **bold face** refer to primary discussions of a given topic in the text when multiple page numbers are listed. Page numbers appearing in *italics* refer to a topic's presentation as a displayed item (figure or table).